Gerald J. Gray
Maia J. Enzer
Jonathan Kusel
Editors

Understanding Community-Based Forest Ecosystem Management

Understanding Community-Based Forest Ecosystem Management has been co-published simultaneously as *Journal of Sustainable Forestry*, Volume 12, Numbers 3/4 and Volume 13, Numbers 1/2 2001.

Pre-publication REVIEWS, COMMENTARIES, EVALUATIONS . . .

"Community-based stewardship is not a panacea–leadership and decisiveness are integral to protecting our best last places. Collaborative approaches to restoring waters and healing the land, however, will ultimately determine whether we can live in productive harmony with the resources that sustain us. This informative and insightful book shows the way to a more sutainable future."

Mike Dombeck
Chief, USDA Forest Service

More pre-publication
REVIEWS, COMMENTARIES, EVALUATIONS . . .

"There is a great deal of discussion and debate today about the various citizen initiatives grouped here as 'community based ecosystem management.' Unfortunately, many perceptions are based on limited knowledge of real, on-the-ground experiments, and even fewer are related to the theoretical framework within which these experiments are occurring. This book offers a detailed theoretical framework as a starting point for understanding the many examples of community based ecosystem management. It should elevate the level of understanding among all who are interested in this growing and management movement."

Sarah B. Van de Wetering, JD
Editor, Chronicle of Community

FPP

Food Products Press
An Imprint of The Haworth Press, Inc.

Understanding Community-Based Forest Ecosystem Management

Understanding Community-Based Forest Ecosystem Management has been co-published simultaneously as *Journal of Sustainable Forestry*, Volume 12, Numbers 3/4 and Volume 13, Numbers 1/2 2001.

The *Journal of Sustainable Forestry* Monographic "Separates"

Below is a list of "separates," which in serials librarianship means a special issue simultaneously published as a special journal issue or double-issue *and* as a "separate" hardbound monograph. (This is a format which we also call a "DocuSerial.")

"Separates" are published because specialized libraries or professionals may wish to purchase a specific thematic issue by itself in a format which can be separately cataloged and shelved, as opposed to purchasing the journal on an on-going basis. Faculty members may also more easily consider a "separate" for classroom adoption.

"Separates" are carefully classified separately with the major book jobbers so that the journal tie-in can be noted on new book order slips to avoid duplicate purchasing.

You may wish to visit Haworth's website at . . .

http://www.HaworthPress.com

. . . to search our online catalog for complete tables of contents of these separates and related publications.

You may also call 1-800-HAWORTH (outside US/Canada: 607-722-5857), or Fax 1-800-895-0582 (outside US/Canada: 607-771-0012), or e-mail at:

getinfo@haworthpressinc.com

Understanding Community-Based Forest Ecosystem Management, edited by Gerald J. Gray, Maia J. Enzer, and Jonathan Kusel (Vol. 12, No. 3/4 & Vol. 13, No. 1/2, 2001). *Here is a state-of-the-art reference and information source for scientists, community groups and their leaders, resource managers, and ecosystem management practitioners. Healthy ecosystems and community well-being go hand in hand, and the interdependence between the two is the focal point of community-based ecosystem management. The information you'll find in* ***Understanding Community-Based Forest Ecosystem Management*** *will be invaluable in your effort to manage and maintain the ecosystems in your community.* ***Understanding Community-Based Forest Ecosystem Management*** *examines the emergence of community-based ecosystem management (CBEM) in the United States. This comprehensive book blends diverse perspectives, enabling you to draw on the experience and expertise of forest-based practitioners, researchers, and leaders in community-based efforts in the ecosystem management situations that you deal with in your community.*

Climate Change and Forest Management in the Western Hemisphere, edited by Mohammed H. I. Dore (Vol. 12, No. 1/2, 2001). *This valuable book examines integrated forest management in the Americas, covering important global issues including global climate change and the conservation of biodiversity. Here you will find case studies from representative forests in North, Central, and South America. The book also explores the role of the Brazilian rainforest in the global carbon cycle and implications for sustainable use of rainforests, as well as the carbon cycle and the valuation of forests for carbon sequestration.*

Mapping Wildfire Hazards and Risks, edited by R. Neil Sampson, R. Dwight Atkinson, and Joe W. Lewis (Vol. 11, No. 1/2, 2000). *Based on the October 1996 workshop at Pingree Park in Colorado,* ***Mapping Wildfire Hazards and Risks*** *is a compilation of the ideas of federal and state agencies, universities, and non-governmental organizations on how to rank and prioritize forested watershed areas that are in need of prescribed fire. This book explains the vital importance of fire for the health and sustainability of a watershed forest and how the past acceptance of fire suspension has consequently led to increased fuel loadings in these landscapes that may lead to more severe future wildfires. Complete with geographic maps, charts, diagrams, and a list of locations where there is the greatest risk of future wildfires,* ***Mapping Wildfire Hazards and Risks*** *will assist you in deciding how to set priorities for land treatment that might reduce the risk of land damage.*

Frontiers of Forest Biology: Proceedings of the 1998 Joint Meeting of the North American Forest Biology Workshop and the Western Forest Genetics Association, edited by Alan K. Mitchell, Pasi Puttonen, Michael Stoehr, and Barbara J. Hawkins (Vol. 10, No. 1/2 & 3/4, 2000). *Based on the 1998 Joint Meeting of the North American Forest Biology Workshop and the Western Forest Genetics Association, Frontiers of Forest Biology addresses changing priorities in forest resource management.*

You will explore how the emphasis of forest research has shifted from productivity-based goals to goals related to sustainable development of forest resources. This important book contains fascinating research studies, complete with tables and diagrams, on topics such as biodiversity research and the productivity of commercial species that seek criteria and indicators of ecological integrity.

"There is clear emphasis on the genetics, genecology, and physiology of trees, particularly temperate trees. . . . These proceedings are also testimony to what does or should distinguish forest biology from other sciences: a focus on intra- and inter-specific interactions between forest organisms and their environment, over scales of both time and place." (Robert D. Guy, PhD, Associate Professor, Department of Forest Sciences, University of British Columbia, Vancouver, Canada)

Contested Issues of Ecosystem Management, edited by Piermaria Corona and Boris Zeide (Vol. 9, No. 1/2, 1999). *Provides park rangers, forestry students and personnel with a unique discussion of the premise, goals, and concepts of ecosystem management. You will discover the need for you to maintain and enhance the quality of the environment on a global scale while meeting the current and future needs of an increasing human population. This unique book includes ways to tackle the fundamental causes of environmental degradation so you will be able to respond to the problem and not merely the symptoms.*

Protecting Watershed Areas: Case of the Panama Canal, edited by Mark S. Ashton, Jennifer L. O'Hara, and Robert D. Hauff (Vol. 8, No. 3/4, 1999). *"This book makes a valuable contribution to the literature on conservation and development in the neo-tropics. . . . These writings provide a fresh yet realistic account of the Panama landscape." (Raymond P. Guries, Professor of Forestry, Department of Forestry, University of Wisconsin at Madison, Wisconsin)*

Sustainable Forests: Global Challenges and Local Solutions, edited by O. Thomas Bouman and David G. Brand (Vol. 4, No. 3/4 & Vol. 5, No. 1/2, 1997). *"Presents visions and hopes and the challenges and frustrations in utilization of our forests to meet the economical and social needs of communities, without irreversibly damaging the renewal capacities of the world's forests." (Dvoralai Wulfsohn, PhD, PEng, Associate Professor, Department of Agricultural and Bioresource Engineering, University of Saskatchewan)*

Assessing Forest Ecosystem Health in the Inland West, edited by R. Neil Sampson and David L. Adams (Vol. 2, No. 1/2/3/4, 1994). *"A compendium of research findings on a variety of forest issues. Useful for both scientists and policymakers since it represents the combined knowledge of both." (Abstracts of Public Administration, Development, and Environment)*

∞ ALL FOOD PRODUCTS PRESS BOOKS AND JOURNALS ARE PRINTED ON CERTIFIED ACID-FREE PAPER

Understanding Community-Based Forest Ecosystem Management

Gerald J. Gray
Maia J. Enzer
Jonathan Kusel
Editors

Understanding Community-Based Forest Ecosystem Management has been co-published simultaneously as *Journal of Sustainable Forestry*, Volume 12, Numbers 3/4 and Volume 13, Numbers 1/2 2001.

Food Products Press
An Imprint of
The Haworth Press, Inc.
New York • London • Oxford

Published by

Food Products Press®, 10 Alice Street, Binghamton, NY 13904-1580 USA

Food Products Press® is an imprint of The Haworth Press, Inc., 10 Alice Street, Binghamton, NY 13904-1580 USA.

Understanding Community-Based Forest Ecosystem Management has been co-published simultaneously as *Journal of Sustainable Forestry*, Volume 12, Numbers 3/4 and Volume 13, Numbers 1/2 2001.

Cover design by Thomas J. Mayshock Jr.

Library of Congress Cataloging-in-Publication Data

Understanding community-based forest ecosystem management/Gerald J. Gray, Maia J. Enzer, Jonathan Kusel, editors.

p. cm.

"Co-published simultaneously as Journal of sustainable forestry, volume 12, numbers 3/4 and volume 13, numbers 1/2."

Includes bibliographical references (p.).

ISBN 1-56022-080-5 (hardcover: alk. paper)–ISBN 1-56022-081-3 (pbk.: alk. paper)

1. Forest management–Citizen participation–Congresses. 2. Forest management–United States–Citizen participation–Congresses. I. Gray, Gerald J. II. Enzer, Maia. J. III. Kusel, Jonathan. IV. Journal of sustainable forestry.

SD387.C47 U53 2000
333.75′152′0973–dc21 00-065437

Indexing, Abstracting & Website/Internet Coverage

This section provides you with a list of major indexing & abstracting services. That is to say, each service began covering this periodical during the year noted in the right column. Most Websites which are listed below have indicated that they will either post, disseminate, compile, archive, cite or alert their own Website users with research-based content from this work. (This list is as current as the copyright date of this publication.)

Abstracting, Website/Indexing Coverage	Year When Coverage Began
• *Abstract Bulletin*	**1993**
• *Abstracts in Anthropology*	**1993**
• *Abstracts on Rural Development in the Tropics (RURAL)*	**1993**
• *AGRICOLA Database*	**1993**
• *Biology Digest (in print & online)*	**2000**
• *Biostatistica*	**1993**
• *BUBL Information Service, an Internet-based Information Service for the UK higher education community <URL: http://bubl.ac.uk/>*	**1995**
• *CNPIEC Reference Guide: Chinese National Directory of Foreign Periodicals*	**1996**
• *Engineering Information (PAGE ONE)*	**1999**
• *Environment Abstracts <www.cispubs.com>*	**1993**
• *Environmental Periodicals Bibliography (EPB)*	**1993**
• *FINDEX <www.publist.com>*	**1999**

(continued)

- *Forestry Abstracts (c/o CAB Intl/CAB ACCESS) <www.cabi.org>* **1993**
- *GEO Abstracts (GEO Abstracts/GEOBASE) <URL: http://www.elsevier.nl>* **2000**
- *Human Resources Abstracts (HRA)* **1993**
- *Journal of Planning Literature/Incorporating the CPL Bibliographics* **1995**
- *Referativnyi Zhurnal (Abstracts Journal of the All-Russian Institute of Scientific and Technical Information-in Russian)* . . **1993**
- *Sage Public Administration Abstracts (SPAA)* **1993**
- *Sage Urban Studies Abstracts (SUSA)* **1993**
- *Wildlife Review Abstracts <www.nisc.com>* **1993**

Special Bibliographic Notes related to special journal issues (separates) and indexing/abstracting:

- indexing/abstracting services in this list will also cover material in any "separate" that is co-published simultaneously with Haworth's special thematic journal issue or DocuSerial. Indexing/abstracting usually covers material at the article/chapter level.
- monographic co-editions are intended for either non-subscribers or libraries which intend to purchase a second copy for their circulating collections.
- monographic co-editions are reported to all jobbers/wholesalers/approval plans. The source journal is listed as the "series" to assist the prevention of duplicate purchasing in the same manner utilized for books-in-series.
- to facilitate user/access services all indexing/abstracting services are encouraged to utilize the co-indexing entry note indicated at the bottom of the first page of each article/chapter/contribution.
- this is intended to assist a library user of any reference tool (whether print, electronic, online, or CD-ROM) to locate the monographic version if the library has purchased this version but not a subscription to the source journal.
- individual articles/chapters in any Haworth publication are also available through the Haworth Document Delivery Service (HDDS).

Understanding Community-Based Forest Ecosystem Management

CONTENTS

ABOUT THE EDITORS

Gerald J. Gray, PhD, is Vice President for Policy at AMERICAN FORESTS and directs the Forest Policy Center. He has directed resource policy activities at AMERICAN FORESTS since 1988. In that capacity, he has worked closely with Congressional staff, federal agencies, private conservation groups, and forest industry on issues related to the management of federal, state, and private forest lands. His current activities focus on ecosystem restoration and maintenance, community-based forestry, and global climate change. Prior to joining the AMERICAN FORESTS, he worked with the Division of Forestry, Minnesota Department of Natural Resources and the U.S. Forest Service, Northeastern Area, State and Private Forestry. He holds a Doctor of Forestry degree from the University of Minnesota (1987), a Master of Forest Science from the Yale School of Forestry and Environmental Studies (1982), and a Bachelor of Arts degree in English from Yale University (1978). He serves on the Steering Committee of the Communities Committee of the Seventh American Forest Congress. He is a member of the National Council on Private Forests (Chair, 1995) and the Society of American Foresters (National Capital Chapter, Policy Committee Chair, 1997).

Maia J. Enzer, MA, is Director of Forest Policy at AMERICAN FORESTS and has been with the organization since 1993. She oversees public land and community-based forestry issues. She has worked with community-based forestry activists from around the country since 1995. Prior to establishing the community-based policy program, she worked on various issues including forest ecosystem health, ecosystem management in mixed ownership landscapes, and climate change. She received a Masters degree in Environmental Studies from the Yale School of Forestry and Environmental Studies and has a B.A. in Political Science from Union College in Schenectady, NY. Previously, she worked as a community organizer dealing with state and local environmental issues for Kentuckians For The Commonwealth and for the Public Interest Research Groups (PIRG) in Massachusetts and New York. She serves on the Executive Committee of the Communities Committee of the 7th American Forest Congress and co-chairs the Policy Task Group. She is also a member of the Society of American Foresters.

Jonathan Kusel, PhD, is Executive Director of Forest Community Research, a research and educational organization based in the Northern Sierra of California. His research focuses on community well-being and its intersection with resource use. He received a Masters in Forest Science from Yale School of Forestry and Environmental Studies and a Ph.D. from the Department of Forestry and Resource Management at the University of California at Berkeley–where he was a visiting scholar and instructor from 1992 through 1996. Since 1993, he has facilitated the Lead Partnership Group, a consortium of fifteen leading watershed and community-based groups from Northern California and Southern Oregon working to make resource management responsive to both community well-being and ecosystem health. He served on the community assessment team for the President's Forest Ecosystem Management Assessment Team (FEMAT), and coordinated the community assessment and public participation teams for the Sierra Nevada Ecosystem Project (1994-96), the latter of which brought together the public and scientists, local and formal knowledge, in an independent science team project.

AMERICAN FORESTS' Workshop

Understanding and Exploring Community-Based Ecosystem Management

As of November 23, 1998

STEERING COMMITTEE

Ron Abraham, State Forester
Ohio Division of Forestry
1855 Fountain Square Court H-1
Columbus, OH 43224
(614) 265-6690 Fax (614) 447-9231
Email: Ron.Abraham@dnr.state.oh.us

Greg Aplet
The Wilderness Society
7475 Dakin Street, Suite 410
Denver, CO 80221
(303) 650-9379 Fax (303) 650-5942
Email: greg_aplet@tws.org

Victor Benavides
Alliance of Forest Workers
and Harvesters
P.O. Box 1814
800 NE Vine Street
White Salmon, WA 98672
(509) 493-5013
Fax (509) 493-5000
Email: vicmosk@gorge.net

Geraldine Bower
Ecosystem Management,
Staff Coordination Planning
USDA Forest Service
201 14th Street SW
Washington, DC 20250
(202) 205-1022 Fax (202) 205-1012
Email: gbower/wo@fs.fed.us

Lance Clark
Resource Planning Analyst
Oregon Department of Forestry
2600 State Street
Salem, OR 97310
(503) 945-7410 Fax (503) 945-7314
Email: lance.r.clark@state.or.us

Doug Crandall
American Forest and Paper Association
1111 Nineteenth Street NW, Suite 800
Washington, DC 20036
(202) 463-2752 Fax (202) 463-2708
Email: Doug_Crandall@afandpa.org

Maia Enzer
America Forests
910 17th Street NW, Suite 600
Washington, DC 20006
(202) 955-4500 ext. 237
Fax (202) 955-4588
Email: menzer@amfor.org

Larry Fisher
Center for the Environment
112 Rice Hall
Ithaca, NY 14853
607/255-4523 Fax 607/255-8207
607/273-9681 (hm)
Email: laf2@cornell.edu

Martin Goebel
Sustainable Northwest
620 SW Main, Suite 112
Portland, OR 97205
(503) 221-6911 Fax (503) 221-4495
Email: mgoebel@sustainablenorthwest.org

Gerry Gray
American Forests
910 17th Street NW, Suite 600
Washington, DC 20006
(202) 955-4500 ext. 217
Fax (202) 955-4588
Email: ggray@amfor.org

Peggy Harwood, Policy Analyst
Bureau of Land Management, Planning,
Assessment, & Community Support
1849 C Street, NW
Washington, DC 20240
(202) 452-0362 Fax (202) 452-5112
Email: pharwood@wo.blm.gov

Lynn Jungwirth
Communities Committee
of the Seventh American
Forest Congress
Box 356
Hayfork, CA 96041
(530) 628-4206 Fax (530) 628-5100
Ernail: lynnj@tcoe.trinity.K12.ca.us

Fay Knox
Newton County Resource Council
HC62 Box 692
Deer, AR 72628
(870) 446-5898 Fax (870) 446-2701
Email: 2077974@mcimail.com

Jonathan Kusel
Forest Community Research
P.O. Box 11
Taylorsville, CA 95983
(530) 284-1022 Fax (530) 284-1023
Email: kusel@fcresearch.org

Bill Painter
U.S. Environmental Protection Agency
Office of Sustainable Ecosystems
and Communities
401 M Street SW
Washington, DC 20460
(202) 260-5498 Fax (202) 260-1935
Email: Painter.Bill@epamail.epa.gov

Jeff Romm
College of Natural Resources
RIPM/ESPM
University of California-Berkeley
Berkeley, CA 94720-3114
(510) 642-6499 Fax (510) 643-5438
Email: jeffromm@nature.berkeley.edu

Leah Wills
Plumas Corporation
P.O. Box 3880
Quincy, CA 95971
(530) 283-3739 Fax (530) 283-5465
Email: plumasco@psln.com

Elisa Adler
Forest Community Research
P.O. Box 11
Taylorsville, CA 95983
(530) 284-1022 Fax (530) 284-1023
Email: eadler@fcresearch.org

Tahnee Robertson
Center for the Environment
112 Rice Hall
Ithaca, NY 14853
607/255-4523 Fax 607/255-820
Email: tmr5@cornell.edu

WORK GROUPS

PROCESS: PARTICIPATION AND DECISION MAKING

Ann Moote, Chair
University of Arizona
Udall Center for Studies in Public Policy
803/811 East 1 Street
Tucson, AZ 85721
(520) 621-7189 Fax (520) 621-9234
Email: moote@u.arizona.edu

Eric Kingsley
Executive Director
New Hampshire
Timberland Owners Association
54 Portsmouth Street
Concord, NH 03301
(603) 224-9699 Fax (603) 225-5898
Email: eknhtoa@aol.com

Gregg Walker
Oregon State University
Department of Speech Communication
Corvallis, OR 97331-6199
(541) 737-2461 Fax (541) 731-4443
Email: gwalker@orst.edu

Steve Marshall
USDA Forest Service-
Cooperative Forestry
201 14th Street SW
Washington, DC 20250
202/205-1 389 Fax (202) 205-1271
Email: Steve.Marshalllwo@fs.fed.us

Donald E. Voth
Professor of Rural Sociology School
of Human Environmental Science
HOEC 118
University of Arkansas
Fayetteville, AR 72701
501-575-2409
Email: dvoth@comp.uark.edu

Beverly Brown
The Jefferson Center
P.O. Box 278
Wolf Creek, OR 97497
(541) 955-9705
Email: jeffctr@internetcds.com

Steve Lee
Heritage Museum/
Gwynns Falls Conservancy
4509 Prospect Circle
Baltimore, MD 21216
(410) 664-6711 Fax: (410) 825-3305
Email: heritage.museum@erols.com

GLOBAL LINKAGES

Nels Johnson, Chair
World Resources Institute
10 G Street NE
Washington, DC 20002
(202) 729-7624 Fax (202) 729-7620
Email: nels@wri.org

Jill Belsky
Department of Sociology
1404 Stanley Street
University of Montana
Missoula, MT 59812
(406) 243-4868 Fax (406) 243-4858
Email: belsky@selway.umt.edu

Ann Hawkins
University of California at Berkeley
c/o College of Natural Resources
4 Giannini Hall #3100
Berkeley, CA 94720-3100
(510) 642-3431 Fax (510) 643-1934
Email: fscholar@nature.berkeley.edu

Sissel Waage
233 South Barrington Avenue
Apt. 214
Los Angeles, CA 90049
Email: swaage@nature.berkeley.edu

Steering Committee Liaisons:
Victor Benavides
Martin Goebel

INSTITUTIONS AND GOVERNANCE

Hanna Cortner, Chair
University of Arizona
School of Renewable Natural Resources
325 Bio Science East
Tucson, AZ 85721-0043
(520) 621-3501 Fax (520) 621-8801
Email: hcortner@ag.arizona.edu

Sam Burns (Co-Chair)
Director of Research
Fort Lewis College-Colorado
Office of Community Services
1000 Rim Drive
Durango, CO 81301-3999
(970) 247-7193 Fax (970) 247-7032
Email: burns_s@fortlewis.edu

Gus Townes
Director of Training and Research
Federation of Southern Cooperatives
P.O. Box 95
Epes, AL 35460
(205) 652-9676 Fax (205) 652-9678
Email: fscepes@mindspring.com

Martha Twarkins, District Ranger
Finger Lakes National Forest
5218 State Rt. 414
Hector, NY 14841
(607) 546-4470 Fax (607) 546-4474
Email:
Twarkins_Martha/r9_gmfl@fs.fed.us

Wendy Sanders, Executive Director
Great Lakes Forest Alliance
P.O. Box 722
Hayward, WI 54843
71 5/634-2006 Fax 715/634-2006
Email: forestls@lsfa.org

Steering Liaisons: Lance Clark

STEWARDSHIP AS COMMON GROUND

Brad Ack, Chair
Grand Canyon Trust
2601 N. Fort Valley Road
Flagstaff, AZ 86001
(520) 774-7488 Fax (520) 774-7570
Email: ack@grandcanyontrust.org

Carol Daly
Flathead Economic Policy Center
919 Elk Park Road
Columbia Falls, MT 59912
(406) 892-8155 Fax (406) 892-8161
Email: cdaly@digisys.net

Juan Mendoza
Forest Workers Alliance
Willamette Valley Reforestation, Inc.
P.O. Box 137
Molalla, OR 97038
(503) 829-9836 Fax (503) 829-9836
Email: wvrdp@molalla.net

Dr. Yvonne Everett, Assistant Professor
Dept. of Natural Resources Planning
and Interpretation
Humboldt State University
Arcata, CA 95521-8299
(707) 826-4188 Fax (707) 826-4145
Email: yel@axe.humboldt.edu

Mary Mitsos
Pinchot Institute for Conservation
1616 P Street NW
Washington, DC 20036
(202) 797-6582 Fax (202) 797-6583
Email: mmitsos@pinchot.org

Ron Ochs
USDA Forest Service
Cooperative Programs
333 SW 1st Avenue
Portland, OR 97204
(503) 808-2647 Fax (503) 808-2339
E-mail: rgochs/r6pnw@fs.fed.us

Steering Committee Liaison:
Fay Knox

INVESTING IN COMMUNITY-BASED ECOSYSTEM MANAGEMENT

Valerie Luzadis, Chair
SUNY ESF
320 Bray Hall, 1 Forestry Drive
Syracuse, NY 13210
(315) 470-6693 Fax (315) 470-6535
Email: vluzadis@mailbox.syr.edu

Carolyn Alkire, Economist
The Wilderness Society
900 17th Street NW
Washington, DC 20006-2596
(202) 429-2685 Fax (202) 429-3958
Email: calkire@tws.org

William Stewart
Fire and Resource Assessment Program
California Dept. of Forestry
1920 20th Street
Sacramento, CA 95814
(916) 227-2650 Fax (916) 227-2672
Email: Bill_Stewart@fire.ca.gov

Catherine Mater
Vice President
Mater Engineering, Ltd.
101 SW Western Boulevard
P.O. Box O
Corvallis, OR 97339-0880
541/753-7335 Fax 541/752-2952
Email: mater@mater.com

Duane Vaagen
Vaagen Brothers Lumber
585 W. 5th
Colville, WA 99114
(509) 684-5071 Fax (509) 684-2168

Steering Committee Liaisons:
Leah Wills
Jeff Romm

MONITORING AND ACCOUNTABILITY

John Bliss, Chair
Oregon State University
Department of Forest Resources
203 Peavy Hall
Corvallis, OR 97331-5703
(541) 737-4427 Fax (541) 737-3049
Email: blissj@ccmail.orst.edu

Paul Jahnige
Community Resources
5131 Wetheredsville Road
Baltimore, MD 21207
(410) 448-0640 Fax (410) 448-0874
Email: info@communityresources.org

Cate Hartzell
Collaborative Learning Circle
P.O. Box 1139
Ashland, OR 97520
(541) 482-4111 Fax (541) 482-8581
Email: hart@mind.net

Dave Kittredge
Department of Forestry &
Wildlife Management
Holdsworth Hall
University of Massachusetts
Amherst, MA 01003
(413) 545-2943 Fax (413) 545-4358
Email: dbk@forwild.umass.edu

Steve Lewandowski
Canandaigua Lake Watershed Task Force
8 Weld Street
Naples, NY 14512
(716) 396-1450 (work)
Fax (716) 394-0377

Mary-Lou Soscia
Columbia River Coordinator
US EPA–Region 10
Office of Ecosystems and Communities
811 SW Sixth Avenue
Portland, OR 97204
(503) 326-5873 Fax (503) 326-3399
Email: soscia.marylou@epamail.epa.gov

Steering Committee Liaisons:
Greg Aplet
Peggy Harwood

Preface

This publication is the product of a collaborative workshop on community-based ecosystem management that was convened in Bend, Oregon, on June 23-28, 1998. It is difficult to fully appreciate this volume without knowing how the workshop was developed and the papers prepared, and the energy of the event itself. The publication emerged from an intense workgroup process in which a diverse mix of people with experience and expertise in community-based ecosystem management prepared collaborative papers on key topics.

The idea for the workshop was born in 1996 as we were developing our program on community-based ecosystem management at American Forests. We had begun working with many community-based partners through the Communities Committee of the Seventh American Forest Congress and the National Network of Forest Practitioners. We were also developing our role as a "bridge group," trying to convey information and build dialogue among local, regional, and national interests. As we talked with local and national partners, it became clear that one of their highest priorities was a means of clearly describing "community-based approaches to ecosystem management" and explaining how such approaches served the public interest. We designed the workshop to meet these specific needs.

As a model, we used workshops American Forests has organized previously to develop information on emerging policy concerns. For this one, we modified the process to address concerns such as inclusiveness and transparency, critical issues to community-based approaches. Working with a few partners who became members of the workshop's Executive Committee, we developed a Steering Committee of 17 people from around the country, including community-based practitioners, scientists, agency officials, and interest group representatives. Our first Steering Committee meeting, convened on January 16-17, 1998, sought to identity a set of topics to explore at

[Haworth co-indexing entry note]: "Preface." Gray, Gerald J., and Maia J. Enzer. Co-published simultaneously in *Journal of Sustainable Forestry* (Food Products Press, an imprint of The Haworth Press, Inc.) Vol. 12, No. 3/4, 2001, pp. xxv-xxvii; and: *Understanding Community-Based Forest Ecosystem Management* (ed: Gerald J. Gray, Maia J. Enzer, and Jonathan Kusel) Food Products Press, an imprint of The Haworth Press, Inc., 2001, pp. xxi-xxiii. Single or multiple copies of this article are available for a fee from The Haworth Document Delivery Service [1-800-342-9678, 9:00 a.m. - 5:00 p.m. (EST). E-mail address: getinfo@haworthpressinc.com].

the workshop and a set of participants to bring the right mix of expertise, experience, and perspective to work-group efforts on each topic. The meeting involved some challenging and fast-paced facilitation as we attempted to synthesize six topics from a wide range of possibilities and to identity, from a long list of possible names, potential chairs and participants for each work group. The great results we achieved throughout the process are due to the experience of the people on the Steering Committee and their commitment to the collaborative spirit of the workshop.

The six work-group chairs were integrated into the final development stages of the workshop during the Steering Committee's second meeting, held March 27-28, 1998. Since the chairs would each play an instrumental role in the workshop, Steering Committee members worked with them to refine their topics, finalize a list of participants who would be invited onto the work groups, and help them fully understand the process challenges they would face at the workshop. The chairs' goal was quite daunting: use a collaborative process to prepare on-site in three days a draft 30-page paper on a broad, complex topic, integrating the diverse values and knowledge systems of the work-group's members.

The six work-group topics were intended to provide a conceptual framework to openly explore community-based ecosystem management in the United States. The Steering Committee saw the framework as helping build a common understanding of how community-based approaches can meet public policy goals and achieve on-the-ground success in restoring and maintaining healthy ecosystems and communities. Each work-group addressed one of the topics–Process, Stewardship, Investment, Monitoring, Institutions and Governance, and Global linkages. Because there was substantial overlap among the topics, a Steering Committee liaison joined each work group to help share information and assure consistency among the groups efforts. The Steering Committee encouraged the work groups to consider how their topics affect both public and private lands and to draw from real examples and case studies in diverse settings and geographic areas. Work-group papers were to be prepared as "disciplined inquiries" into the topics, raising questions and exploring different perspectives. They were intended to stimulate open, constructive dialogue, rather than provide definitive information or final answers. Steering Committee members modeled the workshop process after key principles of community-based approaches: inclusiveness, openness to learning, willingness to adapt, and creative persistence in working through difficult and unanticipated issues.

The work groups were also asked to address several "nested" issues in their papers, issues seen as parts of a conceptual thread that would tie the work-group papers together. Groups were asked to consider: *scale*–the effects of geographic size or scope; *urban and rural*–relevance to both urban and rural communities; *community capacity*–the relative ability of communities to respond to resource management issues; and *diversity*–the interests and involvement of nontraditional and under-represented groups, based on ethnicity, culture, gender and age.

The work-group papers represent the heart of the workshop, but the Steering Committee invited workshop participants to also prepare individual papers on closely related topics in which they had specific expertise. Most of these were written prior to the workshop and peer-reviewed during the event. Several individuals prepared papers immediately after the workshop to round out the publication and fill in gaps in information and perspective. Finally, the workshop's Executive Committee prepared an introductory paper, and the book editors took on the daunting yet exhilarating challenge of preparing a synthesis of all the papers.

The workshop was an extraordinary event that gathered a diverse group of people to share their knowledge, experience, and passion for community-based ecosystem management. The location allowed participants to enjoy the beautiful surroundings while they engaged in the intense workshop process. The central dynamic was the work-groups' ability to build trust and open rapport among their members, a critical factor as they grappled with complex and controversial issues in the course of preparing a paper that captured each of their voices. The emotions over these five days ranged from joyful moments of playing or relaxing together to moments of frustration, even anger, over difficult issues. It was not unlike a final exams week in which groups had to work together under extreme pressure to finish their papers by a deadline. As that final deadline approached, the groups delivered their papers one by one, seeming exhausted but satisfied–even exuberant–for having met their challenge.

It is impossible to convey the energy and emotion of the workshop, or the participants' enthusiasm for sharing ideas and experiences about community-based ecosystem management. For us, it was a special event–the successful culmination of a unique process designed and implemented with a wonderful group of people. We hope this publication effectively captures some of that excitement, stimulates interest and discussion, and encourages more innovative efforts to implement community-based approaches to ecosystem management.

Gerald J. Gray
Maia J. Enzer

Acknowledgments

This publication is a reflection of many collaborative efforts and could never have been written without the support and dedication of our partners. The idea for the workshop was conceived in 1996 and we worked for more than a year to bring the funding and partners together to make it happen. We are indebted to the people who helped us to create and implement this project.

First, we want to thank Dr. Graeme P. Berlyn, Professor of Plant Physiology at the Yale School of Forestry and Environmental Studies and Editor of the *Journal of Sustainable Forestry*. This is the third time Dr. Berlyn has allowed American Forests to publish workshop proceedings in the *Journal of Sustainable Forestry*. We believe the *Journal* is the ideal forum for this topic, as sustainability is at the core of what communities across America are trying to achieve. Without Dr. Berlyn's support and generosity, this project would not have been possible. We will also mention that all three editors of this volume are graduates of the Yale School of Forestry and Environmental Studies. While we graduated at different times, each of us believes the experience and training there enabled us to explore the community and forest connections reflected in this publication.

The Communities Committee of the Seventh American Forest Congress has been an essential partner for American Forests in the development of our Community-Based Ecosystem Management program. We have learned a great deal and benefitted enormously through our relationship with the Communities Committee. The Committee also provided seed funding for this project, which helped open doors to many donors. We want to thank all of the Communities Committee's Steering Committee members for their ongoing support and leadership, and for their early commitment to help us develop this workshop. We are also grateful that a number of the Steering Committee members agreed to participate in the workshop.

Certain individuals have helped to guide and nurture our Community-Based Ecosystem Management program from its very beginnings and have been instrumental in getting this workshop off the ground. Lynn Jungwirth has been a source of inspiration, intellectual stimulation, and constancy. Lynn lives community-based forestry every day and helps keep our program grounded in reality. It is nearly impossible to articulate our gratitude for her leadership and knowledge. Jonathan Kusel has been a friend and trusted partner with whom we have shared much in our efforts to understand and explain community-based forestry. Jonathan's experience with community-based groups in the West and his skills as a social scientist made him the obvious choice as science editor for this volume. We are extremely grateful for his many contributions to this project. Fran Korten, formerly of the Ford Foundation, has fostered a vision for community-based forestry in the U.S., helping many local and national groups develop projects and building a network of community-based forestry organizations. She challenged us to shape our Community-Based Ecosystem Management program and this workshop in ways that would advance the efforts of local groups and define a unique niche in the national policy arena. She provided an essential base of financial support to help us carry out this vision for our program and develop this workshop.

Many other colleagues and friends have contributed to the success of our Community-Based Ecosystem Management program over the last five years. We couldn't possibly acknowledge them all, but we want to take this opportunity to recognize some of the people to whom we have looked for guidance and support: Greg Aplet, Ann Moote, Hanna Cortner, Carol Daly, Thomas Brendler, Mary Mitsos, Michael Georgen, Bill Imbergamo, Fay Knox, Steve Marshall, Jerilyn Levi, Leah Wills, Lance Clark, Peter Titcomb, Kick Weyerhaeuser, Martin Goebel, Dan'l Markham, Steve Daniels, Kevin Smith, Dwight Fielder, and Jerry Filbin. Most of these people also played important roles in helping us bring this workshop to fruition.

This workshop would never have been possible without direct project support from donors, as well as general support for our Community-Based Ecosystem Management program. We are indebted to the following donors for their vision, trust, and patience in providing support: American Forest and Paper Association, American Pulpwood Association, California Department of Forestry, the Communities Committee of the Seventh American Forest Congress, the Conservation and Development Forum, Michigan Department of Natural Resources, Minnesota Department of Natural Resources, Oregon Department of Forestry, the Surdna Foundation, the Ford Foundation, the Sequoia Foundation, the Weyerhaeuser Company Foundation, the Weyerhaeuser Family Foundation, the U.S. Environmental Protection Agency, the Bureau of Land Management and the U.S. Forest Service.

Several interns and project assistants dedicated long hours to this project during its planning and implementation. We are grateful for their commitment and assistance with countless logistical details, manuscript revisions, grant proposals and reports, and communications with a myriad of authors. However, we are most thankful for their passion, fresh perspective, and good humor. We want to formally recognize each of them for their contributions to this book; we couldn't have completed it without them. Thank you Mary Tess

O'Sullivan, Richard Corral, Jessica Taylor, Perrine Punwani, Maggie Parks, and Michelle Weinberg.

This workshop was guided by a small group of dedicated individuals, an Executive Committee, who joined us in committing long hours of planning and preparation to the workshop, as well as working tirelessly during the workshop itself. We can neither give them adequate recognition for their contributions, nor thank them enough. We extend our heartfelt appreciation to Lynn Jungwirth, Jonathan Kusel, and Larry Fisher. In addition, Tahnee Robertson-Fisher, Elisa Adler, and Matthew Buck contributed their time, energy, and talent to help make planning meetings and the workshop itself a series of successful events.

The Executive Committee was part of a 17-member Steering Committee that participated in countless hours of discussion at planning meetings and served as work-group liaisons to ensure a coordinated process during the workshop. Most Steering Committee members became intimately involved in the substance of work-group discussions and ended up as co-authors of group papers. The topics and issues explored in this publication are a reflection of the intellect and passion for forests and communities that Steering Committee members brought to the workshop. The publication cannot convey the high level of emotion and energy expressed during many of the workshop discussions. The Steering Committee was one of the finest groups of people with whom we have ever worked. We salute each of the members for their contributions.

One of the most important Steering Committee tasks was to select individuals who could serve as chairs for each work group. These six chairs each took on the daunting task of facilitating a four-to-seven member work group, ensuring that the Steering Committee's guidance on process and the topic outline were followed, and that the ideas, writing styles, and priorities of each work-group member were included in their paper. The workshop experience was incredibly intense for most chairs, who held primary responsibility for preparing a 30-page draft work-group paper on site in less than a week. Beyond the workshop, the chairs spent weeks on revisions and conference calls with work-group members, ensuring that final papers were truly a product of the group. The work-group chairs are the real stars of this book. Our immense gratitude goes out to them for their commitment and extraordinary effort. They are: Ann Moote (Process); Hanna Cortner (Governance & Institutions); Brad Ack (Stewardship); Nels Johnson (Global Linkages); John Bliss (Monitoring); and Valerie Luzadis (Investment).

The workshop and publication could not have been a success without a diverse mix of participants who actively engaged in the intense process and openly shared their ideas and feelings about communities and forests. All workshop participants were selected by the Steering Committee for their knowledge and experience on issues related to community-based ecosystem management. Participants brought a wide array of perspectives to the workshop as practitioners, researchers, agency officials, and interest group representatives. Although process challenges and differing views proved difficult in a number of the work groups, the quality of the individual participants and their commitment to a collaborative process enabled each work group to complete its task. We are extremely thankful for the effort of each and every workshop participant.

Back at American Forests we want to recognize the support and expertise of Michelle Robbins, who copy-edited all the papers in this volume. Thank you, Michelle, for your diligence and patience. As with most large writing projects, this one didn't always hold to our schedule, but Michelle always greeted us with a smile and rapidly responded to our requests.

Maia J. Enzer
Gerald J. Gray

PART ONE: EDITORIAL SYNTHESIS

Understanding Community-Based Forest Ecosystem Management: An Editorial Synthesis

Gerald J. Gray
Maia J. Enzer
Jonathan Kusel

SUMMARY. This paper synthesizes key issues identified and explored by a mix of practitioners, researchers, and resource managers who participated in a workshop designed to understand community-based ecosystem management. The interdependence between healthy ecosystems and community well-being lies at the heart of community-based ecosystem management. Core ideas focus on linking stewardship to the

---

Gerald J. Gray is Vice President, Policy, and Maia J. Enzer is Director of Forest Policy, American Forests, 910 17th Street NW, Suite 600, Washington, DC 20006.

Jonathan Kusel is Executive Director, Forest Community Research, P.O. Box 11, Taylorsville, CA 95983.

[Haworth co-indexing entry note]: "Understanding Community-Based Forest Ecosystem Management: An Editorial Synthesis." Gray, Gerald J., Maia J. Enzer, and Jonathan Kusel. Co-published simultaneously in *Journal of Sustainable Forestry* (Food Products Press, an imprint of The Haworth Press, Inc.) Vol. 12, No. 3/4, 2001, pp. 1-23; and: *Understanding Community-Based Forest Ecosystem Management* (ed: Gerald J. Gray, Maia J. Enzer, and Jonathan Kusel) Food Products Press, an imprint of The Haworth Press, Inc., 2001, pp. 1-23. Single or multiple copies of this article are available for a fee from The Haworth Document Delivery Service [1-800-342-9678, 9:00 a.m. - 5:00 p.m. (EST). E-mail address: getinfo@haworthpressinc.com].

unique role that local communities and workers can play in its implementation; developing social and institutional processes that are more open, democratic and civil; monitoring ecosystem, social, and economic subsystems and modifying actions based on what is learned; empowering communities to strengthen their participation in natural resource management and address power imbalances; and developing a new understanding and valuation of the broad array of ecosystem products and services. The paper concludes with a discussion of ways community-based ecosystem management can be advanced, including: emphasizing the essential role of communities and workers in sustainable resource management; maintaining and building community capacity; improving understanding of worker issues and promoting an ecosystem workforce; establishing local and institutional processes that are inclusive, accessible and transparent; creating a civic science that respects and engages diverse people and knowledge systems; establishing new public and private approaches to long-term investment in ecosystem restoration and maintenance and in community-based stewardship; and advancing laws and policies that promote collaboration and coordination needed for long-term stewardship.

KEYWORDS. Community-based, forest(ry), ecosystem management, stewardship, collaboration

The interdependence between healthy ecosystems and community well-being lies at the heart of community-based ecosystem management. Community-based practitioners want to heighten awareness of this interdependence and demonstrate how communities and ecosystems can be mutually sustained through its principles and activities. This paper builds on the efforts of a diverse mix of practitioners, researchers, and resource managers who participated in a national workshop designed to understand community-based ecosystem management in the United States. It does not reflect a consensus view, but rather a synthesis of key issues identified and explored by the participants. Our attempt to frame these issues focuses on two questions: What is community-based ecosystem management? and, What pathways will encourage its use and development?

WHAT IS COMMUNITY-BASED ECOSYSTEM MANAGEMENT?

Community-based ecosystem management begins with the conviction that human communities and natural ecosystems are interdependent and builds on two premises of sustainable development theory (PCSD Sustainable Communities Task Force Report 1997). First, communities are key to sustainability because they represent a scale and set of interactions with natural ecosystems that people can comprehend and affect. Second, no community can be sustainable in and of itself; it must work with other communities on ecological, social, and economic issues that transcend socially constructed and administrative boundaries. The second premise raises challenges and opportunities for community-based practitioners: How can communities of place–rural and urban–both work together and work in ways that involve communities of interest?

Community-based ecosystem management starts with the principle of stewardship, a commitment to maintaining and restoring the health of the land. Ecological health is not viewed in a vacuum; it is understood to be linked to community health, economically, socially, and culturally. Although economic return from resource harvest and extraction informs decisionmaking processes, it does not trump ecological health concerns in community-based ecosystem management. Two major investment challenges facing community-based ecosystem management are: increasing funding for ecosystem restoration and maintenance, and developing investment mechanisms that link ecosystem restoration and maintenance activities to the economic vitality of communities.

Community-based ecosystem management requires a shift toward more open, democratic processes and community empowerment. Local communities–including the broad array of public institutions, private businesses, interest groups, landowners, and local residents–need to participate more meaningfully in the planning and decisionmaking processes that affect them and the ecosystems in which they reside. A shift toward more meaningful processes will require restructuring relationships between public and private institutions and among individuals through civil processes that recognize and respect the legitimacy of diverse perspectives, cultures, and knowledge systems.

Community-based ecosystem management demands a firm commitment to monitoring and learning. Monitoring assesses the social, economic, and ecological conditions that inform future planning and management. Monitoring that is responsive to stakeholder concerns–involving diverse parties that are both local and distant–builds accountability and trust and is a precursor to learning. However, to realize the potential that exists for long-term effectiveness and trust-building, it is essential that future actions are based on what is learned through monitoring.

Finally, community-based ecosystem management has historical roots around the world as well as here in the U.S. Many other countries, particularly developing countries, have encountered similar issues, generally earlier than in the U.S., and have responded with community-based efforts. We can learn from the rich experience that exists in other countries as well as share our skills and knowledge with communities globally seeking ways to restore and maintain healthy ecosystems and communities.

Stewardship

Stewardship lies at the heart of community-based ecosystem management. It is the concept of stewardship that draws many people into community-based efforts, guides their actions, and provides a focus for holding groups together. With a long-term emphasis, stewardship is designed to protect, restore, conserve, and prudently use natural ecosystems and all they sustain (Ack et al. 2001). As such, it reflects a unique respect for the land and the human and non-human species it supports.

Community-based ecosystem management involves a fundamental shift in resource management away from output-based or product-based approaches to one focused on the condition of the land. Though resource products or outputs are important, the health and sustainability of the land are emphasized. System outputs, whether ecological services or products, are by-products of stewardship, must be viewed as part of the whole system, and can be used or produced only within the limits of the land and its dependent species.

This emphasis on respect for the land and non-human species is paralleled by a respect for people, their communities, and their relationship to the land. Good land stewardship involves civic processes that embrace diverse perspectives, including but not limited to those held by ethnic and cultural groups, nonresident workers who have a direct relationship with the land, and communities of interests. This allows the land to be managed in ways that meet the needs of the traditionally disenfranchised, address the interests of those that hold power, and, at the same time, maintain ecosystem health.

Where diverse values and interests are involved, there are bound to be differences of opinion. The success of community-based processes, therefore, requires an emphasis on individual respect and civility, which allows difficult issues to be resolved in ways that strengthen the fabric of a community. A basic premise of land stewardship is that it is best done at the community level. At this scale people can know and experience the ecosystem, have a common interest in the community's sustainability and overall well-being, and begin to understand the real connections between ecological, economic, and social considerations, while also being responsive to critical relationships and interdependencies beyond the community level.

Because it is so new, community-based stewardship is often implemented through pilot projects that test new technologies and management practices at smaller scales. Small-scale pilot projects require less investment, are less risky, and, generally, are easier to monitor. The technologies being tested in pilots are often lighter on the land than current industrial-scale technologies. They also tend to be labor-intensive, valuing the role of on-the ground workers, both respecting and responding to the idea that workers are a critical link between communities and ecosystems. As pilot projects are completed and lessons learned, community-based stewardship efforts will likely move to larger scales and longer time frames, requiring greater commitment and investment by participants. The question remains: Can new stewardship practices and technologies attract public and private investment or begin to generate sustainable flows of revenue through market niches?

Monitoring and Learning

Well-designed community-based stewardship projects require a commitment to track or monitor work so successes and failures can be identified, and actions modified–or halted–if necessary. Diverse interests are more likely to support each other when they have common objectives and safeguards to protect their interests. Monitoring is the first step, but it is incomplete by itself. It must be accompanied by reporting mechanisms and the establishment of processes that ensure prompt corrective actions when necessary. These steps are part of adaptive management, which views every management action as an experiment and acknowledges the uncertainty associated with each action.

For many years monitoring has been seen as an important activity in public policy and management models but done in a limited fashion, if at all. Looking forward and putting new projects on the ground has always been more captivating than looking back and monitoring what has been done. Yet, monitoring is the linchpin in efforts to understand and learn from our actions, as well as to begin to build accountability for them. Building accountability requires, first, engaging diverse local and distant stakeholders in monitoring processes and practices, and, second, developing ways that monitoring and learning inform and even obligate subsequent actions. The emergence of community-based ecosystem management, both globally and in the U.S., has brought a renewed emphasis on monitoring and spurred some innovative approaches.

Multi-party monitoring, one approach identified and modeled by community-based groups, draws together diverse stakeholders to assess a project or system. It follows the principle of inclusiveness, a central tenet of community-based ecosystem management, by bringing together parties that care about outcomes. It involves people who live and work in the area of concern and may include members of local businesses, churches and schools, and labor and social service organizations, as well as traditional stakeholders in forest management issues, such as landowners and representatives of public agencies, forest industry, and environmental groups. Community-based forestry practitioners recognize that community members are often uniquely qualified to participate in monitoring because of their long-term knowledge of an area and their community, as well as their individual experiences with activities such as tree planting, logging, nature viewing, or harvesting nontimber forest products. Recognizing the interdependence between healthy ecosystems and community well-being and seeing firsthand the effects of resource harvesting and extraction, practitioners insist that social as well as ecological conditions be monitored (Bliss et al. 2001).

Multi-party monitoring seeks to ensure broad and diverse participation, especially among participants who have particular project or system concerns. Parties do not have to reside locally or be directly involved in monitoring to play a role. They may make recommendations and track measures from afar. Multi-party monitoring responds to the need for both national and local interests to be involved, particularly in efforts relating to public land management. Participants in a number of community-based initiatives are trying to address legitimate concerns about "local control" of information and decisionmaking processes through innovative efforts to involve distant stakeholders.

How to assess or ensure the legitimacy of community-based efforts is a challenge to community-based ecosystem management overall. It requires involving groups from both communities of place and interest and developing an appropriate system of checks and balances for participants wary of sharing power. The national environmental community has expressed concerns about–and on a number of occasions opposed–community-based projects. How can those constituents be confident that what happens in diverse community-based efforts is really about ecosystem stewardship?

Many in the environmental community are looking for "standards" or guidelines for evaluating community-based efforts. Community-based practitioners differ on whether developing such standards would prove appropriate or useful. Some think that standards, such as those developed by the Forest Stewardship Council for certifying sustainable forest management, are necessary (Forest Stewardship Council 1999), particularly for any activities that affect the land. But the question remains: How can a single set of rules or standards address the complexity of issues across a diverse social and ecological landscape? Community-based practitioners have long opposed blanket statements or "one-size-fits-all" solutions. Others suggest that a more appropriate, workable way to facilitate community-based ecosystem management would be through flexible policy mandates (nonprescriptive and non-site-specific)–so long as they included strong monitoring components to ensure accountability. Multi-party monitoring approaches are generally seen as the most promising because they engage diverse stakeholders, both local and national, in efforts to assess and learn from stewardship projects.

Community-based ecosystem management encourages adaptive innovation and respects both practical and scientific knowledge. It seeks to be inclusive and to integrate diverse knowledge, perspective, and ways of understanding with formal knowledge systems. There are many ways to know and understand a landscape beyond the more conceptually based and formalized knowledge generated by agencies and academia. Community-based practitioners may possess knowledge based on historical site-specific experience and observation, remembrances and understandings shared by others, cultural and historic traditions, practical on-the-land experience and learning, and spiritual/aesthetic experiences and relationships, as well as formal classroom instruction. All these sources of knowledge have a place in the practice of community-based stewardship. Multi-party monitoring can offer a forum for these different ways of knowing to mutually inform and enhance each other, but only if an explicit effort is made to do so.

A number of technical and political challenges face emerging community-based monitoring efforts: funding, tension over the level of scientific rigor required to achieve objectives, differing perceptions and expectations, inclusiveness in multi-party monitoring, and integrating social and ecological factors into the monitoring process.

Power and Relationships

The need to empower communities of place to become meaningful participants in the natural resource policy and management decisions that affect them is a basic premise of community-based ecosystem management. Its emergence has been stimulated in large part by the policy gridlock resulting from intense interest group battles at the national level. In other words, "power" in natural resource policy has become concentrated at the national level, among institutions operating in adversarial fashion through the legislatures and the courts (Moote et al. 2001 and Johnson et al. 2001).

Community-based ecosystem management raises complex and controversial issues about the redistribution and sharing of power. While empowering communities raises expectations and offers opportunities to build local capacity, it also affects the status quo and may prove threatening to organizations or interests currently holding power. Community-based ecosystem management represents an experiment in participatory democracy in which it is better to put thorny issues of power and control of resources squarely on the table and work through them, rather than to avoid them because they are too hard, complex, or threatening (Cortner et al. 2001).

Communities of place–the public and private institutions, the citizens and private landowners, and the ecosystem workers–are in a unique position to develop a long-term stewardship relationship with natural ecosystems. Increased power-sharing does not mean these local communities will control natural resource decisions, but rather that they will be able to strengthen their participation in the management and decisionmaking process. The challenges of community-based ecosystem management transcend local communities, requiring active involvement by organizations and interest groups at the regional and national levels, as well. Innovative mechanisms that allow regional and national organizations to participate are needed to address larger-scale ecological, economic, and social issues, and to recognize the legitimate interests of all society in public lands issues. In some cases, open communication about local processes may be sufficient; in others, more direct involvement may be needed to help build community capacity, consider appropriate technology, assess and build accountability, or ensure learning.

Global considerations present further challenges and opportunities. No matter how large, small, or remote, communities across the U.S. are increasingly affected by global trends–environmental, economic, social, and political. Through networks and coalitions, communities in the U.S. and around the world may share information and work together to address complex issues of local and global relevance (Johnson et al. 2001).

Meeting key objectives of community-based ecosystem management–such as strengthening local community participation, ensuring inclusive processes, and promoting collaboration–requires us to restructure relationships and create new interdependencies among public and private organizations. Our country's unique form of government, which combines a federal republic and representative democracy, exhibits contrasting forces of centralization and decentralization. While community-based ecosystem management reflects the Jeffersonian spirit of direct democracy, it cannot be separated from the political checks and balances built into our federal system (Cortner et al. 2001). Our system ensures that neither centralization nor decentralization, national nor community interest, governmental nor individual interest will consistently prevail. Therefore, as local communities are empowered through community-based ecosystem management, local decisionmaking processes will need to either develop response mechanisms or include a system of checks and balances that will provide transparent information and opportunity to review and question actions.

Many community-based ecosystem management efforts involve community sectors and perspectives traditionally excluded from the decisionmaking process. This suggests why some of the governance and process challenges are so significant, and why these efforts appear challenging to institutions and interests empowered in the status quo (Moote et al. 2001). It also suggests why many of these efforts have been so inventive and adaptable–to overcome the many barriers and challenges they encounter.

Community-based ecosystem management recognizes ecosystem workers' unique role as the people who implement stewardship activities and have a direct relationship with the land. Whether they live in close-by communities or elsewhere, workers perform the on-the-ground stewardship practices–or land-service activities such as reforestation, brush disposal, and thinning–planned and funded by public and private land managers. Ultimately, they have a significant role in determining whether stewardship practices succeed or fail. Despite this important role, many of these workers in the U.S. today face difficult conditions: minimum wage, seasonal jobs with few if any benefits, and high rates of poverty and unemployment. These conditions are neither conducive to effective, long-term stewardship nor healthy, productive lives for workers, their families, and their communities. Community-based ecosystem management envisions the creation of an ecosystem workforce characterized by trained workers with the skills and incentives to effectively implement stewardship practices, workers who are respected, justly treated, encouraged to inform land management decisions, paid a living wage, and able to contribute to the well-being of their communities.

A key tenet of community-based ecosystem management is that the types of changes and new relationships contemplated cannot be achieved without an emphasis on civility. To succeed, we must acknowledge the importance of each person and each person's capacity to learn and contribute. Citizens must teach and model the concepts of respect and dignity, trust, goodwill, and exploration in both their individual and collective actions. A respect for diversity within the human community is essential; it reflects our capacity to respect other species (Moote et al. 2001).

Investment

Economic activities related to natural ecosystems and rural communities have long been characterized by patterns of "disinvestment"–inadequate funding for resource use or development which neither covers the full cost nor sustains the long-term capacity of the resource to continue providing the same level of value. This lack of sufficient investment has diminished the capacity of ecosystems to sustain ecological services and the capacity of communities to care for themselves and steward the land. Driven by market values and short-term financial returns, public and private ecosystem investments have focused on resource outputs, often in high volumes requiring capital-intensive, large-scale operations. In many rural communities, these short-term investments have led to economic boom-bust cycles and environmental degradation. Community-based ecosystem management encourages new thinking about investment to restore and maintain the capacity of ecosystems and communities and to achieve the overarching goal of sustainability.

The understanding that we must take care of the land if it is to take care of us and continue providing ecological services and products is a basic community-based ecosystem management principle, but it also represents enlightened self-interest (Ack et al. 2001). Communities should think about stewarding the land the same way a trustee thinks about the management of a financial trust. The land, or an ecosystem, is an asset comprised of natural capital that can sustain its beneficiaries in perpetuity. Beneficiaries in the case of an ecosystem trust include not only people and communities, but also the diverse species dependent on the ecosystem. Since the first duty of a financial trustee is to protect principal so that it can generate flows of income for beneficiaries, an ecosystem trustee will manage the land to protect natural capital and assure a sustainable flow of ecological services and products for all beneficiaries. Participants in community-based ecosystem management processes, whether they are local residents, companies with economic interests, or groups with environmental concerns, need to think of themselves as ecosystem trustees. A major challenge is finding ways to increase investment in activities that restore and maintain healthy ecosystems. Although public concerns about the degradation of natural ecosystems and the loss of benefits such as water quality and species habitat have spurred significant interest in restoration, they have yet to result in an increased investment in long-term ecosystem health. A short-term focus by the general public, policymakers, and corporate investors has resulted in ecosystem investments that are insufficient to restore or maintain the health and productive capacity of the ecosystem.

A lack of understanding about ecosystems–the economic values they provide and the long-term consequences of current uses and activities–is at least partially to blame for insufficient investment (Luzadis et al. 2001). To improve ecosystem investment, we need better information about (Wills and Gray 2001):

1. The economic values of the ecological services and products that ecosystems provide to society;
2. Who benefits from the ecological services and products, to what degree, and their ability and willingness to pay their fair share for ecosystem restoration and maintenance activities;
3. Institutional barriers that constrain efforts to have those who benefit pay their fair share for needed ecosystem investments;
4. Priority activities and long-term maintenance regimes that will help restore and maintain ecosystem capacity; and
5. The cost–or investment needed–to implement ecosystem restoration and maintenance activities.

A second investment challenge is the need to increase funding for activities that link ecosystem restoration and maintenance to the social and economic vitality of communities. From a long-term sustainability perspective, this issue may be more difficult and compelling than the first challenge: How do we build a lasting stewardship that sustains both ecosystems and human communities?

Community-based ecosystem management emphasizes the important role of workers as the people implementing stewardship activities. Their training, skills, and knowledge will contribute significantly to the process of adaptive management, of learning how to care for the land and strengthen the relationship between that land and the community. One investment strategy focuses on training, apprenticeship, and certification programs that will help create an ecosystem workforce.

Another important economic link between ecosystems and communities emphasizes not simply output or quantity, but the quality and value of ecological services and products–particularly those that have been undervalued and underutilized. Community groups are using innovative, often small-scale technologies to capture added product values and the benefits of this economic activity, rather than exporting them elsewhere. Many rural communities have suffered as industry practices extracted natural resources in high volume, processed them to a minimal extent, and shipped the raw materials elsewhere, often closer to urban-based markets for value-added processing. These practices take investment and jobs away from rural areas. Much of the value of the ecological services and products from an ecosystem are neither captured by, nor returned to, the rural communities within the ecosystem, hence new approaches are needed. New ways must be developed to account for and capture the market value of ecological services, such as clean water and air, and the values associated with the quality of small-diameter trees, underappreciated tree species, and a variety of mushrooms, herbs, medicinals, decorative greens, and other special forest products. In many cases, these new approaches will be accompanied by new technologies, often small-scale and labor-intensive, in order to increase product values and market those products from rural communities.

PATHWAYS TOWARD COMMUNITY-BASED ECOSYSTEM MANAGEMENT

To advance community-based ecosystem management in the U.S., change is needed in laws and policies and in the behavior of public and private organizations. We need a new emphasis on communities–on their critical role in developing approaches to sustainability, their participation in planning and decisionmaking, and their capacity to implement stewardship projects while ensuring that checks-and-balances safeguard regional and national interests. Public and private organizations need to focus on policy tools and resources to help build communities' capacity to implement projects and sustain long-term stewardship relationships with the land. Federal, state, and local agencies, as well as private interest groups and organizations, need to develop new mechanisms for collaboration and coordination, engaging in efforts to learn together and share power with each other and with communities. Finally, the public and private sectors need to test and establish new investment mechanisms for restoring and maintaining the health of ecosystems and communities.

Focusing on Process

Process considerations are central to the working of community-based ecosystem management. They encompass principles that must be kept in mind to guide these management efforts, principles based on governance through open, inclusive democratic process. Community-based ecosystem management efforts may address these considerations through different mechanisms; there is no one right, or proven, approach. The principles represent ideals to which practitioners aspire. They acknowledge the complexity of trying to achieve them, but they also appreciate the importance of having the principles as guides.

- *Inclusiveness* suggests the need to proactively engage all interested or affected community groups in the process, regardless of their relative size or influence, and to develop creative opportunities that encourage their voice and participation. It is crucial that community members from diverse backgrounds actively participate from the beginning in the planning, design, decisionmaking, and implementation of any community-based ecosystem management action or program. It is also important to seek ways to engage national and other interests beyond community boundaries. Good process accommodates not only multiple and diverse sources of information, but also multiple kinds of knowledge, whether it is local, cultural, and experiential knowledge or more formal science-based knowledge.
- *Accessibility* involves tailoring the process to match diverse needs and communication styles. Considerations include the cultural manner in which a group is accustomed to meeting and making decisions, the language (professional/scientific or not, English or not, advanced literacy or not), and the history of the group's experience of inclusion or exclusion. Community-based ecosystem management means using innovative and situation-specific participation methods to communicate that each group has a legitimate voice in the process.
- *Transparency* reflects the degree of openness and clarity with which information is provided about and throughout the process. The process itself needs to be clearly and fully understood by all participants–who the participants are, what the schedule for decision making is, who has

decision making authority, and how input from participants will be used. Transparency in process can restore trust between communities and institutions that generate and maintain information and make decisions, both public and private. It means openly sharing information from agency officials and technical experts, making it understandable or accessible to community groups, and explaining how informal information and input from communities will be considered with other information.

- *Mutual Learning* implies that participants in community-based ecosystem management are open and committed to a process of learning with and from one another through ongoing efforts. Theory and practice suggest that such learning best takes place through joint action, such as field work (e.g., monitoring, training, planting trees) and interactive tasks (e.g., problem solving, planning, and analysis). A role change is required for professionals and scientists in these processes, shifting from technical experts and primary presenters of information to participants who share information, facilitate learning processes, serve as resource people, and are learners open to other forms of information (Cortner et al. 2001 and Moote et al. 2001).
- *Adaptability* recognizes the importance of maintaining flexibility in guidelines or management prescriptions for community-based ecosystem management. The amount of uncertainty, risk, and change, and the sheer complexity associated with ecological, social, and economic issues that communities face require approaches that are flexible and able to respond to change and new information. Adaptive learning and management–an ongoing process of planning, implementing actions or plans, monitoring outcomes, and making adjustments based on what is learned–is an approach that allows community-based ecosystem management to move forward in the face of uncertainty.

Building Community Capacity

The concept of "community capacity" is a critical building block of community-based ecosystem management. It refers to a community's collective ability to address local social and economic needs and take on the challenges of stewardship (Kusel 2001). Just as a community with high capacity is more likely to be an effective land steward, good land stewardship can improve community capacity. The practice of stewardship requires many skills, including community organizing, project development and management, monitoring, science and research, and communication. Some communities lack skills and may have limited access to training, technical services, and funding opportunities, or they may not know how to take advantage of opportunities for assistance, if they do exist. Leadership is needed within community-based efforts to marshal the diverse resources needed. This leadership might come from individuals within a community who have seen the need for civil discourse and collaboration. It can come from government officials, private business people, nongovernmental groups, or civic activists who have come to understand the links between the well-being of their community and ecosystem health. Local leaders need to be visionary and creative, able to bridge between different interests and perspectives, and willing to reach out to other communities, organizations, and networks at different levels for assistance. This ability to reach beyond community boundaries to seek resources and assistance and to partner with external organizations is an important dimension of community capacity–and critical to the success of community-based ecosystem management.

Improved communication technologies now enable communities around the world to share information, build capacity, and develop networks and coalitions. Examples of innovative forms of local government are emerging globally that give communities more authority in natural resource management, and new markets are developing around the world for sustainably produced goods and services from small community-based enterprises (Johnson et al. 2001). While a number of communities in the U.S. are connected to these global initiatives, many more need to be aware of the potential contributions of lessons and tools developed internationally. They must consider opportunities to share information–and even work collaboratively–through the global network of communities involved in similar conservation and community development efforts.

Creating a Civic Science

A good example of how institutions can work with communities–showing respect for the informal knowledge of local people and helping to build community capacity–is the creation of a "civic science." Scientists and scientific institutions are all too often aloof, viewing their professional science as superior to informal knowledge systems, such as practical experience or cultural traditions (Cortner et al. 2001). The goal of a civic science, and of community-based ecosystem management, is for formal science to be better informed by local practitioner understanding and vice versa. This requires a research process that involves and respects local people and their diverse knowledge systems and a continual process of mutual informing and learning. Two steps that will help achieve this integration of formal and informal knowledge are:

- Scientists at universities and public agencies need to include community-based practitioners as partners in the study of the environmental, social, and economic issues related to communities and ecosystems, and to do this through a continuous, adaptive learning process. Community practitioners are not passive receptacles for information. They can help identify issues and questions to guide research, produce information used in research, and even directly participate in research endeavors.

- The results of scientific inquiry need to be made more widely available in understandable forms to enable and encourage the participation of community practitioners. The challenge for scientists is to learn how to better translate their work into meaningful, pertinent, and useful information for community practitioners, information that stimulates engagement and further learning.

The civic science needed to advance community-based ecosystem management suggests a new role for university extension programs. In this two-way role, Extension would disseminate information to communities in useful forms and serve as a conduit for bringing local knowledge and issues back to the university for consideration by scientists. Civic science also suggests a role for non-governmental groups in filling the gap between traditional research processes and emerging, community-based research.

Understanding Worker Issues and Promoting an Ecosystem Workforce

The participation of workers and the incorporation of traditional environmental knowledge in planning and decisionmaking are central to community-based ecosystem management. People who work in the woods, watersheds, and elsewhere on the land have an understanding and a perspective about a place that can play an essential role in the practice of stewardship. Workers' knowledge and skills, however, have been highly underutilized. In a somewhat similar vein, Native Americans hold traditional ecosystem knowledge that has been inadequately recognized, respected, and incorporated into resource management.

Worker issues–who they are, how they are treated, and the challenges they face–have only recently received attention. They remain poorly understood. Migration and cultural diversity among ecosystem workers are two examples that illustrate the complexity of these issues. Ethnic minorities make up a large percentage of a migrant forest and farm workforce in the Pacific West and the South. Because they do not live locally, migrant workers are often treated as outsiders, and cultural differences can hamper efforts to integrate them into community-based projects. In some locales, ethnic minorities are perceived as threatening to established ways of life. At the same time, these workers have brought significant cultural and practical knowledge about emerging forest activities rooted in developing countries in South America and Southeast Asia. Recognizing uses and markets for such things as salal, matsutaki mushrooms, huckleberry, mosses, and ferns, immigrant forest and farm workers in the Pacific Northwest have been instrumental in the development of a regional nontimber forest industry now valued at more than $200 million annually. Ethnic, cultural, and migrant-worker differences can slow the acceptance of contributions–such as new knowledge and perspectives and labor–that a diverse workforce can bring to communities.

Ecosystem workers represent an important link between ecosystem health and community well-being. A number of training programs have been created by federal and state governments in the Northwest to help displaced forest workers acquire the diverse array of skills necessary to become ecosystem workers. These programs, such as the federal Jobs in the Woods and Washington State's Jobs for the Environment, are beginning to build an ecosystem workforce capable of doing stewardship. While the supply of workers has been growing, a parallel increase in the demand for workers skilled in ecosystem restoration and maintenance has not yet developed. Although there is growing public concern about restoring threatened watersheds and fish and wildlife habitat, these concerns have not been tied to corresponding increases in funding and jobs that require the skills of trained ecosystem workers.

Encouraging Public and Private Investment

If community-based ecosystem management is to succeed in the long run, new forms of public and private investment need to be developed. Investment strategies and tools need to focus on maintaining ecosystems' long-term capacity to produce ecological services and products and on strengthening communities' capacity to develop and implement stewardship, rather than continuing to emphasize a short-term flow of market-based products from ecosystems. Community groups are experimenting with a number of innovative public and private investment tools to support stewardship projects and create new sources of revenue for maintaining ecosystem and community capacity over the long term. Some of these investment tools might succeed through marketplace demonstration, but making inroads into the marketplace presents significant challenges. Private sector initiatives need tremendous creativity and persistence to overcome financial, institutional, and policy barriers as they try to develop market niches for new technologies or products. This generally requires significant financial support from government, private foundations, or companies. Public sector initiatives must also overcome the political and institutional barriers that exist for longer-term investment proposals, since many may have less tangible objectives and limited prospects for immediate financial returns. Overall, there is a great need to direct more public attention to the lack of investment in ecosystems and communities, the values being lost or threatened by this lack of investment, and public and private investment tools to alleviate the situation.

The public and private investment tools discussed below offer one way to begin to not only recognize the economic value of ecological resources, but also to create financial resources from that value. Those financial resources can then be used as investment funds to help restore and maintain ecosystem capacity. The tools also identify means for public policies and private market-based mechanisms to direct investment toward community-based activities that can strengthen the link between ecosystem restoration and maintenance and community vitality. These investment tools can be grouped into three strategic areas: innovations in private-sector investment, cross-jurisdictional approaches, and government funding of ecosystem management.

Private sector investment tools–Innovative private-sector investment tools are based on market forces, such as anticipated consumer demand for environmentally certified products, entrepreneurs and financial institutions seeking to profit financially from investments in ecosystem management activities, and private nonprofit organizations seeking to demonstrate new methods of financing conservation and ecosystem management. Some private-sector investment tools include:

sustainability

- independent, third-party certification of sustainable forest management;
- leveraging compatible activities that occur in both sustainable forestry certification and carbon sequestration verification in order to reduce costs;
- banking the value of ecological resources so that a landowner has liquid assets while the resource is being managed on a sustainable basis;
- partnership financing by companies seeking to promote the use of new technologies and products, such as low-impact logging equipment; and
- combining new technologies and product marketing initiatives, such as log sort-and-sales yards, and new processing technologies that add value to small-diameter materials and produce high-quality, specialty forest products such as flooring and furniture.

Cross-jurisdictional investment approaches–A set of innovative cross-jurisdictional approaches has emerged that creates mechanisms for beneficiaries to pay for the ecosystem services they want and need. Political and social jurisdictional boundaries, created during the early years of settlement, often hinder ecosystem health and restoration efforts. Some new approaches seek to identify beneficiaries of ecosystem services and develop systems of exchange that increase ecosystem investment to the long-term advantage of both ecosystem managers and beneficiaries. Others use cooperation across boundaries to leverage resources and planning activities, benefitting both cooperating organizations and the larger-scale ecosystem. Examples of cross-jurisdictional approaches include:

- the use of ecological zoning to capture the financial value of open-space for investment in activities that sustain it;
- the development of watershed organizations that identify upstream and downstream interests and facilitate negotiations that lead to new mechanisms for investment in watershed restoration and maintenance;
- new patterns of cooperation across forest economic regions in which public and private landowners choose to make decisions on the basis of ecological and community processes rather than through traditional, independent approaches; and
- increased cooperation and planning among counties to address issues like development patterns and endangered species habitat on a regional basis.

Government funding–Innovations in government funding seek to change traditional approaches to ecosystem finance, which are generally inflexible, short-term measures. The annual appropriations process tends to emphasize short-term operations and focuses on maximizing revenues and reducing expenses. This approach is not conducive to the long-term maintenance or restoration of the land, or the capacity of ecosystems to continue providing ecological services. A congressional commitment to consistent funding levels for ecosystem management over a period of years would give public land managers a much greater ability to plan, implement, and monitor projects, such as multi-year ecosystem restoration projects.

Another policy tool that could help finance community-based ecosystem management would be congressional authorization of new contracting mechanisms that allow contractors to perform multiple stewardship activities on a site over a period of time. Currently, agencies are restricted from consolidating stewardship activities into one "bundled" contract if the activities involve a mix of revenue-generating and land-service activities. Ecosystem restoration projects require multiple tasks at each site; combining them into one contract could improve the agency's administrative efficiency while providing the workers flexibility to schedule and implement a variety of tasks over a longer period of time. Congressional authority and direction may also be needed to test other innovative contracting mechanisms that connect investments in ecosystem restoration and community well-being, such as (1) contracts that favor small, locally based businesses in performing ecosystem restoration work on nearby lands, and (2) contracts that favor highly skilled workers who have completed ecosystem restoration training programs.

State governments also have often-overlooked financing mechanisms for community-based ecosystem management. All states have departments of economic development that leverage surprising amounts of state and federal dollars for economic development projects. Leveraging these finance options using the state as both funding coordinator and one of the investors can entice private investment. An example might be combining investment in private or public land restoration and certification using an investment pool created by watershed beneficiaries including downstream water users.

Collaboration and Coordination

Currently, many laws and policies reflect an interest-based model of government in which a community or region is represented by a specific number of representatives from interest types, such as commodity groups, environmental groups, and local elected officials. Such a model usually does not adequately represent the complex array of interests that make up a community. The actions that result, often based on self-interest, competition, and adversarial relationships, do not effectively address the complexity associated with an ecosystem, nor the best long-term interest of the community or environment. We need laws and policies that promote an open, inclusive, community-based model of deliberation rather than an adversarial, interest-based model. While community-based practitioners understand that interest-based approaches are at times appropriate, advancing democratic principles of civil discourse, mutual respect, and social justice require a new model as well.

The issues of inclusiveness and social justice are among the most complex faced by community-based ecosystem management. Current legal, planning, and budgetary processes tend to exclude low-income and minority groups. For example, land-use support programs reveal a history of indifference toward or blatant discrimination against low-income landowners and minority groups, while similar disinterest or treatment of migrant workers and minorities has been evident in a variety of forest community settings (Ack et al. 2001, Cortner et al. 2001, and Moote et al. 2001). Community-based ecosystem management requires attention to these issues, getting them out in the open and giving voice to groups traditionally excluded. The issues remain equally hard, but a sound community-based process offers the potential for a greater degree of legitimacy and for solutions to be discovered through its energy and creativity.

Role of government–Federal and state governments hold many keys to success for collaborative, community-based efforts. They can help build community capacity by providing assistance to local governments and institutions through a variety of policy tools and resources. These tools, such as the USDA Forest Service's Rural Community Assistance Program and its Stewardship Incentives Program, can provide flexible financial assistance to communities and incentives to individual landowners for forest restoration and maintenance. Technical assistance programs can reach out to a diverse set of private landowners, including low-income and minority groups, but a broader array of educational activities is needed to inform and encourage a future citizenry committed to participating in community-based efforts.

Community-based organizations often seek to base their efforts with agencies on ongoing stewardship activities and collaborative learning among partners, rather than on single-issue emphases and in-depth analyses by external experts. This type of relationship requires change and a greater flexibility and responsiveness in agency decisionmaking processes. "Shared decisionmaking" presents a fundamental challenge for both government agencies and community groups (Ack et al. 2001). While agencies will retain authority as prescribed by law, they will need to broaden the basis for their decisions by improving participation of communities of place and interest in idea generation, planning, monitoring, and learning. Agency officials from top to bottom will need to understand the benefits of working collaboratively. Local area and district managers will need strong leadership from top agency officials and training in collaborative-interactive skills. Public officials will need to turn their agencies outward, being more inclusive of broad-based groups in communities, including low-income and minority groups, and more focused on ongoing partnership efforts (Cortner et al. 2001).

Effective community-based collaboration also requires a citizenry that is willing and well informed enough to work with government. Private nonprofit organizations, for example, can help bridge this gap between communities and public agencies, helping to build public understanding and clarify roles and niches for public agencies, private businesses, landowners, and other community groups (Johnson et al. 2001).

Developing ongoing collaborative relationships between government agencies and community groups will also help build an understanding of institutional issues that have been barriers in the past. In the Forest Service, for example, the separation between functions such as planning, research, land management, and programs that assist landowners and communities, has made it extremely difficult for community-based groups to work with agency officials toward integrated responses to resource management and community issues. A more outward-looking agency and a more involved citizenry promise innovative efforts to effectively integrate agency functions.

Cross-jurisdictional efforts–Cross-jurisdictional efforts are critical to community-based ecosystem management, since activities at a watershed or larger scale inevitably involve multiple ownerships and jurisdictions. Many agencies and institutions are reluctant to partner or share power with other institutions because they are concerned about change in the status quo and fearful of losing decisionmaking authority, administrative discretion, economic benefits, status, market niches, or competitive advantage. Even when it can be effectively shown that cross-jurisdictional efforts will benefit the mission of all organizations, perceived risks often outweigh promised benefits.

The ways in which public and private institutions respond and adapt to change will determine the potential for cross-jurisdictional efforts' success. Institutional change seldom comes easily. Cross-jurisdictional collaboration will require public and private institutions to reflect on their organizational cultures, adapt their missions, and reprioritize resources to encourage collaboration and partnership. For these efforts to succeed, all parties must see the benefits of their involvement and be willing to establish shared criteria and objectives. Both formal and informal mechanisms will be needed, the issue of power must be resolved early on through trust building and shared governance, and sufficient resources must be made available for follow-through. Although cross-jurisdictional efforts are sometimes generated by innovative institutional partners at the local level, cooperation often begins at the highest institutional levels, providing essential leadership to enable local institutions and employees to develop open and collaborative community-based partnerships.

One federal policy tool now being tested in the Northwest is a new authority that allows the Forest Service and Bureau of Land Management to use appropriated funds for watershed restoration work on private lands–so long as the activities pursue shared public and private objectives. This tool provides considerable flexibility and opportunity for cross-jurisdictional partnership, leveraging the resources of the various partners.

CONCLUSION

Community-based ecosystem management is an increasing phenomenon in the U.S. It has emerged as a result of frustration with interest-based, conflict-ridden approaches to resource management and a sense of powerlessness within communities about resource management decisions that affect their well-being. It has emerged from a common desire to move beyond output-based resource management, which has little regard for the condition of the land, to demonstrate the concept of stewardship–restoring and maintaining the health of ecosystems so they can continue to sustain diverse species and provide ecological services and products. It builds on the premise that communities and ecosystems are interdependent and that communities, based on their proximity and vested interest, must play a key role in planning and implementing resource management activities, if those activities are to be sustainable on an ecological, social, and economic basis.

Critical dimensions of community-based ecosystem management are:

- stewardship activities that emphasize ecosystem health but clarify and strengthen the links to the social and economic well-being of communities;

- empowerment of communities through inclusive resource planning and decisionmaking processes that actively engage diverse interests, including forest workers and under-represented groups, and that promote collaboration between public and private organizations by restructuring relationships;
- long-term investment in activities that restore and maintain healthy ecosystems and develop lasting stewardship between ecosystems and communities; and
- monitoring activities that gather and share information in ways that build trust, promote learning, and ensure accountability, including taking immediate corrective measures to inform future actions.

Pathways to advance community-based ecosystem management include:

- a continuous focus on process to ensure inclusiveness, accessibility, transparency, mutual learning, and adaptability;
- technical and financial resources and training to build community capacity;
- increased understanding of ecosystem worker issues, and promotion of an ecosystem workforce;
- innovative public- and private-sector tools and approaches to promote increased investment in ecosystem restoration and maintenance and in community-based stewardship;
- a civic science that allows local practitioner knowledge to inform and become engaged with more formal science; and
- laws and policies that promote collaboration and coordination among public and private organizations, rather than self-interest, competition, and adversarial relationships.

Community-based ecosystem management initiatives will take different approaches, encounter many obstacles, and achieve varying degrees of success and failure. Our hope is that the dimensions of and pathways toward community-based ecosystem management identified in this and other papers in this collection will be helpful and will promote further sharing and learning about this exciting approach to sustaining ecosystems and communities.

REFERENCES

Ack, Bradley L., Carol Daly, Yvonne Everett, Juan Mendoza, Mary Mitsos and Ron Ochs. 2001. The Practice of Stewardship: Caring for and Healing Ecosystems and Communities. In Gray, G.J., Enzer, M.J., and Kusel, J. Understanding Community-Based Forest Ecosystem Management, pp. 117-141. The Haworth Press, Inc., NY.

Cortner, Hanna J., Sam Burns, Lance R. Clark, Wendy Hinrichs Sanders, Gus Townes and Martha Twarkins. 2001. Governance and Institutions: Opportunities and Challenges. In Gray, G.J., Enzer, M.J., and Kusel, J. Understanding Community-Based Forest Ecosystem Management, pp. 65-96. The Haworth Press, Inc., NY.

Forest Stewardship Council. 1999. FSC Principles and Criteria. Version 1.2. Forest Stewarship Council, Oaxaca, Mexico.

Johnson, Nels, Jill Belsky,Victor Benavides, Martin Goebel, Ann Hawkins and Sissel Waage. 2001. Global Linkages to Community-Based Ecosystem Management in the United States. In Gray, G.J., Enzer, M.J., and Kusel, J. Understanding Community-Based Forest Ecosystem Management, pp. 35-63. The Haworth Press, Inc., NY.

Kusel, Jonathan. 2001. Assessing Well-Being in Forest Dependent Communities. In Gray, G.J., Enzer, M.J., and Kusel, J. Understanding Community-Based Forest Ecosystem Management, pp. 359-384. The Haworth Press, Inc., NY.

Luzadis, Valerie A., Carolyn Alkire, Catherine M. Mater, Jeff Romm, William Stewart, Leah Wills and Duane R. Vaagen. 2001. Investing in Ecosystems and Communities. In Gray, G.J., Enzer, M.J., and Kusel, J. Understanding Community-Based Forest Ecosystem Management, pp. 169-194. The Haworth Press, Inc., NY.

Moote, Margaret A., Beverly A. Brown, Eric Kingsley, Steven X. Lee, Steven Marshall, Donald E. Voth and Gregg B. Walker. 2001. Process: Redefining Relationships. In Gray, G.J., Enzer, M.J., and Kusel, J. Understanding Community-Based Forest Ecosystem Management, pp. 97-116. The Haworth Press, Inc., NY.

President's Council on Sustainable Development. 1997. Sustainable Communities Task Force Report. *President's Council on Sustainable Development*: Washington, DC.

Wills, Leah and Gerald J. Gray. 2001. Exploring Reinvestment from a Community-Based Watershed Perspective. In Gray, G.J., Enzer, M.J., and Kusel, J. Understanding Community-Based Forest Ecosystem Management, pp. 385-399. The Haworth Press, Inc., NY.

PART TWO: INTRODUCTION AND WORK GROUP PAPERS

An Introduction to Community-Based Ecosystem Management

Gerald J. Gray
Larry Fisher
Lynn Jungwirth

SUMMARY. This paper introduces the concept of community-based ecosystem management and explores cases and trends reflecting this emerging approach to natural resource management. It presents the workshop tone as a disciplined inquiry and addresses several questions

Gerald J. Gray is Vice President, American Forests, 910 17th Street NW, Suite 600, Washington, DC 20006.

Larry Fisher is Director, Cornell Program on Environment and Community, 112 Rice Hall, Ithaca, NY 14853.

Lynn Jungwirth is Executive Director, The Watershed Research and Training Center, Box 356, Hayfork, CA 96041.

[Haworth co-indexing entry note]: "An Introduction to Community-Based Ecosystem Management." Gray, Gerald J., Larry Fisher, and Lynn Jungwirth. Co-published simultaneously in *Journal of Sustainable Forestry* (Food Products Press, an imprint of The Haworth Press, Inc.) Vol. 12, No. 3/4, 2001, pp. 25-34; and: *Understanding Community-Based Forest Ecosystem Management* (ed: Gerald J. Gray, Maia J. Enzer, and Jonathan Kusel) Food Products Press, an imprint of The Haworth Press, Inc., 2001, pp. 25-34. Single or multiple copies of this article are available for a fee from The Haworth Document Delivery Service [1-800-342-9678, 9:00 a.m. - 5:00 p.m. (EST). E-mail address: getinfo@haworthpressinc.com].

such as where community-based ecosystem management came from, what it looks like, its potential benefits, and its key challenges. *[Article copies available for a fee from The Haworth Document Delivery Service: 1-800-342-9678. E-mail address: <getinfo@haworthpressinc.com> Website: <http://www.HaworthPress.com>]*

KEYWORDS. Community-based, forest(ry), ecosystem management, stewardship, collaboration

> All ethics so far evolved rest upon a single premise: that the individual is a member of a community of interdependent parts. His instincts prompt him to compete for his place in that community, but his ethics prompt him also to co-operate (perhaps in order that there may be a place to compete for). The land ethic simply enlarges the boundaries of the community to include the soils, waters, plants, and animals, or collectively the land. (Aldo Leopold 1949)

> No longer is community . . . the refuge within which tradition lurks to shackle progressive social trends. Instead it has become the focus of thinking on devolution of power, meaningful participation, and cultural autonomy–all of these believed, in turn, to be critical to successful resource management and development. (Arun Agrawal 1998)

INTRODUCTION

There is change afoot in the way we think about and relate to our living environment. There are also valuable new ways of thinking about and involving communities in the important business of managing our natural resources. Emerging innovative efforts link us to the environment in ways that expand our understanding of our relationship to the land; they challenge our basic assumptions, behaviors, and our most established institutions. There is innovation and energy where conflict, frustration, and apathy have held sway.

These changes are built upon basic principles such as increased respect for inclusiveness and collaboration as being central to democratic and sustainable systems and on the interdependence of human communities and natural ecosystems. Emerging efforts respond to a variety of conditions and trends in environmental, social, and economic systems as well as the failure of current policies, institutions, and market mechanisms to effectively deal with problems and conflict.

These approaches point to a global movement toward increased community participation in natural resources management (Western et al. 1994; Wells et al. 1992). At the same time we see an enhanced appreciation for the concept of ecosystems as a basis for policy decisionmaking (GAO 1994; Keystone Center 1996). During the past three decades, conservation and community development practitioners have collaborated in a plethora of community-based conservation programs around the globe:

- In India, joint forest management provides an alternative to traditional custodial policing systems, along with a fundamental shift from timber extraction to nontimber forest products. These changes also represent a shift from centralized forest planning and management to local planning and control. Thousands of forest protection committees have formed over the past decade, leading to the development of federations of forest protection committees throughout the subcontinent.
- In the Philippines, the integrated social forestry program has addressed land-tenure uncertainty for upland villagers, providing them with new rights to forest protection and use. The project has integrated agroforestry technologies and worked to develop new opportunities for product processing and marketing. The program is estimated to affect more than 60 percent of the land area under government jurisdiction (Ford Foundation 1997).
- In Africa, the CAMPFIRE program (Zimbabwe) and the Amboseli National Park (Tanzania), among others, offer new strategies for involving communities in planning and managing protected areas and biodiversity conservation. These programs have engendered important changes in the way benefits and incentives help strengthen community capacity and economic development (Western 1994; Metcalfe 1994).

Case studies of individual programs have been documented recently in numerous publications and discussed in several international meetings on community-based approaches to ecosystem management. Collectively, they indicate the breadth and vision of these innovative approaches around the world (Western et al. 1994; Wells et al. 1992; Fox et al. 1997; World Bank, 1998).

In the United States, these issues are a recent addition to our national policy debate, but they are rapidly becoming central to the public discourse on natural resource management. Ironically, much of the impetus for this debate is found in urban settings where poverty, deterioration of the resource base, and new concepts of community mirror the conditions under which these approaches have emerged internationally. Urban case studies have been documented across the U.S.–from New York to Los Angeles, Seattle to Atlanta, and Chicago to Austin (American Forests 1997, 1999; Gray in press). At the same time, there is a growing recognition of America's historical traditions–such as documentation of Native American (Menominee, Choctaw) and

European-American experiences–and the cultural traditions of growing minority groups, such as Latinos and Southeast Asians (Forest Community Research in press). Compelling examples of collaborative resource management can now be found across the U.S., from Monhegan Island, Maine, to Plumas County, California and have been documented in recent publications (EPA 1997; Bernard and Young 1997; Yaffee et al. 1998).

What's going on here? What do these emerging innovations and their collective lessons tell us about our relationships, institutions, and policies? What do they tell us about our notions of community and the emergence of civil society? Are these approaches really new or simply old, common-sense approaches we are beginning to see in new ways?

The sheer number and range of experience in community-based ecosystem management suggests these programs are no longer isolated, "fringe" phenomena that can be dismissed by policymakers, scientists, and activists. The momentum of this change is at least a trend, at most a fundamental reorientation. The growing volume of case analyses, the wrestling with new terms and vocabulary, the emerging networks, and the pace of institutional policy reform suggest the latter.

We are seeing a maturing of this experience into new concepts, paradigms, and policies, an institutionalization through changed policies and new organizational configurations, the establishment of innovative mechanisms for monitoring and assessment, and the implementation of creative fiscal arrangements. Authors have begun to raise second-generation concerns of equity, linkages, and institutional change (Saigal 1997; Daly 2001).

Community-based ecosystem management approaches are rooted in process–quite simply, the way we make environmental management decisions given the diversity of interests and perspectives. Process is central to the shift toward community-based approaches, and many new tools and strategies enable more informed and inclusive decisionmaking. On-the-ground innovations offer communities new models of environmental stewardship and capacity-building. New investment tools provide insight into the critical question of how to make this work pay its way. We see changes in the shape and function of our public and private institutions. And emerging new alliances and partnerships transcend conventional governance systems to address the challenges of scaling-up local experience to broader landscapes and social systems.

This workshop, *Understanding Community-Based Ecosystem Management in the United States*, provides the setting for a "disciplined inquiry" into these approaches, allowing us to explore whether they represent a sea change in the way we go about the business of relating to our social and physical surroundings. We started with some very basic questions: Where did "community-based ecosystem management" come from? What does it look like? What are some potential benefits? And, what are some key challenges facing these approaches?

WHERE DID COMMUNITY-BASED ECOSYSTEM MANAGEMENT COME FROM?

There has been a wealth of international experience in community-based natural resource management over the past three decades. This is particularly true in developing countries where institutions collectively have tried to respond to the continued destruction and degradation of forests and the needs and rights of indigenous communities dispersed throughout the landscape. Many of the community-based efforts in India, Africa, Southeast Asia, and South America have been driven by governments, international donor agencies, and nongovernmental organizations (NGOs) seeking to strengthen community participation in decisionmaking and management. In contrast, U.S. efforts generally have emerged at the community level in response to ecological, social, and environmental concerns. These efforts continue to reside with local community-based groups, although state and federal governments are increasingly interested in partnering with and nurturing these efforts.

Urban and rural communities in the U.S. have been developing innovative socially and politically acceptable approaches to natural resource management since the late 1970s. Community-based approaches have emerged in diverse settings and for various purposes. In urban areas, community groups often organize around the need to restore and maintain such ecosystem values as open space and water quality or to build the capacity to partner with public and private institutions to implement projects. In rural areas, most notably in the West, community-based groups have formed in response to recent conflicts over natural resource management issues, such as the protection of old-growth forests and endangered species. As the negative social, economic, and ecological effects of continued conflict become clear, diverse local interests come together to seek common ground and work collaboratively. In other rural areas, such as in the Northeast where private farms and forests dominate the landscape, community-based efforts have long been part of the traditional conservation movement. More recently, however, innovative approaches have dealt with broader ecological issues such as water quality protection and forest fragmentation. These issues require information and conservation strategies at a landscape scale.

WHY COMMUNITY-BASED ECOSYSTEM MANAGEMENT?

A variety of terms internationally and in the U.S. describe innovative efforts that more fully engage communities in natural resource management and decisionmaking processes. These include community forestry, social forestry, urban and community forestry, sustainable forestry, community-based natural resource management, community-based conservation, and integrated conservation and development. We have chosen the term "community-based ecosystem management" for this workshop for two reasons. First, "community-based" captures the essential characteristic that these initiatives are place-based. They involve a range of stakeholders from "communities of place" (groups of people sharing an interest in a common geographic area). In addition, many initiatives are seeking ways to involve stakeholders beyond the community of place. Second, "ecosystem management" implies an approach that is broader than forest, natural resource, or watershed management. Ecosystem management looks at whole systems and requires thinking that integrates ecological, social, and economic considerations at various scales across the landscape and over time.

Principles and Features

Community-based ecosystem management initiatives take place in diverse community contexts, involving unusual combinations of partners, a variety of processes, and unique land conditions. They can occur on communal or tribal lands, mixed-ownership landscapes dominated by private nonindustrial forest owners, areas dominated by federal or state lands, and in urban and suburban neighborhoods with multiple political and institutional overlays. While it is difficult to generalize about community-based ecosystem management, we have identified four principles common to many efforts:

1. *The land is treated as part of the community.* Hearkening back to Leopold's land ethic, participants acknowledge the health of the ecosystem–the land and all its components–as a critical part of the community.
2. *Land decisions are made through inclusive, open, and transparent community-based processes.* These processes strive to involve people who know the land, care about it, and are affected by the consequences of the decisions–whether they live there or not.
3. *When benefits flow from the land to the greater community, they flow through local communities.* Processes try to estimate values provided by the land–values such as ecological services, resource use, and processing–and try to ensure that a fair share of the values are cycled through communities to sustain local capacity.
4. *The community is acknowledged as part of the landscape.* People and the communities in which they live are recognized as integral parts of forest ecosystems.

A number of features commonly emerge through community-based ecosystem management efforts, including:

- Planning and design efforts openly recognize the interdependence of community well-being and ecosystem health.

- Planning and decisionmaking processes are unusually inclusive of all stakeholders, including traditionally disenfranchised groups such as forest workers and ethnic minorities.
- Collaboration among interests and organizations that traditionally have not worked together transcends social, cultural, economic, and political barriers, often to create common ground.
- New and various scales of analyses are used to understand ecosystem functions and the potential impact of human activities. These scales cover geographic, institutional, social, and temporal parameters. One important scale is often referred to as "human" and seems to reflect a group's ability to think together on the same scale.
- Assessment and information-gathering efforts consciously include local knowledge as well as scientific expertise.
- An emphasis is placed on monitoring. Groups see monitoring as a means of addressing accountability, building trust, and promoting learning, innovation, and adaptive responses.
- Groups create incentives for investment in a community's capacity to develop and use mechanisms to restore and sustain forest ecosystems. Stewardship is common ground.

These guiding principles and common features suggest and lead to new mechanisms for integrating competing values. Deliberative attempts to "integrate" as opposed to "balance" competing factions can make an important contribution to healing degraded landscapes and communities; they also have the potential to prevent degradation.

POTENTIAL BENEFITS

Community-based ecosystem management has brought a new sense of excitement, creativity, and hope into the realm of natural resource management and decisionmaking. Scientists, resource managers, and practitioners involved in these efforts see many benefits from community-based processes and projects. These benefits are not easily categorized because they tend to integrate and reflect the interrelationships of ecological, social, and economic concerns inherent in the concept of sustainable ecosystems and communities. For example, community-based processes are often recognized as viable alternatives to adversarial, interest-based politics; this is a social or political benefit. Although these processes involve conflict, they seek to reconcile diverse interests and identify common ground on which to move forward, rather than just determine winners and losers.

Community-based ecosystem management projects are also a source of innovation for new technologies and on-the-ground practices to implement ecosystem stewardship, providing ecological and economic benefits. Community-based ecosystem management's commitment to monitoring seeks to safeguard against unintentional harm to ecosystems while providing essential information for learning and adaptive responses. And, community-based ecosystem management projects identify and test new mechanisms for attracting investment to restore and maintain the capacity of ecosystems and communities to sustain themselves. Finally, many proponents recognize that community-based ecosystem management can develop a workforce that is skilled in the concepts, technologies, and practices of ecosystem management; is committed to the land; and is part of the community's social fabric–a high-skill, living-wage, ecosystem workforce rooted in communities of place.

ISSUES AND CHALLENGES

While generating excitement, community-based ecosystem management has also raised questions among those observing specific community-based efforts and the momentum of the broader trend. The workshop identified a number of issues and challenges, reflecting a variety of those perspectives. Workshop participants grappled with different types of challenges while preparing collaborative papers for this publication:

- How can community groups find new approaches that ensure inclusivity? These approaches must involve ethnic, social, economic, and philosophical diversity, as well as stakeholders from beyond communities of place.
- How can community groups strengthen their participation in natural resource decision making and management without appearing to challenge the authority of federal, state, and local authorities?
- How will community-based ecosystem management address governance, institutional jurisdiction, and landowners' rights and responsibilities in large-scale, mixed-ownership settings?
- What types of public and private investment mechanisms can help community-based initiatives gain access to and sustain financial resources over time, particularly where forest resources and local economies have been depleted?
- What constitutes a credible community-based approach? How do we assess a community-based initiative's strengths and weaknesses? What lessons have we learned from existing efforts? How are initiatives monitored and who is held accountable?

CONCLUSION

Community-based efforts worldwide are emerging in a variety of contexts, exploring innovative ways to strengthen communities' participation in natural resource decision making and management. The excitement and momentum have captured the interest of scientists, practitioners, and policy makers. Many believe community-based initiatives could create processes and mechanisms to move us toward a vision of sustainable ecosystems and communities that integrates ecological, social, and economic concerns. At the same time, however, these efforts raise many questions about necessary changes and the implications for institutions, policies, markets, and actions on the land.

The workshop papers that follow expand on the information and questions this introduction presents about community-based ecosystem management. They bring together the knowledge and perspectives of a diverse set of participants from across the U.S. Many came with international experience; all were asked to consider global trends and forces, although their primary challenge remained to explore community-based ecosystem management as it is uniquely developing in the U.S. And so participants examined community-based ecosystem management issues and case studies throughout the country, from different regions, from both urban and rural areas, at different spatial scales, and involving a variety of institutions and ethnic groups. That workshop participants shared an enthusiasm for community-based ecosystem management comes through in many of the papers. The collaborative work-group papers are open and inquisitive, inviting readers to join the dialogue and help build a broadly shared understanding of community-based ecosystem management in the United States.

REFERENCES

Agrawal, A. 1998. Communities and conservation: Beyond enchantment and disenchantment. Gainesville, FL: Conservation Development Forum.

American Forests. 1997. Cities by Nature's Design. Proceedings of the 8th National Urban Forest Conference, Atlanta, Georgia. Washington, DC: American Forests.

American Forests. 1999. Building Cities of Green. Proceedings of the 1999 National Urban Forest Conference, Seattle, Washington. Washington, DC: American Forests.

Bernard, T. and J. Young. 1997. *The Ecology of Hope: Communities Collaborate for Sustainability*. Gabriola Island, B.C.: New Society Publishers.

Daly, Carol. 2001. Surviving the Transition: Issues for Maturing Community Forestry Groups. Journal of Sustainable Forestry 13(1/2):321-334.

EPA. 1997. *Community-Based Environmental Protection: A Resource Book for Protecting Ecosystems and Communities*. Washington, DC: United States Environmental Protection Agency.

Ford Foundation. 1997. Forestry for sustainable rural development: A review of Ford Foundation-supported community forestry programs in Asia. New York: Ford Foundation.

Forest Community Research Technical Report. In press. *Forest Communities, Community Forests.* A collection of case studies for the Communities Committee of the Seventh American Forest Congress. Forest Community Research.

Fox, J., L. Fisher, and C. Cook (eds.). 1997. *Conflict and Collaboration: Eighth Workshop on Community Management of Forest Lands.* Honolulu: East-West Center.

GAO, 1994. Ecosystem Management: Additional Actions Needed to Adequately Test a Promising Approach. GAO/RCED-94-111. Washington, DC: U.S. General Accounting Office.

Gray, Gerald J. The New York City Watershed Agreement: A Grand Experiment. In *Forest Communities, Community Forests.* A case study for the Communities Committee of the Seventh American Forest Congress. In press.

Keystone Center. 1996. The Keystone National Policy Dialogue on Ecosystem Management. Colorado: The Keystone Center.

Leopold, A. 1949. *A Sand County Almanac.* New York: Ballantine Books.

Metcalfe, S. 1994. The Zimbabwe communal areas management programme for indigenous resources (CAMPFIRE). In D. Western, R. M. Wright, and S. Strum (eds.), *Natural Connections: Perspectives in Community-Based Conservation.* Covelo, CA: Island Press.

Saigal, S. 1997. Beyond Experimentation: Emerging Issues and Conflicts Over Joint Forest Management in India. In J. Fox, L. Fisher, and C. Cook (eds.), *Conflict and Collaboration: Eighth Workshop on Community Management of Forest Lands.* Honolulu: East-West Center.

Wells, M., K. Brandon, and B. Hannah. 1992. *People and Parks: Linking Protected Area Management with Local Communities.* Washington, DC: World Bank, World Wildlife Fund, USAID.

Western, D., R.M. Wright, and S. Strum (eds.). 1994. *Natural Connections: Perspectives in Community-Based Conservation.* Covelo, CA: Island Press.

Western, D. 1994. Ecosystem Conservation and Rural Development: The Case of Amboseli. In D. Western, R.M. Wright, and S. Strum (eds.), *Natural Connections: Perspectives in Community-Based Conservation.* Covelo, CA: Island Press.

World Bank. 1998. Workshop materials. International workshop on community-based natural resources management, Washington, DC.

Yaffee, Steven L. et al. 1998. *Ecosystem Management in the United States: An Assessment of Current Experience.* Washington, DC: Island Press.

Global Linkages to Community-Based Ecosystem Management in the United States

Nels Johnson
Jill Belsky
Victor Benavides
Martin Goebel
Ann Hawkins
Sissel Waage

Nels Johnson is affiliated with the World Resources Institute, 10 G Street NE, Washington, DC 20002.

Jill Belsky is affiliated with the Department of Sociology, University of Montana, Missoula, MT 59812.

Victor Benavides is affiliated with Alliance of Forest Workers and Harvesters, P.O. Box 1814, White Salmon, WA 98672.

Martin Goebel is affiliated with the Sustainable Northwest, 620 SW Main, Suite 112, Portland, OR 97205.

Ann Hawkins is affiliated with the University of California-Berkeley, College of Natural Resources, 4 Giannini Hall #3100, Berkeley, CA 94720-3100.

Sissel Waage is affiliated with the University of California-Berkeley, 233 South Barrington Avenue, Los Angeles, CA 90049.

The authors express their appreciation for the hard work of Maia Enzer, Gerry Gray, the rest of the team from American Forests, and the Steering Committee Members for the Workshop on Community-Based Ecosystem Management that made this paper possible. They are especially grateful to Melanie McDermott Hughes and David Hughes McDermott of the University of California at Berkeley for their text on linking community-based resource management in California, Zimbabwe and the Philippines which appears in Box 8. Finally, Ronnie Yimsut of the U.S. Forest Service contributed important insights on migrant workers and the non-timber forest products economy of the Pacific Northwest.

[Haworth co-indexing entry note]: "Global Linkages to Community-Based Ecosystem Management in the United States." Johnson, Nels et al. Co-published simultaneously in *Journal of Sustainable Forestry* (Food Products Press, an imprint of The Haworth Press, Inc.) Vol. 12, No. 3/4, 2001, pp. 35-63; and: *Understanding Community-Based Forest Ecosystem Management* (ed: Gerald J. Gray, Maia J. Enzer, and Jonathan Kusel) Food Products Press, an imprint of The Haworth Press, Inc., 2001, pp. 35-63. Single or multiple copies of this article are available for a fee from The Haworth Document Delivery Service [1-800-342-9678, 9:00 a.m. - 5:00 p.m. (EST). E-mail address: getinfo@haworthpressinc.com].

SUMMARY. Community-based ecosystem management (CBEM) in the United States is closely tied to global processes. Increasing and shifting international market demands for ecosystem products and services together with global trends in climate change and biodiversity loss have tangible impacts in communities in every region of the country. Meanwhile, community-based natural resource management efforts in other parts of the world, particularly in developing countries, have a longer history than in the United States. This history, and the tools and techniques developed elsewhere for community-based resource management, can help to inform North American advocates and practitioners of CBEM. This paper has four objectives. The first is to identify key global trends that affect communities in the United States. These trends include demand and supply relationships (especially of forest resources) and environmental changes that will shape economic and policy choices in coming decades. Second, the paper examines countervailing forces of globalization and decentralization. How are shifting patterns of governance and control around the world affecting the prospects for sustainable community-based resource management? Third, the paper seeks to understand migration as a growing feature of many communities. What challenges and opportunities does migration pose for sustainable resource management? Finally, the paper summarizes a few of the tools and techniques used internationally that might have relevance in the United States. *[Article copies available for a fee from The Haworth Document Delivery Service: 1-800-342-9678. E-mail address: <getinfo@haworthpressinc.com> Website: <http://www.HaworthPress.com>]*

KEYWORDS. Community-based, forest(ry), ecosystem management, global, sustainable, migration

Community-based ecosystem management (CBEM) in the United States is closely tied to global processes. These include trends in natural resource demand, migration, and international efforts to integrate natural resource management with economic development. In this chapter we discuss these "global linkages" and some of their implications for community-based ecosystem management in the United States.

Considering global linkages in community-based ecosystem management is worthwhile for several reasons. First, the fate of people and places across the American landscape is tied to accelerating flows of trade, investment capital, labor, and environmental changes that circle the globe. Second, the rise of CBEM in the United States parallels efforts in other countries to give local communities a larger voice in natural resource management. And third, despite important cultural, economic, and environmental differences between communities in the U.S. and elsewhere, there are important lessons to be learned from community-based conservation efforts in other parts of the world.

In other countries, community-based natural resource management falls under a variety of names. The most broad-based approaches today are "community-based conservation" and "community-based resource management." These build on earlier efforts known as "integrated conservation and development projects," which considered peoples' economic and cultural concerns (Brown and Wyckoff-Baird 1993). Today's approaches build on more than two decades of international experience involving communities in managing resources and sharing the benefits that flow from them. These programs use different terminology reflecting their own social and historical contexts. For example, in the Philippines the focus has been on "upland development," in Thailand on "social forestry," and in India on "joint forest management." Other terms used include "co-management," "rural development forestry," and "community-based forestry" (Ford Foundation 1998). These programs' experiences are relevant to emerging community-based ecosystem management efforts in the United States.

It's worth noting that "ecosystem management" means different things in different places. In the United States, it has its origins in legal and policy developments as well as in science. For example, federal legislation such as the National Environmental Protection Act, the Endangered Species Act, and the National Forest Management Act are often cited as catalysts for ecosystem management in the United States (Keystone 1996). Community-based ecosystem approaches are sometimes colored by perceptions of their legal or policy motives. Some national environmental groups in the United States are skeptical about community-based ecosystem management because they view legislative approaches as more representative of public interests, and more effective in dealing with industry (McCloskey 1996). Rural community residents, on the other hand, may see community-based ecosystem management as another effort by bureaucrats and academics to increase their influence. Meanwhile, at least as defined in practice, ecosystem management in the United States often attempts to foster collaborative partnerships for both ecological and socioeconomic objectives (Yaffe et al. 1996; Keystone 1996).

In Europe, ecosystem management often refers to a highly technical scientific approach to resource management along multiple temporal and spatial scales. Outside Europe, "bioregional management" and "integrated conservation and development projects" may be used in place of "ecosystem management." These efforts generally seek a regional approach to conservation and resource management that necessitates cooperation across diverse peoples, property regimes, and communities (Miller 1996). Conservationists working in developing countries have generally recognized that the fate of the forest and other habitats lies with the man or woman with the shovel and the machete. In developing countries, some observers suggest understanding resident people and the factors influencing their resource use may be more important than either science or policy in promoting sound natural resource management (Wells and Brandon 1992; Western and Wright 1994).

It's also important to note differences in how the term "community" is used. For example, in contrast to the United States, in many parts of the world communal forms of tenure and resource management figure prominently in "community" forestry and resource management. Community resource management, therefore, may refer specifically to social processes for managing common lands but not necessarily to individual actions on private lands. Research around the world suggests that differences within communities are as important as differences between them. Individual differences based on age, gender, class, race, ethnicity, and being "newcomer" or "old timer," among others, affect views on forest management goals and processes. In the tropics, these differences shape land and resource tenure, define household economic strategies, and help explain why different individuals and communities are willing or not to manage trees and forests (Ford Foundation 1998). In the United States, differences within and between communities are often not clearly defined (Box 1).

This chapter has four goals. The first is to identify key global trends that affect communities in the United States. These trends include demand and supply relationships (especially of forest resources) and environmental changes that will shape economic and policy choices in coming decades. Second, the chapter examines countervailing forces of globalization and decentralization. How are shifting patterns of governance and control around the world affecting the prospects for more sustainable community-based

BOX 1. Who's in a Community, Anyway?

In resolving resource management conflicts in the United States, two types of communities are often recognized–"communities of place" (geographically bounded areas) and "communities of interest" (a group of people with shared beliefs or concerns). Some researchers, drawing on international experience, believe these categories neither sufficiently define "communities" in the context of natural resource management nor do they provide insight into how the two communities are related (Agarwal 1997; Brosius et al. 1998; Belsky in press). Migratory peoples, such as shifting cultivators in the tropics and migrant forest workers in the Pacific Northwest, are often left out of both categories and may not figure at all in discussions about community-based forest management. The categories are also too simplistic; they assume that members of one community or the other have similar access to resources and power. Within most communities–defined by geographic region, special interest, or any other definition–a range of assets, values, and capacity to collaborate will exist. These differences need to be recognized and dealt with early in the CBEM process.

resource management? Third, the chapter seeks to understand migration as a growing feature of many communities. What challenges and opportunities does migration pose for sustainable resource management? Finally, the chapter summarizes a few tools and techniques used internationally that might have relevance to community-based ecosystem management in the United States.

GLOBAL RESOURCE TRENDS AND COMMUNITIES

No matter how remote they are, communities across the United States are increasingly affected by global trends and international events. Shifting supply and demand for natural resources in other parts of the world together with global signs of environmental distress, such as climate change and biodiversity loss, have tangible impacts on every region of the country. International trends can increase opportunities for community-based ecosystem management. For example, new markets in Europe for sustainably produced wood products can provide incentives for community-based ecosystem management in the United States. Communities aware of these demands can expand options for economic diversification and sustainable natural resource management. In other cases, international trends can frustrate or complicate these opportunities. For example, the rapid growth of international demand for nontimber forest products, such as ginseng in Virginia or mushrooms in Oregon, can overwhelm efforts to put sustainable harvest regimes in place (Duke 1998).

This section briefly examines those international resource and environmental trends most likely to intersect with community-based ecosystem management in the United States. We focus on forests because they are better documented than other natural resources.

Global Supply and Demand of Forest Resources

The world's supply of forest resources–for both timber and nontimber products–is declining even as demand for those products is increasing. Although there are enormous regional variations in the supply and demand of forest resources, the rapid rise of international trade and the movement of investment capital increasingly means communities from Ghana to Siberia to Tennessee can feel the effects of growing demands and shrinking supplies in other parts of the world. Measuring the supply of resources, especially nontimber forest products and ecosystem services, is extremely difficult. Our best surrogate is to look at overall forest cover.

About a quarter of the earth's land area is forested today, with most countries having at least some forest cover (WRI 1998). However, just seven countries–Russia, Canada, Brazil, the United States, China, Indonesia, and the Republic of the Congo–have more than half the world's forests. Slightly more than half the world's forests are found in the tropics, the remainder in temperate and boreal regions. Before the advent of agriculture 8,000 years ago, it's estimated that global forest cover may have been nearly double what it is today (Bryant et al. 1997). Deforestation and forest degradation have occurred in some regions for thousands of years (e.g., the Mediterranean and China), but the rate and scale of forest loss today is unprecedented. Between 1980 and 1995, the world's forest area declined by an estimated 160 million hectares (or just over 3 percent of total forest area)–a number just slightly smaller than Mexico's total land area (WRI 1996). While most of the recent forest loss has been in the tropics, many natural forest ecosystems in temperate and boreal regions have been damaged by pollution, conversion to tree plantations, and fragmentation by extensive logging and road building (Dudley 1992).

Deforestation is the result of complex forces. Agricultural expansion–for subsistence and for production of export commodities–is widely recognized as the leading cause of deforestation, followed by destructive logging, and infrastructure development (Sharma 1992). However, the root causes of deforestation are extremely complex and synergistic–lack of secure tenure, widespread corruption, population growth and migration, tax and economic policies, subsidies for agricultural development and logging, inadequate investment in forest management, etc. (Johnson and Cabarle 1993).

As they convert their natural forests, a growing number of countries have moved from the ranks of timber exporters to timber importers. For example, the Philippines, Thailand, Sri Lanka, Nigeria, Togo, El Salvador, Haiti–all once extensively forested–now have virtually no natural forest and must import wood. Many other countries, particularly in Southeast Asia (traditionally the source of most internationally traded tropical timber) and Central America, have logged or allocated concessions on all nonprotected natural forest areas. Partially offsetting the loss of natural tropical forest area is the expansion of tree plantations, especially in Brazil, Chile, New Zealand, Indonesia, and China (FAO 1996). Still, the area planted worldwide is but a fraction of the natural forest area lost and is often at the expense of natural forests.

Even more than population growth, economic growth determines forest resource demand. For example, the United States, with a 1995 per capita GDP of $26,026, consumes 700 pounds of paper per capita (EDF 1995). By contrast, per capita paper consumption averages 66 pounds in Latin America, where per capita GDP is typically less than $2,000, and 50 pounds in Asia, where per capita GDP (outside Japan, Korea, and Taiwan) is typically less than $1,000. The United States consumes paper and other wood products like it thirsts for oil, claiming nearly 26 percent of the world's industrial wood products for less than 4 percent of the world's population (FAO 1996).

Population growth also drives demand for forest resources. In 1998, the world population stood at 5.9 billion with a 1.5 percent annual growth rate. Most of this growth was in developing countries, although population growth rates are declining. Still, middle-range projections place the world's population at 8.3 billion in 2025 and between 9 billion and 10 billion in 2050 (WRI 1998). With a 1998 population of 273 million, the United States has the highest annual growth rate of any industrialized nation (1 percent). By the middle of the next century, the country's population will surge to at least 350 million (U.S. Bureau of the Census 1997).

Most of the growth in wood consumption–on a per capita basis–has been in developed countries, especially the United States, Canada, Germany, and Japan, where growth rates have been in excess of 1 percent annually (FAO 1993). Per capita consumption has remained relatively stable in developing countries, where most of the wood is used as fuel for cooking. Global demand for paper and wood-based panels (the two fastest growing wood products sectors) are projected to grow steadily at 3 percent annually over the next decade (FAO 1993). Much of this growth is expected from emerging economies in Asia and to a lesser extent in Latin America and Eastern Europe. In China, paper consumption has been growing at 10 percent annually since 1980 (FAO 1996). China, which now ranks only 11th or 12th in importing of wood products, is likely to top that catagory by 2010.

Still, it's important to keep in mind that the growth from emerging economies starts at a very low base per capita consumption level. The vast majority of internationally traded forest products are consumed in developed countries. It is their appetite that drives the global search for fiber–a search that is expanding to relatively unexploited natural forests in parts of the Amazon Basin and Guyana Shield in South America, Central Africa, the Russian Far East, and Canada (Bryant et al. 1997). This demand is also driving plantation development in Chile, New Zealand, and Brazil and a new round of harvesting in secondary forests in the eastern United States. For the first time in its history, the United States is becoming a net importer of wood products (Box 2).

International demand is also growing for nontimber forest products and services. For example, the majority of wild mushrooms and floral display materials (e.g., ferns, and moss) collected in the Pacific Northwest are exported to restaurants and markets in Europe and Japan (Cohn 1995). Nature-based tourism, called ecotourism, is now estimated to be worth as much as $50 billion worldwide and is increasing at a rate of 7 percent annually (Filion et al. 1992). Forests are often the focus; for example, Indonesia recently issued a 100,000-hectare forest concession for the express purpose of nature-based tourism development rather than for logging (Barber et al. 1995).

BOX 2. Logs Passing in the Night

In the mid-1990s, the Brightwood Corporation, a Bend, Oregon, manufacturer of edge-glued wood products, was faced with a timber supply crisis. Declining supplies of high quality local ponderosa pine and rising timber prices meant the company faced layoffs, curtailed operations, and possible closure. The solution came by importing radiata pine (a species native to California) from Chilean plantations more than 5,000 miles south. Even with the distance, this made economic sense since the aggressively marketed Chilean log exports were relatively cheap and the value added by Brightwood's highly skilled workforce left room for profits (The Oregonian 1998). In Chile such exports generate jobs and revenue but also social and environmental conflicts as natural forests and agricultural lands are converted to plantations. Meanwhile, along the Tennessee River and the Tennessee-Tombigbee Canal in Alabama and Tennessee, barges bulging with woodchips head to ships on the Gulf of Mexico destined for Asian ports. The woodchip export market has exploded in recent years due to wood shortages in Asia, maturing second-growth forests in the southeastern U.S., and technologies that now allow virtually any tree species to be used in engineered wood products. Although the region benefits economically, this widespread clearcutting alters one of most biologically diverse parts of the country and causes a shortage of sawlogs for local mills. So even as the U.S. becomes a net importer of wood, it continues to export large quantities of wood to other regions, especially Asia.

Global Environmental Trends

Ecosystems around the world, and the people and communities that depend on them, are experiencing rapid change. Climate change, biodiversity loss, air pollution, and introduced species are widespread and are increasingly felt in communities everywhere, including across the United States.

Climate Change–Forests are sensitive to climate change. Average temperature changes of as little as 1 degree C can affect the reproduction and growth of many tree species (IPCC 1996). There is significant risk that climate warming, as projected by the International Panel on Climate Change (IPCC), will adversely impact forest ecosystems through changes in species composition, increased prevalence of insect pests and microbial pathogens, and more frequent and severe fires in some areas and more severe storms and flooding in others (IPCC 1996). While temperate forests are expected to be less affected by climate warming than boreal and tropical forests, U.S. forests in the next century could undergo significant changes in species composition, pest and disease outbreaks, and fire frequency and severity. These changes could substantially alter the mix of forest fiber available to industry and irreversibly deplete forest biodiversity.

Biodiversity Loss–The accelerating loss of biodiversity is a global phenomenon (UNEP 1995). Significant losses are being felt by the United States, itself home to more species than all but a few tropical countries such as

Brazil, Colombia, and Indonesia (Eisener et al. 1995). For example, The Nature Conservancy estimates that nearly a third of all species in the U.S. are at some risk of extinction during the next century (TNC 1996). Biodiversity losses in other parts of the world can directly affect U.S. citizens. First, since virtually all crops in the United States originated elsewhere, today's very high losses of "agrobiodiversity" (e.g., wild relatives of crop species) limit our options to adapt modern crops to new diseases and climate change (Raeburn 1995). Second, biodiversity losses preclude the discovery and use of genetic and biological materials that may have enormous industrial and medical value. For example, bacteria collected from hotsprings in Yellowstone National Park have yielded the enzyme taq polymerase, an essential tool that literally opened the way for biotechnology (Research and Development 1998).

Air Pollution–Local areas can be adversely impacted by pollutants that originate beyond state and international boundaries. Acid rain that comes principally from coal-fired power plants and automobiles has been implicated in widespread forest declines in Europe, Asia, and the eastern United States (Likens et al. 1996; MacKenzie and El-Ashry 1989). Although there is scientific debate about the specific damage caused by acid rain, research indicates forest, agricultural, and aquatic ecosystems are being affected in parts of Canada, the United States, Europe, Asia, and Latin America. At an experimental forest in West Virginia, research links substantial tree mortality and growth declines to car exhaust hundreds of miles away in the Midwest (Gilliam et al. 1996). Communities can be adversely and measurably affected by emissions even thousands of miles away.

Introduced Species–One of the by-products of international trade is the spread of invasive species. Examples are numerous and recognizable throughout the United States–from zebra mussels in the Great Lakes region, kudzu vine in the Southeast, leafy spurge on the northern Great Plains, and gypsy moths in the Northeast. In some cases, introduced diseases have wiped out native species of tremendous ecological, social, and economic importance. For example, the American chestnut–one of the most widespread and valuable species in eastern hardwood forests until chestnut blight was introduced from Asia–has disappeared from forests over its entire range. The spread of introduced (or "exotic") species is believed to be a close second to habitat loss as a factor in species extinctions and considerably more important than pollution and overharvesting (UNEP 1995). Climate change and rapidly expanding international trade could substantially increase the vulnerability of U.S. forests to pest and disease outbreaks in coming decades (IPCC 1996).

In short, communities across the United States are increasingly linked–for better or worse–to resource and environmental trends in other parts of the world.

SHIFTING PATTERNS OF GOVERNANCE AND CONTROL

We live in a world where the private sector is assuming a larger role in many aspects of life, government is assuming less, and civil society is organizing itself to take on new roles and responsibilities. These trends are global in scope and particularly evident with respect to natural resources. The countervailing forces of globalization and decentralization are shifting the control and governance of natural resources in various ways. Many of these changes will profoundly alter the relationship between individuals and communities and their environment.

The Private Sector Swells While Government Shrinks

Energy, agriculture, and forest products are increasingly dominated by a small group of multinational corporations whose operations span the globe. Malaysian logging companies, having run out of forests at home, are now investing in such far-flung countries as Guyana, Suriname, Gabon, Papua New Guinea, and Russia (Sizer 1996). Pulp and paper companies in North America, Europe, Japan, and Brazil are hoping to survive in increasingly competitive international markets and become one of a small handful of global players (IIED 1996). Mining around the world is dominated by a relatively few large companies, especially from Canada, Australia, and the United States, that can afford the huge capital investments required (Miranda et al. 1998). These companies invest wherever the returns are relatively high and the risk can be managed (e.g., publicly financed insurance from home country governments or the World Bank). For example, international oil company investments in a pipeline across Burma provide otherwise scarce revenue for a military government widely engaged in human rights abuses and destructive logging schemes (Brunner et al. 1998). Rapid influxes of capital and new absentee landlords can irreparably disrupt or destroy local resource practices and communities (Box 3). International economic and political forces can quickly render community-based efforts meaningless.

National governments from Brasilia to Jakarta and from Washington to Wellington are liberalizing economic policies and scaling back their commitment to social programs, including the environment. Governments have increasingly turned to multilateral institutions and instruments to govern trade and investment policies. In some cases, such as in the United States, these trends are actively promoted by those in business and government who seek increased access to other countries' markets and resources. In other cases, such as in Indonesia, Brazil, and Russia, countries reluctantly embrace multilateral institutions and their economic policies because they feel they have no choice. Many countries being pushed to open their economies and scale back government expenditures are dependent on natural resources (Box 4).

BOX 3. Cultural Collisions

Few cultures are as disparate as those of multinational corporations and communities of indigenous forest peoples. While multinational corporations are good neighbors in some communities, relations between outside companies and indigenous peoples are frequently tense and often damaging for local residents. In Irian Jaya, Indonesia, the world's largest gold and copper mine has produced both great wealth for investors and enormous environmental impacts for local residents. Mine tailings have plugged streams flooding farms and villages and local residents assert that toxic chemicals are poisoning their drinking water. Social and cultural effects have been immense as the Amungme and other tribal peoples have been displaced from customary lands and resources (Press 1995). In addition to being shorn from historic resource use patterns and lifestyles, local indigenous groups have seldom been employed by the new companies. In Ecuador, the Chevron Oil company brought in settlers and workers from other parts of the country as it explored for oil in indigenous territories (Sawyer 1996). The indigenous peoples were displaced more by the newcomers and their land claims, diseases, alcoholism, and violence than by the exploration itself. From Canada and Venezuela to Indonesia and the Philippines, indigenous groups increasingly are fighting back and blockading access for mining, logging, and oil drilling.

BOX 4. Multilateral Organizations: Fish or Fowl?

The 1930s and 1940s brought widespread economic and social upheaval as the Great Depression dried up investment capital and jobs, and then the Second World War laid waste to millions of lives and thousands of towns and cities. In the wake of this devastation, several new global institutions–including the United Nations, International Monetary Fund (IMF), World Bank, and General Agreement on Trade and Tariffs (GATT)–were created to help rebuild shattered economies and stabilize the international financial system. Today, these organizations play a central role in shaping international economic and trade policies. They have attracted intense criticism for the environmental and social impacts of their policies, although improvements have been made recently. Still, policies in multi-lateral agencies are typically developed within small circles closed to outside scrutiny. Highly trained economists and trade specialists are influential; virtually everyone else is not.

By some measures, the influence of these organizations is declining. For example, rapidly growing private investment dwarfs World Bank lending. And, the highly secretive proposed Multilateral Agreement on Investments (MAI), which sought to eliminate regulatory obstacles to investment, was recently shelved after intense opposition from nongovernmental organizations (NGOs) and local governments around the world. Still, these agencies retain enormous power to influence trade and investment around the world. For example, the World Trade Organization (WTO) recently ruled the United States could not prohibit shrimp imports from countries that have failed to reduce incidental catches of endangered marine turtles in their shrimping fleets. The IMF and the World Bank have tremendous sway in economically distressed countries, and frequently push governments to expand foreign trade and investment in forestry, agriculture, fisheries, oil and minerals to remedy their debt problems. Multilateral institutions continue to play a role that is quite distinct from either government agencies or private business. They are less accountable than either the government or a private business since a citizen cannot vote for them and consumers can't express their opinion in the marketplace.

Meanwhile, even committed social welfare states such as Canada and Sweden have slashed government expenditures. In Canada for example, budgets for national forest inventory and monitoring have dropped drastically (Canadian Council of Forest Ministers 1997). Forests Canada, a federal agency, has turned over to the provinces a substantial portion of its responsibilities for collecting forest information. In Sweden, as in many other developed countries, budgets and staff for international development agencies have dropped significantly in recent years.

New Forms of Governance for Natural Resources

Citizens and non-governmental organizations (NGOs) have reacted to the growing power of corporations and the declining capacity of government in a number of ways. One way is by international consumer boycotts of multinational corporations, such as that of Shell following publicity about environmental damages caused by its Nigerian operations. The successful NGO and citizen protest against the Multilateral Agreement on Investment is another. And, independent certification of sustainable forest practices shows how consumers are bypassing conventional market institutions to obtain environmental and economic change. Civic society has found an astonishing array of ways to fill the vacuum in governance left by inadequate and declining government capacities.

Social forestry in India, catalyzed by support from the Ford Foundation during the early 1970s, became one of the early modern experiments in community-oriented approaches to natural resource management. Among the key motivations for social forestry was the lack of trust that often existed between state forest agencies and local people. The goals of social forestry included not just better forest management, but increased incomes for poor rural people and new institutional mechanisms at the village level to address various community social needs (Ford Foundation 1998). Social forestry programs were also initiated during the 1970s and 1980s in the Philippines, Thailand, Indonesia, Mexico, and other countries.

By the early 1990s, social forestry programs in some parts of Asia had evolved into "joint forest management" projects (Poffenberger and McGean 1994). Now, public forest lands are jointly managed by both village committees and state forest departments to improve soil and water conservation and villagers' incomes. This evolution reflected increasing sophistication on the part of both government and communities. Government recognized the important role of local people in managing forests, and villagers recognized the need for technical and financial assistance from government.

Many of these programs and changes would not have come about without the growth and evolution of NGOs. In Indonesia, the number of NGOs supporting rural development and conservation grew from a small handful in the mid-1970s to hundreds by the early 1980s. So many organizations developed that an environmental umbrella organization was formed (Wahana Lingkungan Hidup or WALHI) to coordinate advocacy in support of community efforts to conserve natural resources. The phenomenal growth of national and regional NGOs has been fueled by donors–such as the MacArthur and Ford Foundations, U.S. Agency for International Development, and German GTZ–seeking more effective and responsive mechanisms to improve social and environmental conditions. Not surprisingly, NGOs are often most numerous in areas and sectors where government capabilities are relatively weak (Meyer 1997).

Problems and Opportunities Posed by New Patterns of Governance and Control

Globalization and decentralization pose both problems and opportunities for communities seeking to improve natural resource management. Among the problems caused by new forms of governance and ways to control natural resources:

- Liberalized economic policies that undermine community forms of natural resource governance. For example, the Mexican government weakened long standing communal forest management systems to favor private forest land ownership consistent with provisions of the North American Free Trade Agreement (NAFTA).
- Rapid influxes of money and people to develop resources that overwhelm government capacities to plan, monitor, and regulate environmental and social impacts.
- A lack of authority and legitimacy for civic organizations and NGOs to negotiate or enforce regulations to alter unsustainable natural resource practices by individuals, businesses, and government agencies.
- The rapid movement of capital which complicates efforts to find long-term solutions to environmental and social problems.

Among the opportunities globalization and decentralization present to communities:

- Growing public and institutional support to increase the capacity of local, national, and international NGOs to assess, advocate, and implement community-based ecosystem management.
- More accessible communication technologies (e.g., fax and internet) that enable communities to share information, experiences, and ideas, and to coordinate actions on common concerns.

- The growing capacity of indigenous peoples, networked through the internet and coordinated by umbrella groups, to address land claims and resource management policies at local, national, and international levels.
- New markets around the world, which enable small enterprises and communities to sell sustainably produced goods and services to buyers who were out of reach just a few years ago.
- More responsive and innovative forms of local government that sometimes enable communities to have more authority and responsibility to govern natural resources.

In short, globalization and decentralization are shaping new opportunities and constraints for community-based ecosystem management in the United States.

MIGRATION: THE TIES THAT BIND COMMUNITIES AND ECOSYSTEMS

The United States is a nation of migrants. Migration ties rural and urban residents in this country to diverse cultures around the world. We cannot avoid tracing our origins to one or another wave of voluntary or forced migration from Europe and Africa and more recently from Asia and Latin America. That heritage has probably contributed to the propensity of Americans to move much more frequently within their own country than do the residents of other countries. Migration is a fundamental part of our cultural fabric.

The migration and trade patterns of the earliest Asian migrants and their Native American descendants also established biological bonds we cannot break. Most of the staple foods we take for granted in the United States, including corn, potatoes, tomatoes, wheat, and barley, came from somewhere else, especially Central and South America and Asia. Much of the genetic material we require for continuous crop improvement still originates in the mountains and valleys of countries far from the agricultural fields in the United States (Raeburn 1995). Only recently have we developed a deeper understanding of the connectivity of nature. From tiny hummingbirds to migratory whales, thousands of species migrate across the hemisphere irrespective of political boundaries. Ecological health cannot be maintained within borders–local or international.

"Knowledge" has migrated into and out of this country as well. Physics, chemistry, and now biotechnology and computer science have benefited enormously from the migration of individuals and their knowledge. Indigenous knowledge of tropical plants for medicinal purposes is now helping to spark a renaissance in the field of ethnopharmacology in the United States and other developed countries (Peppall 1996). The importance of U.S. information technology in promoting global integration and understanding global trends–including ecological ones–cannot be underestimated. By the same token, we cannot underestimate the contributions that foreign knowledge and human resources are making to a variety of ecological and resource management issues in the United States. One example is the rapid growth of the nontimber forest products economy in the western United States, driven in large measure by immigrants from Asia and Latin America. This economy has provided new incentives to manage forests for more than timber.

Underlying Factors That Drive Human Migration

Today few of us take time to understand the underlying reasons for human migration. When we do, we are typically drawn to simple explanations–overpopulation, crowding, and a lack of jobs. Many of us don't consider possible political and environmental connections. The migration of Central Americans to the United States during the last two decades has been strongly motivated by synergistic interactions between human rights abuses, political strife, and environmental degradation. So too, Irish immigration to the United States during the 19th century–often blamed simply on the potato blight–was the result of a web linking environmental degradation, inequitable land tenure, and economic domination by a very small elite (Coffey and Golway 1998). Recent studies on Mexican migration show desertification to be one of the most overlooked factors driving this (Schwartz 1997). Addressing these underlying causes over the long term is likely to be far more effective than focusing on short-term, simplistic solutions (Box 5).

Migration: Challenges and Opportunities to Community-Based Ecosystem Management

Migration–internal as well as external–poses both challenges and opportunities for community-based ecosystem management in the United States. Newcomers seldom are integrated rapidly into existing communities. Yet, migrants often bring new ideas, new skills, and strong work and family ethics that can strengthen community efforts to sustainably manage natural resources.

Migration poses certain challenges for community-based natural resources management including:

1. Often perceived as outsiders who are not part of the "community," migrants often are excluded when it comes to community activities. This is especially problematic for forest and agricultural workers who do not remain in one place for long.

2. Community residents often understand little about migrants and perceive them as threatening established ways of life. Such perceptions often mean community-based efforts overlook migrants' potential contributions of knowledge, perspectives, cooperation, and labor.
3. Some individual migrants and groups of migrants choose to remain detached from the community and from community-based ecosystem management initiatives.

It's worth noting that most migrant populations eventually become vested and rooted in a community. Ultimately they develop influence and interest in the environmental and natural resource problems it faces. Even small and mobile populations of migrants can add their interests and skills to community-based ecosystem management initiatives. The key is to invite their participation and keep the process open to them in a constructive manner. Not doing so may bring challenges that truncate community-based ecosystem management processes, plans, and projects.

Immigrants may be able to contribute to ecosystem management initiatives in positive and unforeseen ways. Two such possibilities are:

Knowledge and Perspectives–Community-based ecosystem management is a new approach that requires uncommon knowledge and perspectives. Ecotourism, agroecology, and biosphere reserves, for example, have roots in developing countries. The development of a nontimber forest products industry in the Pacific Northwest is due in large part to immigrant farm workers who recognized uses and markets for such things as bear grass, salal, matsutaki mushrooms, huckleberry, mosses, and ferns. This industry is now valued at least $200 million in Washington, Oregon, and northern California (Johnson and Ditz 1997).

BOX 5. Pointing the Finger

The recent debate within the Sierra Club over whether the organization should lobby for restrictive immigration policies is a good example of a simplistic prescription to treat symptoms rather than causes. Proponents of restrictive immigration policies conveniently blamed urban sprawl and environmental degradation in their backyards on people from somewhere else. They failed to address our own profligate resource consumption in the United States and how U.S. companies' practices and U.S. government policies might be contributing to environmental degradation overseas. Ironically, that degradation often occurs in the developing countries from which many immigrants come. And, the proponents of restrictive immigration policies overlook the contributions immigrants make to the economy, the enrichment of our cultural fabric, and even to the solution of environmental problems. In the long term, it will be more productive to support sustainable economic development in poor countries and implement policies and economic incentives that reduce our own excessive resource consumption at home (Gwynallen 1998).

Labor–Immigrant labor has been critical to restoring forests and watersheds in some parts of the United States. Reforestation, ecosystem restoration, and other activities essential to restoring ecosystem health in degraded areas often require hard physical labor under demanding environmental conditions. Low wages and few benefits typically mean that most non-immigrant residents are not interested in such work. Migrant workers, therefore, have become an increasingly important part of forest economies in the United States, especially in the Pacific Northwest. Their history and contributions to natural resource management are little understood by either scholars or the public. Sustainability is as important to them as it is to people whose families may have lived in a place for generations (Box 6).

TOOLS AND TECHNIQUES FROM OTHER PARTS OF THE WORLD

Conservation in developing countries most often takes place in a context dominated by poverty, dependence on natural resources, and limited opportunities for public participation in economic and policy decisions. Conservation strategies developed abroad, therefore, have often pivoted around five core principles:

1. All resource users should have an opportunity to participate or contribute to planning and policy processes.
2. Technical and financial resources should target building capacity locally to sustainably manage resources and to reform regional and national institutions so they work more effectively with local institutions.
3. Resource conservation and management should improve community social and economic conditions, particularly for the poor.
4. Conservation efforts should be monitored and evaluated at regular intervals to test underlying ecological and social assumptions.
5. Information on natural resource-related conditions, trends, policies and organizations should be widely accessible to all resource users.

Tools and techniques developed in other parts of the world to address these same principles may also be germane to many situations in the United States. The remainder of this section briefly summarizes a few of these tools and techniques.

Participatory Planning Approaches

Natural resource policy and planning became progressively centralized and specialized in many countries following the Second World War. This

BOX 6. Migrant Forest Workers in the Pacific Northwest

There is a long and rich history of immigrants becoming forest workers in America. The first European-American settlers during the Colonial era were in essence forest workers who transformed forests into houses, ships, and agricultural fields. Later, immigrants from northern Europe and Scandinavia provided much of the itinerant labor force that pushed logging frontiers from the Appalachians to the Great Lakes and on to the Pacific Northwest. Meanwhile, immigrants from Ireland, China, and elsewhere built the railroads, mines, and towns that helped transform America's natural resources into the industrial and economic assets that made it a world power. Like many who would follow, these immigrants were disenfranchised and poor, attracted to the promise of a better life for their children if not for themselves (Brown 1995). Often exploited and sometimes unwelcome, many of these immigrants eventually put down roots, fought for better economic and social conditions, and became important members of their communities.

Half a century later, beginning in the 1940s, Mexican workers were brought to the Pacific Northwest to harvest fruit and vegetables. By the mid-1960s, a small number of Mexican workers were contracted to plant trees on public and private forest lands. Economic hardships in Mexico helped expand the region's migrant workforce and promote conditions where low wages and poor working conditions were commonplace. These conditions pushed local workers into other occupations and by the early 1980s, 85 to 90 percent of the reforestation and precommercial thinning were done by Mexican workers. The recently burgeoning nontimber forest products industry also relies on a pool of migrant workers, especially from Laos and Cambodia, to harvest wild mushrooms, fruit, and floral display materials from the region's forests.

Like other migrants, today's forest workers face extremely difficult conditions and little or no legal protection. For years, reforestation contractors used a pay formula that varied from federal lands (where federal rules set relatively high rates) to much lower "prevailing wages" on county, state, and private lands. Workers were often kept in the dark about whose lands they were planting, thus allowing dishonest contractors to pocket the difference. Meanwhile, contractors often use the "company store" concept, inflating the prices of food, lodging, and equipment for workers who have little choice. Many labor contractors use the "ahorro" or "saving formula"–money deducted from pay for savings or credit that workers can ostensibly access at any time. Unfortunately, abuses of the "ahorro" system are widespread, and workers are often cheated out of substantial earnings by corrupt contractors. Even if workers are not cheated outright, the inequities between workers and contractors are frequently stark. For example, workers may receive as little as 15 to 20 percent of the full value of a reforestation contract while contractors retain 80 percent or more–an exhorbitant sum even after accounting for legitimate overhead costs (Benavides 1998).

Poor working conditions have continued for several reasons. One is that many workers are recent immigrants with limited education from places with very few jobs. The lack of alternatives makes them more willing than others to tolerate low wages and deplorable working conditions. Second, many workers are unaware of their rights and the services available to them because of their limited or nonexistent English language skills and inadequate funding for the public and nonprofit agencies that serve them. This has been exacerbated by recent welfare reform legislation that makes it harder for migrant workers to get access to health care, disability, and other benefits. Third, many workers are undocumented and therefore especially vulnerable to exploitation. Some contractors simply fire workers who seek better conditions because they know undocumented workers fear deportation by the Immigration and Naturalization Service (INS).

The diversity of people working in American forests is more evident today than in any period of history. Unfortunately, the conditions and prospects for migrant forest workers today are no better, and probably worse, than those faced by earlier generations. Better pay and improved working conditions for migrant forest workers should be on the agenda for community-based ecosystem management efforts. Practical dividends are likely since improved working and social conditions can help to draw migrant workers' skills and cooperation into community-based ecosystem management efforts.

occurred for a variety of reasons including the export of management models from industrialized countries, the growing complexity of planning technologies, and the concentration of political and economic power in capital cities. International efforts to slow deforestation and other forms of widespread environmental problems tended to reinforce this tendency. For example, the Tropical Forest Action Plan (TFAP) was designed in the mid-1980s by the U.N. Food and Agriculture Organization, the World Bank, and the World Resources Institute to mobilize international financial resources and national policy reforms to reverse alarming forest losses. In practice, TFAP processes at the national level were dominated by strategies for accelerating donor investments to build national government bureaucracies and to subsidize industrial forest development (Winterbottom 1990). The TFAP process has since been widely viewed as a failure, in part because local stakeholders had little or no influence.

Against this backdrop of centralization a counter current of participatory planning began to develop during the 1980s. These efforts assume that top-down regulation will not change resource use. Rather, involving a wide variety of resource users in planning is more likely to identify important problems and realistic solutions supported by the resource users themselves. Participatory Rural Appraisal (PRA) and Rapid Rural Appraisal (RRA) are related forms of planning now widely used in the tropics in community-based resource management. Both are adaptive methods for working with resource users to systematically assess local ecological conditions, the socio-economic context for resource use, institutional interactions, and changes in practices and policies to reach objectives (Kenya NES et al. 1990; Chambers et al. 1989; Poole 1995; Freudenberger 1996). These exercises are vital departure points for further discussions about resource use because they form a common base of understanding of what is happening across the landscape, both in terms of resource use and ecological change (Box 7).

One factor complicating many planning efforts is a lack of agreement on the validity of data. For example, in some areas environmental groups, the forest industry and the U.S. Forest Service maintain separate Geographic Information Systems (GIS) and contest one another's data. Yet, implementation of watershed and ecosystem-based resource management depends on various parties agreeing about what data is credible. In Zimbabwe's CAMPFIRE program, local communities, international conservation groups, and the government jointly decided what wildlife and related information to collect, divided responsibilities for collecting it, and then came together at regular intervals to assess what the information meant for management and policy (Metcalfe 1994). Similar strategies are being used in the United States. For example, in northeastern Oregon's Wallowa County a community-based group formed to integrate concerns about watershed health and local jobs (Waage

BOX 7. Participatory Rural Appraisal

Participatory Rural Appraisal (PRA), and the closely related Rapid Rural Appraisal (RRA), approach were originally developed to offer an alternative to centrally planned and externally managed development in rural regions of Asia and Africa (Ford et al. 1992). They have since been adapted and used for natural resource management and conservation purposes in many developing countries. They are based on three premises. First, local resource users and residents have significant knowledge about the environment, but it needs to be organized. Second, local individuals and communities have the technical, financial, and human resources to address ecosystem management problems, but they need to be mobilized. Third, outside resources are frequently needed to solve problems, but communities must define their own priorities first. While RRA and PRA are related, they differ in several important respects. RRA relies on an interdisciplinary team of experts interviewing residents while PRA actually engages residents in information gathering and analysis. RRA is intended to quickly identify potential solutions to the most severe problems identified by the community. PRA can take months and its techniques may be used as part of an ongoing community-based monitoring system (Margolius and Salafsky 1998). Both use similar appraisal methodologies. These include sketch maps and transects to build an overview of area resources and problems, seasonal resource availability and use calendars, long-term trend lines, institutional diagrams, and ranking of resource problems and solutions (Kenya NES et al. 1990).

PRA has been used for a wide range of objectives. In the Gambia, farming communities faced with perennial drought, population growth, and little or no prospect of significant help from the government or outside NGOs, used PRA to prioritize communal actions to address the most serious problems (Ford et al. 1992). In Pakistan, farmers in remote mountain valleys identified new crops, practical ways to organize technical assistance within the community, and strategies to market and transport their produce to generate increased family incomes (McCracken et al. 1988). While used principally in developing countries, PRA has been increasingly used in developed countries. For example, local communities in the Scottish highlands used PRA to evaluate forestry proposals for a block of poorly managed public forest lands to assess proposals for their potential contributions to local employment, increased tourism income, and improved wildlife habitat and sheep pastures (Inglis and Lussignea 1995). PRA is not easy. It requires substantial time from many volunteers, technical information that may not be locally available, and it can generate biased results if participation is not carefully structured. Still, PRA techniques can provide effective tools when communities can't wait for or don't trust assessments by outside specialists and government agencies.

1998). Together with the local Soil and Water Conservation District, the Natural Resources Conservation Service, the U.S. Forest Service, and more than 50 private land owners, the group is talking about initiating a participatory, multi-party watershed assessment. The groups plan to develop a mutually agreed-upon protocol for collecting data from both public and private lands for common understanding of what is occurring within the watershed. In coming years the assessment would be used to develop a watershed management plan that ranks management and restoration activities.

Locally-Appropriate Technical Support

The rapid expansion of local and national NGOs has been supported internationally by non-profit counterparts based in North America and Europe, as well as by governmental foreign aid agencies. Lessons in developing locally appropriate resource management and conservation strategies are now being brought home to the United States. For example, the Grand Canyon Trust, Sustainable Northwest, Ecotrust, and the Sonoran Institute all bring tools and staff with international experience to their regional activities in the United States. They use different approaches to convene, facilitate, and build capacity in rural communities to address ecological degradation and economic challenges. Many of the staff in these four organizations worked on social forestry and integrated conservation and development projects in other countries for many years prior to working at home.

Universities also play a role in helping communities become active participants in policy debates that affect their livelihoods. For example, the Centre for Applied Social Science at the University of Zimbabwe is working closely with communities, NGOs, and government agencies to help provide the analytical basis for implementing the CAMPFIRE program. In the United States, the Ford Foundation's Community Forestry Research Fellowship Program is developing reciprocal relationships between universities and rural communities. Graduate students receive research fellowships sponsored by a partner community-based group; the student and the group then jointly assess research and information needs on a topic of mutual interest. This collaborative work brings people on the ground closer to policy debates and researchers closer to community issues and realities. A recently initiated program brings together international researchers and community resource managers to share lessons (Box 8). The premise is that researchers and practioners in the United States can learn from the relatively long record of community-based resource management in other parts of the world.

Locally-Appropriate Financial Support

One universally common challenge facing community-based resource management is insufficient funding. Advocates of community-based conservation and development have to be creative in finding financial support for their efforts. Sustainable enterprise grants and loans and community-controlled revolving funds are just a few of the innovative tools that have been developed during the past decade. For example, Chicago's South Shorebank, in partnership with Ecotrust, helped to create Shorebank Pacific to finance locally based "sustainable" enterprises in the coastal Pacific Northwest. Similar enterprise-based conservation efforts are being tested in Asia to support independently certified sustainable forest management, monitoring and mar-

BOX 8. Linking Community-Based Resource Management in California, Zimbabwe, and the Philippines.

In December 1997, three organizations seeking to learn from each other's experiences in community-based resource management convened a workshop in the Philippines. From California, the Department of Environmental Science, Policy and Management organized a delegation that included faculty and students, a U.S. Forest Service district forest ranger, a tribal forester working for the Hoopa Valley Tribe, and members of the Watershed Research and Training Center, a small nongovernmental organization (NGO) based in forest-dependent Trinity County. From Zimbabwe, the Centre for Applied Social Science (CASS) brought NGO partners working with the Communal Areas Management Program for Indigenous Resources (CAMPFIRE), southern Africa's first initiative devolving control of wildlife and forests from central to district and local levels. The University of the Philippines School for Environmental Science and Management (SESAM) hosted the workshop and brought in representatives of Philippine NGOs and community organizations actively involved in community-based resource management. To put the workshop into context, participants visited a communal forest stewardship site in the small community of Kalahan in the mountains of central Luzon.

While all the groups have been pioneers in community-based resource management, each came from a distinctly different policy and management setting. Community-based approaches in northern California are relatively recent, growing out of the need to find economic alternatives to a timber industry that began a precipitous decline in the late 1980s. By contrast, nearly two decades of "social forestry" development in the Philippines led President Ramos in 1996 to declare community-based resource management as the foundation for national forest policy. Meanwhile, the CAMPFIRE program has been widely acclaimed as a model for building local capacity to sustainably manage natural resources with active involvement by government agencies and local, national, and international NGOs.

During the week-long workshop, participants presented their research experiences and their issues, problems, and achievements implementing community-based resource management. In each case, workshop participants sought to identify which factors facilitated or constrained community-based resource management. In this sense, differences are as significant as similarities, and failures are as important as successes. Indeed, one of the goals of the collaboration was to frame a critical analysis of community-based efforts, not simply to blindly promote them.

Still, the group found many similarities among rural communities in the three countries. These included economic dependence on natural resources, lack of legal access and control, lack of capital, the export of raw products, and persistent poverty. The discussion also pointed to common issues that demand further investigation. For example, do communities really call the shots in community-based resource management, or do they mostly "participate" in a scenario designed by the state or corporations? Who is accountable to whom in community-based resource management? How much authority can communities exercise when they do not control capital and lack legally guaranteed rights to land and natural resources? How do markets work in remote areas, and can local folks get fair prices in national and international markets? Finally, who is the "community" and who benefits from community-based resource management–indigenous people, long-standing migrants, recent migrants, men, women? Not surprisingly, the international workshop raised more questions than conclusive answers. As a result, participants are developing proposals to continue a joint U.S.-Zimbabwe-Philippines program of research, training and exchange to understand and overcome obstacles to community-led rural development.

keting of nontimber forest products, and locally managed ecotourism (BSP 1997).

Large-scale debt swaps have financed tens of millions in conservation activities in a handful of developing countries during the past decade. In the 1980s, a number of developing countries, including Bolivia, Mexico, and the Philippines, faced staggering debts triggered by ill-advised loans to public sector enterprises from the World Bank and private banks with large international portfolios. Many of these countries were simultaneously facing persistent rural poverty, high deforestation rates, and a lack of funding to invest in rural conservation and development projects. In 1987, Conservation International pioneered an innovative new mechanism for conservation funding in which some of a country's public sector debt would be exchanged for local currency funds from the central bank dedicated to conservation and development activities (Guerin-McManus 1998). The first of these "debt-swaps for nature" took place in Bolivia and Mexico, and others soon followed in the Philippines, Ecuador, and elsewhere. A dollar of debt, often purchased for only 10 or 20 cents, raised a dollar's worth of local currency for conservation. While debt swaps were sometimes controversial (many critics suggested debts incurred in poorly advised loans should be forgiven outright), and some had implementation problems, this innovative concept was extremely successful in turning a problem into an opportunity.

In the United States, the debt-swap concept may have potential in certain situations. For example, the Bonneville Power Administration (BPA) in the Pacific Northwest has responsibilities both to provide the region with an economical and reliable power supply and to ensure that its actions protect fish and wildlife resources. BPA also has incurred a mountain of debt ($16.3 billion) building and maintaining its infrastructure–a factor cited by the agency when it recently cutback its funding for conservation activities. As Goebel (Goebel 1995) points out, the BPA case may be ripe for use of the debt-swap concept. Funding is a major challenge for the recently established community-based watershed councils throughout Oregon, Washington, and Idaho. They are searching for ways to promote new economically and environmentally sustainable forms of development. In exchange for forgiveness of its public debt, BPA could pay into a regional trust fund supporting creation of local trust funds dedicated to watershed health recovery and environmentally compatible forms of economic development. BPA is already experimenting with creative salmon and watershed enhancement through marketing of "green" power, an approach that may be adaptable to developing countries. Thus, creative financing approaches here and abroad can be transferred and adapted to resolve real ecosystem management needs.

Sharing Lessons: Opportunities and Limitations

Although community-based resource management projects have many lessons that can be applied to the United States, there are some significant differences that limit direct, uncritical transfer. Community-based ecosystem management in the United States cannot be based on romantic notions of local residents living in balance and harmony with nature. However, it also should not be assumed that local residents are disinterested in sustainable natural resource management. Viewing resource users as full partners in developing solutions to ecological problems is a prerequisite to any community-based approach to natural resource management.

CONCLUSIONS

Though relatively new to the United States, community-based ecosystem management should be viewed as part of a broader local environmental movement. All over the world community-driven efforts have arisen to balance economic development and protect cultural values and natural resources. They emerge, spontaneously it sometimes seems, as people respond to environmental change and disruptions and to countervailing forces of globalization and devolution. In the 1990s the nation-state is proving too small to manage the transnational global economy, and yet too large and bureaucratic to be responsive to all its constituents, especially rural people in remote places. The private sector, and non-governmental organizations in particular, have stepped in to take up the middle ground. They are providing significant support and co-leadership to local community groups struggling to obtain a voice at the policy table. All across Asia, Africa and Latin America–and increasingly here in the United States–partnerships and collaborative decisionmaking are replacing the old centralized model of resource management.

So what are the lessons for community-based ecosystem management in the United States? First, although community-driven efforts to manage natural resources respond to locally perceived environmental and social problems, the forces behind such transitions–whether liberalized economic policies, rapid demographic change, rapid movement of capital, or escalating consumption rates–are transnational. NGOs and local institutions (including government) usually lack authority and legitimacy to negotiate agreements or enforce them with institutions that operate on national and international levels. Community-based efforts and collaborative processes cannot, in and of themselves, be effective at changing these global forces. They must be strengthened through increased communication and sharing between communities and by finding new political and economic mechanisms to influence global forces.

Second, the diversity that characterizes many communities is both a strength and a weakness for community-based ecosystem management. Diverse interests bring more information and skills to the table. Still, mobilizing different interests and finding common ground is complicated, time-consuming, and sometimes simply not possible. In particular, migratory peoples such as forest workers in the Pacific Northwest add diversity to communities but are often left out of communities of place and interest. Bringing traditionally marginalized groups into the partnership–be it because of race, ethnicity, or gender–still plagues community-based resource management around the world. In countries with longer histories in community-based resource management, there is often a tendency for more economically or politically powerful interests to co-opt others in ways that short-circuit wide participation in resource management. Local elites can frustrate public participation and benefit sharing just as easily–sometimes more so–than national elites. Local control and public participation are not synonymous.

Third, while many tools and techniques have relevance in the United States–participatory planning and appraisal methods, use of indigenous knowledge and experience, creative funding and transnational research and exchange programs–"technology transfers" need to recognize the particulars of place, culture, and geography.

Finally, there are places and issues where community-based ecosystem management is unlikely to work. Despite almost two decades of such efforts overseas, many involved in them still consider them experiments. Better monitoring and evaluation methods, however, are helping to harvest new lessons and insights. Meanwhile, the rise of community-based approaches to natural resources management is a global phenomena, not a collection of isolated and unrelated cases. Community-based resource management may not be perfect, but its rise signals widespread frustration with natural resource management dominated by small groups of specialists and powerful interests.

REFERENCES

Agarwal, A. 1997. Community in conservation: beyond enchantment and disenchantment. *Conservation & Development Forum Discussion Paper.* Conservation and Development Forum. Gainesville: Tropical Research Institute.

Barber, C. V., S. Affif, and A. Purnomo. 1995. *Tiger by the Tail: Reorienting Biodiversity Conservation and Development in Indonesia.* Washington: World Resources Institute.

Belsky, J. M. *In press.* The meaning of the manatee: An examination of community-based ecotourism in Gales Point, Belize. In Charles Zerner (ed.), *People, Plants and Justice.* New York: Columbia University Press.

Benavides, V. 1998. Conversation with Martin Goebel, Sustainable Northwest, 26 June.

Brown, B. 1995. *In Timber Country*. Philadelphia: Temple University Press.

Brown, M. and B. Wyckoff-Baird. 1993. *Designing Integrated Conservation and Development Projects*. Washington, DC: Biodiversity Support Program.

Brosius, J. P., A. L. Tsing, and C. Zerner. 1998. Representing communities: histories and politics of community-based natural resource management. *Society and Natural Resources* 11(3):157-168.

Brunner, J., K. Talbott and C. Elkin. 1998. *Resources and the Regime: Logging Burma's Frontier Forests*. Washington, DC: World Resources Institute.

Bryant, D., D. Nielsen, and L. Tangley. 1997. *The Last Frontier Forests: Ecosystems and Economies on the Edge*. Washington, DC: World Resources Institute.

BSP. 1997. *Stories from the Field and Lessons Learned: Evaluating an Enterprise-Oriented Approach to Community-Based Conservation in the Asia/Pacific Region*. Washington, DC: Biodiversity Support Program.

Canadian Council of Forest Ministers. 1997. *Compendium of Canadian Forest Statistics 1996*. Ottawa: Natural Resources Canada.

Chambers, R., A. Pacey, and L. A. Thrupp. 1989. *Farmer First: Farmer Innovation and Agricultural Research*. London: Intermediate Technology Publications.

Coffey, M. and T. Golway. 1998. *The Irish in America*. New York: Hyperion Books.

Cohn, L. 1995. The growing market for special forest products: an opportunity and challenge. *Forest Perspectives* 4(4): 4-9.

Dudley, N. 1992. *Forests in Trouble: A Review of the Status of Temperate Forests Worldwide*. Gland, Switzerland: WWF-World Wide Fund For Nature.

Duke, J. A. 1998. Phytomedicinal forest harvest in the United States. In: G. Boedker, K. K. S. Bhat, J. Burley, and P. Vantomme (eds.), *Medicinal Plants for Forest Conservation and Health Care*. Rome: UN Food and Agriculture Organization.

EDF. 1995. *Paper Task Force Recommendations for Purchasing and Using Environmentally Preferable Paper*, The Paper Task Force: Duke University, Environmental Defense Fund, Johnson & Johnson, McDonald's, The Prudential Insurance Company of America, Time Inc., New York.

Eisner, T., J. Lubchenco, E. O. Wilson, D. S. Wilcove, and M. J. Bean. 1995. Building a scientifically-sound policy for protecting endangered species. *Science* 268: 1231-1232.

FAO. 1993. *The Challenge of Sustainable Forest Management: What Future for the World's Forests?* United Nations Food and Agriculture Organization, Rome.

FAO. 1996. *FAO Yearbook: Forest Products, 1983-1994*. Rome: United Nations Food and Agriculture Organization.

Filion, F. L., J. P. Foley, and A. J. Jaquemot. 1992. The economics of global ecotourism. Paper presented at the Fourth World Congress on National Parks and Protected Areas, January 1992, Caracas, Venezuela.

Ford Foundation. 1998. *Forestry for Sustainable Rural Development*. New York: Ford Foundation.

Ford, R., C. Kabutha, N. Mageto, and K. Manneh. 1992. *Sustaining Development Through Community Mobilization*. Worcester, MA: Clark University.

Freudenberger, M. S. 1996. *The Future of the Ndongo Lagoon: A Rapid Rural*

Appraisal Case Study of Stakeholders and Conservation in Coastal Gabon. Washington, DC: World Wildlife Fund, West & Central Africa Program and Social Science and Economics Program.

Gilliam, F. S., M. B. Adams, and B. M. Yurish. 1996. Ecosystem nutrient responses to chronic nitrogen inputs at Fernow Experimental Forest, West Virginia. *Canadian Journal of Forest Research*. 26(2): 196-204.

Goebel, M. 1995. Money for salmon recovery could come from debt swaps. *The Oregonian*: August 14, 1995.

Guerin-McManus, M. 1998. The greening of international finance: ten years of debt-for-nature swaps. Issues paper. Washington, DC: Conservation International.

Gwynallen, R. 1998. Immigration restrictions won't save the planet. *The Oregonian:* April 24, 1998.

IIED. 1996. *Towards a Sustainable Paper Cycle: Final Report*. London: International Institute for Environment and Development.

Inglis, A. and A. Lussignea. 1995. Participation in Scotland: The Rural Development Forestry Program. *Participatory Learning and Action Notes* 23: 31-36.

IPCC. 1996. *Working Group II Second Assessment Report*. Geneva: International Panel on Climate Change.

Johnson, N. and B. Cabarle, 1993. *Surviving the Cut: Natural Forest Management in the Humid Tropics*. Washington, DC: World Resources Institute.

Johnson, N. and D. Ditz. 1997. Challenges to sustainability in the U.S. forest sector. In: Dower et. al. (eds.) *Frontiers of Sustainability: Environmentally Sound Agriculture, Forestry, Transportation, and Power Production*. Washington, DC: Island Press.

Kenya NES, Egerton University, Clark University, and WRI. 1990. *Participatory Rural Appraisal Handbook*. Washington, DC: World Resources Institute.

Keystone. 1996. *The Keystone National Policy Dialogue on Ecosystem Management. Final Report*. Keystone, CO: The Keystone Center.

Likens, G. E., C. T. Driscoll, and D. C. Buso, 1996. Long-term effects of acid rain: response and recovery of a forest ecosystem. *Science* 272: 244-246.

MacKenzie, J. J. and M. T. El-Ashry, 1989. *Air Pollution's Toll on Forests and Crops*. New Haven, CT: Yale University Press.

Margolius, R., and N. Salafsky. 1998. *Measures of Success: Designing, Managing, and Monitoring Conservation and Development Projects*. Washington, DC: Island Press.

McCloskey, M. 1996. The skeptic: Collaboration has its limits. *High Country News* 28 (9):7.

McCracken, J. A., J. N. Pretty, and G. R. Conway. 1988. *An Introduction to Rapid Rural Appraisal for Agricultural Development*. London: International Institute for Environment and Development.

Metcalfe, S. 1994. The Zimbabwe Communal Areas Management Programme for Indigenous Resources (CAMPFIRE). In Western, D. and R.M. Wright (eds.). *Natural Connections: Perspectives in Community-Based Conservation*. Washington, DC: Island Press.

Meyer, C: 1997. Public non-profit partnerships and North-South green finance. *Journal of Environment and Development*. 6(2):123-130.

Miller, K. R. 1996. *Balancing the Scales: Guidelines for Increasing Biodiversity's Chances Through Bioregional Management*. Washington, DC: World Resources Institute.

Miranda, M., A. Blanco-Uribe, L. Hernandez, J. Ochoa, and G. E. Yerena. 1998. *All That Glitters Is Not Gold: Balancing Conservation and Development in Venzuela's Frontier Forests*. Washington, DC: World Resources Institute.

Peppall, J. 1996. Putting a price on indigenous knowledge. *International Development Research Centre (IDRC) Reports*. October 25, 1996.

Poffenberger, M. and B. McGean (eds.). 1994. *Policy Dialogue on Natural Forest Regeneration and Community Management*. Proceedings of a workshop, Honolulu, HI. The Asia Sustainable Forest Management Network. Honolulu: The East-West Center.

Poffenberger, M. and R. Stone. 1996. Hidden faces in the forest: A twenty-first century challenge for tropical Asia. *SAIS Review* (Winter-Spring). 16:203-214.

Poole, P. 1995. *Indigenous Peoples, Mapping and Biodiversity Conservation: An Analysis of Current Activities and Opportunities for Applying Geomatics Technologies*. Washington, DC: Biodiversity Support Program.

Press, E. 1995. Freeport-McMoran at home and abroad. *Nation* July 31/August 7.

Raeburn, P. 1995. *The Last Harvest: The Genetic Gamble That Threatens to Destroy American Agriculture*. New York: Simon and Shuster.

Research and Development. 1998. Bioprospecting at Yellowstone. *Research and Development* 40(2):19.

Sawyer, S. 1996. Indigenous initiatives and petroleum politics in the Ecuadorian Amazon. *Cultural Survival*. 20:26-30.

Schwartz, M. 1997. *Environmental Degradation and Migration*. San Francisco: Natural Heritage Institute.

Sharma, N. P. 1992. *Managing the World's Forests*. Dubuque, IA: Kendall/Hunt Publishers.

Sizer, N. 1996. *Profit Without Plunder: Reaping Revenue from Guyana's Tropical Forests Without Destroying Them*. Washington, DC: World Resources Institute.

TNC, 1996. *Priorities for Conservation: 1996 Annual Report Card for U.S. Plant and Animal Species*. Arlington, VA: The Nature Conservancy.

UNEP. 1995. *Global Biodiversity Assessment*. Cambridge: Cambridge University Press.

U.S. Bureau of the Census. 1997. *Statistical Abstract of the United States 1997*. Washington, DC: U.S. Department of Commerce.

Waage, S. 1998. Conversation with Nels Johnson, World Resources Institute, 28 August.

Wells, M. and K. Brandon. 1992. *People and Parks: Linking Protected Areas with Local Communities*. Washington, DC: World Bank.

Western, D. and R. M. Wright. 1994. *Natural Connections: Perspectives in Community-Based Conservation*. Washington, DC: Island Press.

Winterbottom, R. 1990. *Taking Stock: The Tropical Forestry Action Plan After Five Years*. Washington: World Resources Institute.

WRI. 1996. *World Resources 1996-97: A Guide to the Global Environment*. New York: Oxford University Press.

WRI. 1998. *World Resources 1998-99: A Guide to the Global Environment*. New York: Oxford University Press.

Yaffee, S. L., A. F. Philips, I. C. Frentz, P. W. Hardy, S. M. Malecki, and B. E. Thorpe. 1996. *Ecosystem Management in the United States: An Assessment of Current Experience*. Washington, DC: Island Press.

Governance and Institutions: Opportunities and Challenges

Hanna J. Cortner
Sam Burns
Lance R. Clark
Wendy Hinrichs Sanders
Gus Townes
Martha Twarkins

SUMMARY. The institutional and governance issues that facilitate or impede community-based ecosystem management are discussed. Community-based ecosystem management has a variety of institutional models that differ in purpose, history, scope of interest, and capabilities. Several governance issues face these institutions, including: the adequacy of existing laws and policies; the role of government and policy tools such as money, messages, and mandates; mechanisms for cross-jurisdictional management; the effects of organizational culture and

Hanna J. Cortner is affiliated with the University of Arizona, School of Renewable Resources, 325 Bio Science East, Tucson, AZ 85721-0043.

Sam Burns is affiliated with the Fort Lewis College, Office of Community Services, 1000 Rim Drive, Durango, CO 81301-3999.

Lance R. Clark is affiliated with the Oregon Department of Forestry, 2600 State Street, Salem, OR 97310.

Wendy Hinrich Sanders is affiliated with the Great Lakes Forest Alliance, P.O. Box 722, Hayward, WI 54843.

Gus Townes is affiliated with the Federation of Southern Cooperatives, P.O. Box 95, Epes, AL 35460.

Martha Twarkins is affiliated with the Finger Lakes National Forest, 5218 State Route 414, Hector, NY 14841.

[Haworth co-indexing entry note]: "Governance and Institutions: Opportunities and Challenges." Cortner, Hanna J. et al. Co-published simultaneously in *Journal of Sustainable Forestry* (Food Products Press, an imprint of The Haworth Press, Inc.) Vol. 12, No. 3/4, 2001, pp. 65-96; and: *Understanding Community-Based Forest Ecosystem Management* (ed: Gerald J. Gray, Maia J. Enzer, and Jonathan Kusel) Food Products Press, an imprint of The Haworth Press, Inc., 2001, pp. 65-96. Single or multiple copies of this article are available for a fee from The Haworth Document Delivery Service [1-800-342-9678, 9:00 a.m. - 5:00 p.m. (EST). E-mail address: getinfo@haworthpressinc.com].

resources in organizations such as agencies and science; and power-distribution concerns that have made community-based ecosystem management politically contentious. For each of these issues a vision is presented for enhancing the precepts of community-based ecosystem management to ensure that it serves traditionally underrepresented populations and embodies democratic principles of open public deliberation, inclusiveness, and social justice. A significant number of success stories and innovations in dealing with these issues are documented as are barriers and challenges. Recommendations for addressing these challenges are made. While community-based ecosystem management reflects the Jeffersonian spirit of direct democracy, it also depends on individual civility and cannot be separated from the political pull and tug and the checks and balances of a complex federal system of representative democracy. The ultimate opportunity and challenge for community-based ecosystem management is to explicitly recognize its political dimensions and to continually experiment and innovate in ways to improve American democracy as well as long-term ecological sustainability. *[Article copies available for a fee from The Haworth Document Delivery Service: 1-800-342-9678. E-mail address: <getinfo@haworthpressinc.com> Website: <http://www.HaworthPress.com>]*

KEYWORDS. Community-based, forest(ry), ecosystem management, institutions, governance, power

> The problem which all federalized nations have to solve is how to secure an efficient central government and preserve national unity, while allowing free scope for the diversities, and free play to the . . . members of the federation. It is . . . to keep the centrifugal and centripetal forces in equilibrium, so that neither the planet States shall fly off into space, nor the sun of the Central government draw them into its consuming fires.
>
> Lord James Bryce
> *The American Commonwealth*

The federal system of government in the United States has long grappled with the basic question of balancing national and local interests and responsibilities. Increasingly, methods of governing in the U.S. are affected by global institutions and a global economy (Johnson et al. this volume) and by heightened public concerns about the need to ensure future generation's long-term ecological sustainability. Within this political context, community-based ecosystem management attempts to make local communities and local people more effective participants in the process of governance. It specifically recognizes the importance of fostering participation by local participants (including elected representatives, individual citizens, and tribal entities) to create strategies that ensure local commitment to natural resource and environmental stewardship. Community-based ecosystem management is characterized by inclusivity. It acknowledges the contributions non-local citizens (e.g., citizens from communities of interests or occupational communities, part-time residents, and recreationists) can make in the decision-making process. Community-based ecosystem management also acknowledges that healthy democratic processes are inclusive ones that bring to the table previously under-represented individuals and groups.

The community-based philosophy also expands traditional notions of ecosystems to specifically acknowledge the importance of humans and their socioeconomic goals and objectives. And it embraces a land ethic, expanding the traditional notion of community to include the land as a partner. It demands individual citizen commitment to civility, trust, and stewardship, a holistic approach that examines relationships between people and natural resources across a broad range of temporal and spatial scales. It accepts that the ecological goal of ecosystem management is long-term sustainability of both human communities and natural resources. In essence, community-based ecosystem management is premised on the belief that healthy democracies and sustainable ecosystems are both intertwined and interdependent.

The purpose of this paper is to discuss how existing institutions and governance issues facilitate or impede the realization of the above principles, and to suggest ways that more inclusive community participation can achieve greater benefits for both people and the environment. After describing the variety of institutional models that exist for collaborative, community-based ecosystem management, the paper examines a number of related institutional issues associated with this approach. For each of these issues, our vision for enhancing the precepts of community-based ecosystem management is presented, along with a discussion of existing opportunities and challenges. Finally, the paper offers a number of recommendations to address the challenges facing community-based ecosystem management and to ensure that the visions can be realized.

DEFINITIONS OF GOVERNANCE AND INSTITUTIONS

Governance is the act, process, or power of governing. Through their chosen institutions citizens make and administer public policy and exercise political authority. Issues surrounding the who, what, where, and how of governing are, by their very nature, highly political. They reach to the very core of the nation's democratic and federal system of government and to the balances struck among levels and branches of government and between citizens and their government.

While formal organizations like Congress or the Federal Bureau of Land Management come to mind when thinking of institutions, the broadest definitions emphasize informal and social-cultural elements as well as formal organizational structures. Institutions can be defined not only in terms of organization and structure, but also as sets of rules and standards of behavior (Ostrom 1986). For Bellah and others (1991), institutions embody a "pattern of expected action of individuals or groups enforced by social sanctions, both positive and negative," and can range from a handshake to highly formal organizations. They are the "normative patterns embedded in and enforced by laws and mores (informal customs and practices)" (pp. 10,11). Under these broad definitions, institutions can include agencies, laws, regulations, religions, marriage, science, universities, cooperatives, and land tenure/property. It must also be kept in mind that not all institutions–e.g., slavery and racism–are good. Because ecosystem management will require changes both in formal organizational structures and in informal customs and practices (Cortner and Moote 1999), this paper adopts a broad definition of institutions.

AN OVERVIEW OF INSTITUTIONAL ARRANGEMENTS

Given the various arrangements and diverse community situations, no single institutional model fits all community-based ecosystem management organizations or groups. They vary by purpose, how they were established, the role of government agencies and the general public, the nature of their networks, and the resources they have available (Natural Resources Law Center n.d.; Carey and Loeks 1995).

Purpose: The primary purpose of organizations ranges from nonprofit forums to profit-making small businesses. Some exist for the sole purpose of exchanging information, while others engage in ecological advocacy or conduct community development or stewardship work.

Formation: Some resource management agencies establish community-based ecosystem management organizations to accomplish a short- or long-term mission. Others develop to solve specific concerns or needs within their community. A few form through legislative action, while others are ad hoc in nature, acting without force of government.

Networking: In some cases the community-based group or organization operates as an interagency council, bringing together agencies, organizations, or authorities to address the needs of a single, often large ecosystem such as the Chesapeake Bay or the Columbia River Basin. In other situations, institutions or organizations undertake collaborative efforts in a loosely affiliated network or association. Because no formal coalition is created, each entity can join and support a common objective without losing or reducing its own identity or autonomy.

Resources: The stability and structure of a community-based ecosystem management organization can be affected by the resources available to it. Community-based ecosystem management groups may have to limit their scope and attention based on the availability of certain types of funding. Many are required "to scramble for resources" (Snow 1998), which may ultimately affect their sustainability.

It matters greatly whether governing is by tightly focused coalitions that seek new and expanded roles in community-based ecosystem management or by loosely connected networks of existing institutions that use traditional or long-standing norms. When, for example, a university, a county government, a timber association, and a Forest Service district partner to initiate a stewardship process, they must align their existing authorities and powers to address collaboratively the goals of community-based ecosystem management. Each might have in place rules to help ensure that participation, decisionmaking, and problem-solving are inclusive, equitable, and in the public interest.

Owner or community-operated cooperatives are an example of an alternative institutional model that deserves more in-depth examination. Cooperatives are owned, operated, and controlled by the people–or members–who use their services and, unlike other businesses, exist to serve members, not to provide goods or services at a profit. Savings after expenses are distributed to member-owners in proportion to their investment and any dividends on capital invested in the cooperative. Voting control is based on membership rather than investment amount, and each member usually has only one vote. Cooperatives play a vital role by maximizing the number of economic transactions that remain within the community (allowing it to retain capital and profits and increasing the economic multiplier effects), harnessing the decisionmaking power and profit-making incentive of individual landowners, placing the responsibility for environmental health back in the hands of local people and communities, and improving and conserving a natural resource for present and future generations. Cooperatives are quite prevalent in Scandinavian countries as well as Japan, and their experiences can do much to inform U.S. experiences.

For example, a forestry cooperative can help members develop and implement management plans and counsel them on proper harvesting and reforestation techniques. Cooperatives can also assist with the conservation of land and water by establishing buffer zones, restoring riparian areas, and implementing other conservation practices, then monitoring each step to ensure that all landowners, or members, become good stewards of the land. As an example, Michigan's Western Upper Peninsula Forest Improvement District, which is facilitated by the county forester, is a landowner member cooperative for forest planning, marketing, sale of timber, and forest stewardship.

In a recent analysis of more than 100 USDA Forest Service Rural Community Assistance-funded projects, Frentz and others (1998) constructed a model that delineates three basic types of organizations: (1) community-oriented, which focus on community or economic development or diversification, e.g., job creation or new career training; (2) stewardship-oriented, which focus on restoration or ecological health projects that require community collaborative efforts in watersheds, forests, or on rangeland; and (3) policy-oriented, which focus on planning, visioning, or management processes, often in concert with an agency or governmental body responsible for public lands. Linkages, interactions, and overlaps often occur among these three primary types. In some instances, organizations begin with a deep concern about their own situation and gravitate toward working arrangements with public land agencies to improve ecosystem conditions. Sometimes this experience leads the organization to take on a policy-oriented role, such as participating in an ecosystem management plan. In turn, organizations that begin with a planning or policy focus can sometimes move into a stewardship responsibility such as visitor education, resource monitoring, or vegetation restoration.

Continuing development and analysis of knowledge about community-based ecosystem management groups and partnerships can suggest linkages between different types of organizations and governance that can be examined empirically. (Such investigations could be organized around a typology similar to that illustrated in Table 1.) While the work group believes there is a need to address more specifically the governance issues associated with different types of community-based ecosystem management entities and processes, our focus in the remainder of this paper is on a series of questions and topics that apply in varying degrees to all sorts of groups.

In sum, several institutional models exist for community-based ecosystem

TABLE 1. A Prototype Typology of Community-Based Ecosystem Management Groups

	Primary Focus		
Organizational Attributes	Community-Oriented	Stewardship-Oriented	Policy-Oriented
Community-initiated			
Coalitions			
Institutional networks			
Owner/Community-cooperatives			

management. All have specific purposes, different histories, and varying interests and capabilities. Regardless what institutional arrangements are formed, myriad questions and issues will result. It is to these questions that we now turn.

INSTITUTIONAL AND GOVERNANCE ISSUES

We have chosen five issues upon which to focus our attention: the adequacy of existing laws and policies, the role of government and policy tools in facilitating community-based ecosystem management, mechanisms for cross-jurisdictional management, the effects of organizational culture and resources, and concerns over power distributions that have made community-based ecosystem management politically contentious. Our vision for each issue area is followed by a discussion of opportunities and challenges and recommendations for realizing the vision.

The Legal Basis for Community-Based Ecosystem Management: Laws and Policies

> *Vision*–Laws and policies will support community-based ecosystem management in an inclusive, diverse, sustainable, equitable, and just manner. There will be adequate implementation and enforcement of existing laws.

Ecosystem management is the official policy of various federal agencies, and federal level, environmental and natural resource laws exist that support implementation of this approach (Keiter et al. 1995). Yet, there appears to be little explicit legal authority for agencies to establish community-based groups or for such groups to enter into working partnerships with federal agencies. Many of the existing laws reflect an interest-based model in which the community or region is represented by commodity groups, local elected officials, or environmental groups; a specific number of representatives is allowed from each interest type. Such groups alone usually do not represent the complex interests of a community, and their actions are sometimes antithetical to it.

Federal land management strategies undertaken recently on a regional scale include citizen advisory groups, although these groups also somewhat reflect the interest-based model. President Clinton's Northwest Forest Plan includes a hierarchy of local, provincial, and regional advisory committees to guide its implementation (Tuchmann et al. 1996), but those committees tend to be dominated by agency, commodity group, and conservation group representatives rather than the general public.

The 1990 Farm Bill authorized the Rural Community Assistance (RCA) Program, which explicitly provides for more community-based management. The RCA offers communities near national forests opportunities to plan, build local leadership and capacity, and diversify their economies, especially where a dependency on forest resources existed or still exists. Through RCA funding, communities are improving their infrastructure, constructing recreational trails, starting new businesses, and operating ecosystem education and visitor information centers. Moreover, the U.S. Forest Service's new strategy for its state and private forestry programs openly advocates more attention to community issues and partnerships. The five priorities it outlines include: sustainable natural resources and communities, watershed issues and conditions, urban forest resources, forest information for landowners and managers, and tribal governmental relations (USDA Forest Service 1998).

At the state level, Oregon provides legislation that specifically fosters community-based approaches. The 1993 law authorizes and encourages creation of state and county-recognized community-based watershed councils. The 1997 Oregon Plan for Salmon and Watersheds explicitly states that these councils will do the bulk of watershed planning and restoration work to restore salmon habitat in the state (Clark this volume).

In addition to laws and programs that facilitate community-based approaches, there are also a number of significant legal barriers. For instance, the Federal Advisory Committee Act (FACA) was passed into law in the 1970s with the worthy goal of ensuring that private interests did not gain unfair access and influence over federal agencies through meetings that excluded open public access. Some citizen groups, usually formed by agency initiative, officially operate in compliance with FACA. For example, the Ouachita National Forest Advisory Committee provides formal recommendations on forest planning and policy for the national forest (Frentz et al. 1997). Yet, FACA's rigid advance notice and nonexclusion requirements can be an onerous obstacle. In some instances, a strict interpretation of FACA has hindered the formation of community groups or discouraged agency participation. In other instances, groups may ignore FACA requirements and simply hope not to be sued.

New legal interpretations and recent changes, such as an amendment that exempts interagency and intergovernmental meetings, may alleviate some of these concerns. Moreover, overly strict interpretations may not be warranted. One legal analysis, for example, maintains that federal agency participation in informal community groups does not violate FACA if the group is not formed or controlled by a federal agency, or if financial assistance by the agency does not result in a high degree of control over a group (Rieke 1997; see also Brendler and Crosman n.d.).

Perhaps a more ominous obstacle to equitable governance, and one related to social justice, is that present legal, planning, and budgetary processes tend to exclude low-income landowners and minority groups from the matrix of participants. Through a history of indifference and blatant discrimination, low-income landowners and minority groups have at best been overlooked, and at worst excluded, from land-use support programs (National Commission on Small Farms 1998, p. 40). Even community-based groups, which are perhaps best suited to address inclusivity and diversity, are sometimes behind the curve on this issue.

Bliss and others (1998) compare the prospects for sustainable development in two forest-dependent communities in Alabama to illustrate that locally driven economic development opportunities appear stronger if the industry is diversified and the entrepreneurial spirit strong. Conversely, the absence of sound local policy, including effective tax and investment strategies, can result in inadequate funding for public education and other social and physical infrastructure needs. In places like the study area this can perpetuate patterns of racial discrimination that limit economic opportunities for African Americans.

Questions of scale can present barriers to community-based ecosystem management. Centralized, large-scale planning processes conducted under the Forest Service's Resources Planning Act or under the National Forest Management Act can present problems for the goal of local participation. Community groups often do not have the institutional mechanisms or financial resources to facilitate interaction with national-level processes to ensure that local decisions integrate well with big-picture criteria and objectives. Decisions made at higher administrative levels or by congressional committees can overturn those crafted in local forest plans. This also holds true at the state level. For example, Oregon's highly regarded land-use planning law allows a state commission to veto county zoning decisions to maintain statewide planning goals.

Other federal laws may impede community-based ecosystem management. For example, a community group may favor a local, small-forest contractor to implement light-on-the-land forest restoration in a nearby national forest, but federal timber contracting laws make it prohibitively expensive. Current law favors large timber sales purchased by timber companies with enough bank credit or cash on hand to put up expensive performance bonds. Within the timber sale boundary, the Forest Service can redirect some of the money to pay for restoration, which is usually done by a separate contractor. This creates an incentive for the agency to cut a lot of timber to do restoration. If the agency could exchange timber for restoration services, it could pay enough money to get a local contractor started, and the contractor could then sell the cut timber. Such a "goods for services" arrangement is just one tool in a larger policy kit that uses land stewardship contracts to restore and maintain resources and promote sustainable economies. While significant questions still surround "perverse" incentives that might encourage agencies or contractors to harvest more timber than they should, these arrangements deserve further investigation. (For a review of existing statutory and administrative authorities to conduct stewardship contracting, see Ringgold 1998; Mitsos and Ringgold 2001; see also Ack et al. 2001.)

Private property issues also present challenges. Some landowners fear that community-based groups attempting to influence land management across ownerships will somehow gain control of private land decisions over and above that currently exercised under federal and state law or county zoning decisions. This concern can sometimes be ameliorated through education, landowner involvement, and assurances that community groups cannot trump private property rights guaranteed by the U.S. Constitution. Other landowners may fear that, by cooperating with community groups, information about resources on private land will be shared with federal agencies that may then use that information to enforce land-use restrictions under laws such as the Endangered Species Act (ESA). This is a valid concern but mitigated somewhat by new policies that protect landowners who willingly enter into conservation agreements with the federal government. In exchange for a landowner's legal agreement to enhance or maintain endangered species habitat, the federal government guarantees their obligations under the ESA will not change over the life of the agreement. Landowners can also be encouraged to participate when incentives, such as tax benefits or dollars for restoration projects, are made available. Voluntary, nonthreatening processes that reward a future of good land stewardship, rather than punish past wrongs, can help integrate private property owners into community-based efforts.

> *Recommendation*–Enabling legislation or revision of existing laws to further community-based ecosystem management should ensure equal access for all citizens. Voluntary, nonthreatening processes should be encouraged.

Governmental Roles and Policy Tools: Money, Messages, and Mandates

> *Vision*–Governments will play a constructive role in facilitating community-based approaches by building community capacity. They will experiment with using the appropriate mix from an array of available policy tools, including economic and stewardship incentives, educational and technical assistance programs, enabling legislation, and regulations that define minimum expectations.

Governments can build community capacity and facilitate meaningful community-based approaches to ecosystem management through three categories of policy tools: messages, money, and mandates. Any number of policy tools can be an asset as groups experiment with doing the best job possible to meet the needs of the environment and the community. The challenge is to determine the government's role and the appropriate mix of its policy tools given the nature of the situation and the scale of problems to be addressed.

One study, which discussed the role of federal and state agencies and the policy tools they could use to promote civic environmentalism, concluded that states have a comparative advantage over the federal government in using nonregulatory tools. That's because they have the capacity to customize initiatives to fit local situations, engage citizens in an active and personal manner, work across professional and agency boundaries, and act quickly and with precision (John 1994, pp. 271-275). The author goes on to suggest that local institutions may have an even greater advantage. The federal government, he concludes, has the advantage both in using regulatory tools when uniform applications of rules and regulations are required and in redistributing resources from a large segment of the population to another, e.g., from the rich to the poor through welfare programs. This study suggests that examining the connection between policy tools and federalism is a useful way to sort out the different policy roles of federal, state,.and local governments. This merits additional research and consideration in the context of community-based ecosystem management.

Money: Economic and Stewardship Incentives–Government institutions and policy can create market conditions that provide incentives for community-based ecosystem management (Ludazis et al. this volume; Ack et al. this volume). For example, development or use fees (e.g, water user fees or fees for recreation or grazing) might be used to support programs and projects, or tax rebates might be given for developing and implementing a stewardship plan. Such market-based policy tools are an alternative to regulation or can be used in combination with legal and regulatory mandates.

Voluntary action by industry and other organizations is another way to achieve stewardship. The American Forest & Paper Association's (AF & PA) Sustainable Forestry Initiative (SFI) attempts to create market visibility for member companies that meet its criteria and indicators. Companies that fail to meet those criteria are dropped as members of AF & PA, demonstrating this major timber industry association's commitment to the initiative. Nonetheless, some still challenge the credibility of voluntary and self-policing programs. In another instance, some companies have made sincere attempts to implement an all-party monitoring process but have had difficulty securing the participation of some environmental groups, who feared involvement would compromise their ability to argue against the process in national arenas. Lack of participation by a full array of interests could detract substantially from the public credibility accorded the monitoring process.

Messages: Education and Technical Assistance Programs–Technical assistance that is customized to community culture and needs can be a valuable tool in building community-based ecosystem management. After conducting workshops around the nation to gather input on the potential for collaborative management across mixed ownerships, American Forests' Forest Policy Center concluded that the single greatest opportunity for improving stewardship and coordination on private lands is through technical assistance. "Most landowners regard themselves as good land stewards but need assistance in identifying key ecological values on their lands and developing management plans that better protect those values" (Sample et al. 1995, p. v).

A variety of technical assistance is available from federal, tribal, state, and local groups, as well as the private sector. The Rural Community Assistance Program, for example, has fostered an effective community-based process in the Ozark-Ouachita Highlands area to stimulate sustainable development of communities, encourage broad-based participation in forest planning, and build capacity for collaborative planning (Voth and Pell 1998). The more established technical assistance delivery systems like the Extension Service and the state foresters also are reorganizing to better assist community-based groups. For example, a county extension agent in Dodge County, Wisconsin, who is skilled in facilitation and understands the social, ecological, and economic issues of natural resources, has successfully helped facilitate a community group to study and develop a plan for the Horicon Marsh. This environmentally sensitive natural marsh is a high profile tourist attraction during the seasonal migration of geese and is surrounded by agricultural land, housing areas, and businesses. In the private sector, several networks (e.g., the National Network of Forest Practitioners and the Communities Committee of the Seventh American Forest Congress) have linked technical assistance to community-based ecosystem management groups. The Lead Partnership Group in northern California has done the same.

Building a future of citizens who are informed and committed enough to participate actively in community-based ecosystem management depends upon sound early education programs. A number of programs provide young people with knowledge and "hands on" experience in community-based ecosystem management, including, for example, Agriculture in the Classroom, Project Learning Tree, forestry youth camps, and 4-H sponsored programs on natural resources and civics. Other programs aim at teachers and mid-career professionals. For example, the Michigan Geographic Alliance, the area's universities, and the Michigan Forest Resource Alliance worked jointly to develop a 10-day teacher training institute that focused elementary teachers on the environmental and economic components of forest management. The group plans to continue the program and has been invited to expand it to nearby states. State rural development councils, such as Wisconsin Rural Partners, also are developing and facilitating leadership programs that focus on civic education. An Extension-designed and offered program, the Wisconsin Rural Leadership Program, offers mid-career rural professionals the chance to study and travel around the country and the world to learn about other approaches to civic engagement.

To assure that community-based ecosystem planning is built upon sound knowledge of environmental, economic, and social fundamentals, a comprehensive, coherent, and accessible system of education and technical assistance must be available to all communities and citizens. Yet, the environmental education curriculums developed by some state education departments can be criticized for focusing solely on the environmental component without addressing the social and economic community elements. Moreover, many groups unfortunately operate and therefore carry out their experiments in various degrees of isolation, and some lack the capacity (technical skills, financial resources, or experience in governmental processes) to develop and sustain efforts without assistance. As discussed above, there exist several noteworthy formal and informal education and technical assistance programs. Federal and state agencies that provide technical support to landowners and forestry professionals have recently begun to provide such support to those practicing civic engagement. But all too frequently these programs are disconnected and noncomprehensive. Some high quality programs are inaccessible to many communities or may be overextended. Moreover, in a climate of conservative public expenditure, educational and technical assistance programs may be the first to go. There has also been a history of insufficient allocation of resources to institutions that serve minorities and other underserved landowners (National Commission on Small Farms 1998, p. 41). Poorly delivered, biased, or inadequate technical assistance can set back community-based ecosystem management groups.

Mandates: Legislation, Rules, and Practices–Legislative mandates specific to community-based ecosystem management might take two forms. Broad, enabling legislation can provide legitimacy for community-based partnerships without prescribing how to accomplish their goals. One example of an enabling act is the Minnesota Sustainable Forest Resources Act of 1995, which established a mix of councils involving diverse forest interest groups and professional agency staff to study, analyze, and suggest forest management rules. In the second form, Congress or state legislatures can authorize or provide appropriations to community group projects on a case-by-case basis, such as occurred in 1998 with the Quincy Library Group's legislation.

While intended to reinforce or support a locally developed collaborative process, legislation that prescriptively singles out a particular community, forest, or bioregion has the potential to short-circuit community-based discussions and solutions. Some stakeholders in community-based efforts may seek legislative remedies when cooperative or collaborative outcomes cannot be achieved. However, such legislative appeals might be looked upon as "end runs" or "special treatment" by other stakeholders who believe that community-based ecosystem management must remain grounded in study and dialogue occurring in face-to-face relationships. Another danger of the case-by-case legislative approach is that it might transform community-based ecosystem management into yet another example of pork-barrel politics that rewards communities with access to politically powerful committees.

Agency policy can facilitate community-based ecosystem management by adding staffing functions for a group within existing job descriptions, involving community groups and the public early-on in planning and decisionmaking, encouraging local community groups to devise their own solutions, and honoring local solutions negotiated at the community level, as long as those agreements comply with the law. Many of these ideas require agency headquarters to provide managers with clear authority to participate in community processes, and more discretion. It will also require broadening the role of the manager to include educator, mediator, catalyst, and scientist as well as decisionmaker.

In sum, there is a rich and diverse pool of policy tools that can be tapped to facilitate community-based ecosystem management. There is a need to frame new laws, reframe existing ones, and explore the appropriate mix of standards and incentives to creative strategies and quality results for the environment and the community. Such strategic thinking will not happen overnight. It must be recognized that creating incentives that complement existing plans and regulations is challenging, complex, and takes time. Technical assistance must be available and accessible to those communities that wish to raise community capacity, reinforce knowledge and expertise at the local level, and adequately address a holistic view of the ecosystem in the context of sustainable economy and community. Many believe that integrating environmental, civic, and economic education, beginning at the elementary level, will help prepare the next generation to face continuing and ever more complex challenges. A key challenge is to create working relationships and partnerships across jurisdictional levels that can address these issues with optimal strategies. Just as the state of biophysical resources needs to be monitored, the use of various policy tools needs to be continuously evaluated and adjustments made based upon accumulated experience.

> *Recommendations*–Institutions should develop, promote, and adequately fund tools to facilitate community-based ecosystem management, including, but not limited to, effective training, technical assistance for all communities, and legislative incentives.

State, federal, and local agencies should initiate and support policies for sharing among themselves educational and technical assistance programs.

Organizations that oversee technical and financial support programs should be monitored to ensure compliance with guidelines for inclusion of low-income and minority landowners.

Laws and policies to facilitate community-based ecosystem management should generally be flexible, nonprescriptive, and non-site-specific, and monitoring mechanisms should ensure that implementation is just and equitable.

Cross-Jurisdictional Coordination

- *Vision*–Effective coordination mechanisms, both formal and informal, will exist across jurisdictional boundaries to facilitate collaboration over time. These mechanisms will reach across public, private, and tribal jurisdictions.

Several mechanisms are designed to facilitate collaborative efforts across jurisdictional boundaries. Formal mechanisms include memoranda of understanding (MOUs), the Challenge Cost Share Program, cooperative agreements, alliances, advisory councils, and cooperative charters or articles of incorporation. Informal mechanisms also exist and oftentimes are very successful. Study groups, informal partnerships, and networks have been used where the formal mechanism has either impeded cooperation or is deemed unnecessary. The mechanism used depends on the purpose the coordination will serve. Oftentimes social capital and trust must be built informally before institutions are willing to participate in formal coordination efforts.

Priester and Kent (1997) discuss formal and informal organization as it relates to how humans organize themselves. Formal organization is characterized by established membership, leadership, chains of command, and written communication. Its functions are to maintain political, economic, and ideological control. Informal organization is characterized by social networks that support individuals in predictable ways. Such networks have flexible goals, informal word-of-mouth communication, and respected leadership. The function of the informal system is survival, maintaining culture, and caretaking (Priester and Kent 1997). Priester and Kent's discussion of how formal and informal organizations interact illustrates why cross-jurisdictional efforts either succeed or fail. If coordinative mechanisms are to successfully facilitate community-based ecosystem management, they must consider the best mix of formal and informal organization for their purposes.

Institutions that wish to cross jurisdictional boundaries for coordination will need a true sense of shared governance among all parties or the effort likely will fail. All parties must also sense that there is something to gain by the association. This premise is most apparent when there is a perceived loss of power by one entity. Many agencies and institutions are reluctant to extend or share power with other institutions for any number of reasons–e.g., fear of losing decisionmaking authority, administrative discretion, economic benefits, or status and prestige–and therefore do not initiate collaborative measures. Even if it can be argued that coordination would benefit the mission of all organizations, the perceived risks and costs often outweigh the promised benefits. Many organizations do not have the resources to devote to coördinative activities; even informal networking cannot occur without scarce resources such as time. These problems are exemplified in the Southeast, where few efforts are made to coordinate ecosystem management among low-income landowners. Another example: when counties fear collaboration that extends beyond their borders because they fear losing influence over local land-use decisions (Press 1994, p. 107).

In the case of Native Americans, it must be remembered that some tribes act as sovereign nations and are under no obligation to join in coordinative efforts. They may or may not choose to participate in community-based ecosystem management, depending on whether they see a benefit to the tribe or how they perceive cooperation will impact tribal sovereignty. One tribal government that has made such a commitment is Wisconsin's Lac Courte Oreillas tribe. It is developing a MOU with the Chequamegon-Nicolet National Forest and the state of Wisconsin for cooperative forest management around the Chippewa Flowage. In another example, in 1993 the Nez Perce tribe and officials in Wallowa County, Oregon, cooperatively developed a recovery plan for the area's threatened Snake River chinook salmon.

For success to occur there must either be a formal mechanism to develop and enforce rules and regulations or all parties must exhibit a high level of trust and integrity. Informal coordination built on trust and integrity works well for many smaller institutions; for example, on Colorado's San Juan National Forest the Ponderosa Pine Forest Partnership (Richard 1995; Larmer 1996) is carrying out a 1,000-acre restoration project within a 180,000-acre ponderosa pine ecosystem. Increased citizen participation in resource planning for the San Juan National Forest (Richard and Burns 1998) as well as the Finger Lakes and Green Mountain national forests (Twarkins et al. this volume) has also been accomplished through informal coordination between institutions and individuals.

While mechanisms may be in place for cross-jurisdictional cooperation, information is oftentimes incompatible. This is especially true in sharing computerized information. Agencies and institutions must develop criteria and standards acceptable to both parties for true information sharing to occur. Both the Oregon Department of Forestry and the Great Lakes Forest Alliance are developing forest criteria and indicators that can be used by federal, state, and local governments as well as private institutions and landowners. This type of coordination prevents needless duplication that tests the public's trust of agencies. In one instance, two separate agencies used public funds to fly the same landscape for aerial photographs because they could not agree on photo resolutions.

Cross-jurisdictional cooperation does have its success stories (Knight and Landres 1998; Yaffee and Wondolleck 1997). The Applegate Partnership in southwestern Oregon has built a participatory program of ecosystem restoration; the Rincon Institute and Saguaro National Park in southern Arizona are grappling with the rural-urban interface. The Finger Lakes National Forest has engaged in educational efforts to build community-based ecosystem management through an informal cross-jurisdictional affiliation between the U.S. Forest Service, multiple county Extension agencies, state and federal watershed agencies, private industry, and local private individuals. Governors of the states of Michigan, Minnesota, and Wisconsin and the Province of Ontario created the Great Lakes Forest Alliance in 1987 as a forum to foster and facilitate cooperative efforts that enhance management and sustainable use of public and forest lands. Alliance trustees represent forest resource agencies at the national and state/province levels as well as citizen interests. The Alliance has been successful in building bridges toward promoting community-based ecosystem management (Sanders this volume).

Oftentimes cross-jurisdictional cooperation occurs at the highest institutional levels, which in turn can enable local governments to empower institutions for successful community-based ecosystem management. The Western Governors' Association, for example, has adopted a resolution, Principles for Environmental Management in the West (Enlibra), outlining a new doctrine to guide natural resource and environmental policy in the West. The resolution is based upon the governors' experience that environmental strategies work best when they have strong gubernatorial commitment, vested local support, and federal collaboration. It envisions local governments and communities working collaboratively with sufficient authority to implement agreed-upon projects. Another example of state-level cross-jurisdictional cooperation designed to facilitate action by lower levels of government is the Oregon Plan for Salmon and Watersheds. This plan brings together all governmental agencies that impact salmon in the state. Under it, actions to conserve and restore salmon must be worked out by communities and landowners, with solutions reflecting local knowledge of problems and ownership (GNRO 1997).

Despite the success stories, many challenges remain before widespread success can be claimed. The issue of power must be resolved early on through trust building and shared governance. All parties must see the benefits of their involvement and be willing to establish shared criteria and objectives. The resources necessary for follow-through must be available. True community-based ecosystem management cannot succeed without cross-jurisdictional cooperation.

> *Recommendations*–Agency directors should support cross-jurisdictional coordination through actions expressed in both budgets and resources.
>
> Agencies should plan for and prioritize funding for cross-jurisdictional coordination early in the budgeting process.

Organizational Resources and Culture

> *Vision*–Organizational resources will effectively facilitate community-based ecosystem management. Organizational cultures, in both public and private institutions, will responsively adapt to changes in ecological and social values, to the need for more understandable scientific information, and to the expectations of workers for greater participation in decision making.

Central to the success of community-based ecosystem management is the way the organizational resources of various institutions, both public and private, respond and adapt to proposed change. That, in turn, is often based on how the organizational culture responds to change. Private institutions, such as chambers of commerce, community leadership development programs, industrial councils, and civic forums, can redirect resources to increase stewardship of ecosystems among diverse interests in a community. Public institutions can examine their mission, make adaptations as needed, and reprioritize resources to encourage rather than impede collaborative community-based efforts. However, institutional change seldom comes easily. Public agencies can be protective of turf and reluctant to look beyond their own mission, values, and traditions, while private for-profit companies can be protective of market niches, subsidies, and competitive advantage. To illustrate the opportunities as well as difficulties inherent in organizational change, we examine two very different institutions–agencies and science–in more detail.

Agencies

Federal and state agencies hold most of the keys to success for many community-based organizations and processes. The staff and line officers of resource management organizations traditionally have been the primary, if not sole, agents of control, planning, research, and monitoring. Their involvement with members of the public or community has been within fairly narrow channels of communication. Contacts with user groups, such as between a Forest Service range conservationist and a range permittee, have been limited to specific obligations and standards on a limited piece of ground, not focused on larger-scale landscapes or ecosystems.

During larger-scale planning processes or project development, contact with local governments has typically been limited and minimally successful. That contact usually takes the form of written correspondence on the impact public land management decisions have on limited fiscal matters, such as county road budgets and law enforcement costs. While the public at large has had the opportunity to become involved in NEPA processes, these procedures have tended to be dominated by interest groups on both sides of a given issue. Public hearings, the primary form of participation, have become opportunities for polarization rather than problem solving, collaborative stewardship, and increased civic responsibility. Citizens vent their frustrations, but little two-way communication occurs.

Many community-based organizations seek continuous rather than episodic collaborative approaches with federal and state agencies, ones based in stewardship ethics rather than single-issue debates and in-depth resource analysis and communal learning rather than external expertise (Walker and Daniels this volume). These types of public engagement will require some fundamental changes in natural resource agencies as they seek and develop broad-based collaborative partnerships with diverse public interest communities.

To accomplish this, agencies will need flexibility and responsiveness in their decision making. While they retain authority as prescribed by federal and state laws, they will need to broaden the basis for their decisions by improving participation in idea generation, planning, learning, and monitoring. Both communities of place and interest need to become involved earlier in long-range, large-scale strategic planning processes, as happened recently on the San Juan and Finger Lakes national forests (see Richard and Burns 1998; Twarkins et al. this volume; Duane 1997).

Agency staff need encouragement and support in involving community groups and publics. Some have few skills in partnership development or collaboration. Some may believe their certain scientific perspective is the only answer, while others may simply feel uncomfortable or ill prepared to interact with community members in an open, learning manner. Many staff need and desire training in collaborative processes, like that offered by the Bureau of Land Management (BLM) Training Center in a course titled, "Community-Based Partnerships and Ecosystems for a Healthy Environment," and in the Pinchot Institute for Conservation's "Open Decision Making" workshop. Unfortunately, the length (one week) and the cost of these collaborative training experiences make them somewhat inaccessible. Consideration should be given to making such training efforts more accessible in terms of time, tuition costs, and travel expenses. Another option to benefit public land agency personnel would be to organize internal collaborative partnership support teams that share lessons learned from model projects and partnership processes (USDA Forest Service Collaborative Stewardship Team 1997).

To achieve flexibility and to develop partnerships and collaborative-interactive skills will require strong vision and leadership from key agency leaders and line officers. Bolstered by a consistent commitment from Washington headquarters, local area and district managers will need to turn their agencies outward, involving the community not merely for short-term input or episodic records of decision, but in ongoing, helpful partnerships.

Resource management agencies will also have to become more inclusive. The so-called "usual suspects" cannot be the only representatives in true community-based processes. While new interests must be brought to the table, low-income landowners in many regions of the country will need special opportunities. For example, technical and financial support from agricultural and forestry programs often is determined by local county committees. In many instances, these committees have few low-income or minority members (National Commission on Small Farms 1998). Such organizations must be mandated, monitored, and sanctioned to ensure that low-income and minority landowners are included in critical programs. Significant outreach and education are required to address this long-term deficit in local community and economic sustainability programs.

In the area of forest management, several studies have addressed ways to redesign the organizational structure of the U.S. Forest Service to improve land stewardship, enhance public respect, replace polarization with cooperation, and reduce the financial burden on taxpayers. Such studies have been conducted by the agency itself (USDA Forest Service Collaborative Stewardship Team Report 1997), by nonprofits (Thoreau Institute 1998), and by academic and scientific groups (National Research Council 1998; Boyle et al. 1994). These reports offer a wealth of ideas for improving linkages between federal and nonfederal entities, fostering public involvement, testing a range of innovative policy tools, and creating a receptive culture for partnerships and collaboration.

Private nonprofit organizations are often effective in bridging the gap between local communities and federal agencies by creating new sources of jobs and other opportunities in rural areas. A collaboration among public resource management agencies, private industry, and nonprofits can allow each institution's strengths and particular capabilities to be used to its utmost effectiveness. For example, nonprofits' specific knowledge of local, social, and economic circumstances combined with the federal agencies' organizational frameworks and assistance capabilities could result in more effective, longer-term support for communities in developing ecosystem management.

Science

Natural resource science has traditionally been the purview of agency staff or university personnel. Typically, institutionalized research processes have focused on single species or on resource elements such as soil or water. In keeping with the increased emphasis on ecosystem management, individual scientists and resource specialists are beginning to work in a more holistic, integrated, and comprehensive ecological framework. While the picture remains somewhat mixed in regard to fully integrated ecosystem assessments, the value of this approach seems assured.

Parallel to this trend is the need to integrate local and practitioner knowledge. In the past, professional science was viewed as superior because it supposedly conforms to positivistic and rational procedures. In fact, professional science has not brought about the promised benefits of ecological health (Nelson 1995); it is overwhelmingly expensive and cannot truly be expected to provide the sole solution for practical resource management. Increased success in integrating scientific knowledge with community and practitioner perspectives could be achieved by involving more representation from community development, civic education, and social science.

The goals of community-based ecosystem management call for the integration of formal and informal knowledge, for rational science to be informed by practitioner understandings, and for good, objective study to be tested by comparison with local wisdom and vice versa. This mutual informing process, leading to integration where appropriate, will require two steps.

First, whether within agencies or universities, scientists need to include local community practitioners, ranchers, loggers, recreationists, bird-watchers, hunters, schoolteachers, and youth in the study and assessment of the biophysical and social components of ecosystems. Bringing this set of knowledge into the totality of the resource information base needs to be a continuous process from field studies, analysis, planning, monitoring, and adaptive management.

Second, pertinent and useful scientific inquiry needs to be made more widely available in understandable forms. Because environmental issues are characterized by a mix of highly technical information and uncertainty, problems become compartmentalized into technocratic organizations that are biased against the information-processing capabilities of highly participatory governance structures (Press 1994, pp. 41-42). The challenge is to balance the technocratic nature of science-informed decision making with participatory governance, which will require scientists to learn how to better translate their work into meaningful presentations. Collaborative study processes, field trips, and advisory groups are needed to fit the requirements of the local ecosystem management situation or objective. Enabling policies may be required for these forms of scientific integration and collaborative participation to move forward unimpeded, but leadership among scientists and land managers is also essential.

There is likely a new role here for universities, especially for land grant universities and their USDA and state cooperative Extension programs, to participate in providing and integrating scientific and local knowledge. This can take many forms but would include partnering with federal and state agencies during research and planning, hosting community workshops, and designing new means of public communications that are more accessible to and inclusive of marginalized community groups. Colleges and universities, especially the 1860s and 1890s land grant institutions, can be instrumental in fostering community-based ecosystem management among underserved landowners.

> *Recommendations*–Education and training about the economic risks/benefits of sustainable ecosystem priorities should be provided to low-income, underserved landowners to increase their opportunities to participate economically and socially.
>
> Agriculture, forestry, and other natural resource agencies should be mandated to provide outreach services and to support developing underserved communities', local organizations', and low-income citizens' capacity to contribute to ecosystem management.
>
> Federal and state natural resource agencies should utilize private nonprofit organizations to build capacity for ecosystem stewardship within local communities and to bridge the communication gap between agencies and citizens.
>
> Natural resource agencies should operate in an actively inclusive manner to infuse new and emerging leaders and broader community audiences with community-based ecosystem management principles and practices.
>
> Federal, state, and local government agencies should provide employees with training in community collaboration using the BLM National Training Center model and provide resources in more places for more participants.

Agencies should be encouraged to organize internal collaborative support teams that share lessons learned from model projects and partnership processes.

Scientists should encourage the development of "civic science," involving citizens in study design and implementation, creating opportunities to integrate practitioner knowledge into the scientific process and making scientific knowledge more understandable and accessible.

The Distribution of Power and Political Feasibility

> *Vision*–The distribution of political power among levels of government and among interests will be perceived as equitable by all. The political feasibility of community-based ecosystem management will be ensured because where society (through our representative democracy) has identified a national interest, that interest is respected. The interests of the poor, the politically weak, and future generations are protected. Civic dialogue occurs in an atmosphere of trust and civility.

Many of the governance issues surrounding community-based ecosystem management reflect the democratic debate that first took place in the decades following the revolutionary war–a debate that is most directly manifest in the tensions between the political thought of Thomas Jefferson and James Madison. Madison's primary concern was that one group of citizens could control government (the "tyranny of the majority") with no protections for the interests of minorities. According to Madison, a direct democracy would not solve this problem, because it would not have controls preventing individuals without virtue from usurping control. Madison, who is often called the Father of the Constitution, instead believed a large republic with a system of constitutional checks and balances would diffuse power among many parts so that no one part dominated the whole. Reflecting the Madisonian view of human nature, the Constitution divides power among three branches of government and between national and state governments.

For Thomas Jefferson the ultimate concern was that government be placed in the hands of the people: "I know of no safe depository of the ultimate powers of the society but the people themselves; and if we think them not enlightened enough to exercise their control with a wholesome discretion, the remedy is not to take it from them, but to inform their discretion by education" (Jefferson in a letter to William Charles Jarvis, 1821, reproduced in Padover 1939, pp. 89-90). Jefferson believed democratic ideals could best be realized by public participation in government. While Madison's views about democracy prevailed in the Constitution, Jefferson's belief in the value of people and their right to participate fully in political decisionmaking remain ideals toward which democratic governance continues to strive. Jefferson is often called the American spirit. American democracy continues to struggle with the question of how much power should reside directly with the people and how much with their elected representatives, as well as the appropriate balance of national and state responsibilities.

The history of the environmental movement in this country is one in which the federal government assumed increasing responsibilities, responding to the perceived failure of state and local governments to take action and to citizen demands for remedy. Distrust of the ability of federal agencies to incorporate environmental values into their decisionmaking processes led to more and more prescriptive legislation. This gave new participants the leverage to take action judicially as well as legislatively and administratively. Today there is considerable criticism that this situation has created a federal behemoth, interest-group polarization, and gridlock. Each side portrays the other's motives in extreme terms, creating a climate of intense distrust and agency dysfunctionalism. To escape the rigidities of centralized policy prescription, there is increasing support for shifting more governmental authority to the state and local level in all areas of public policy, including natural resources and the environment.

Support for this decentralization has taken many forms. Many cities have passed policy initiatives to build the institutional capacity for sustainable and liveable cities. Urban communities are working to redevelop brownfield sites and to address environmental justice concerns (President's Council on Sustainable Development 1997). The National Association of Counties and the U.S. Conference of Mayors have joined together to create the Joint Center for Sustainable Communities, which has the potential to foster urban-rural linkages. As noted above, governors are taking a more active role.

In the West, the tree of decentralization has many branches. One includes the wise use, property rights, and counties movements that emphasize "resource development, enmity toward environmentalists, and a desire to confront federal land managers for control over public lands" (Davis 1997, p. 23). Another represents the policy declarations of the Western Governors' Association, such as the Park City Principles for water management and the Principles for Environmental Management in the West. These declarations call for a greater decision-making role for state governments while acknowledging the role of the federal government in setting environmental standards. They explicitly embrace inclusive, collaborative processes and market-based solutions along with regulation to achieve a synergistic policy effect.

As a movement that advocates more direct and inclusive citizen participation, more opportunities to access local knowledge, and more policy flexibility to accommodate both regional variations and new knowledge, the governance issues surrounding community-based ecosystem management involve a redistribution and sharing of political power. Redistributions of power are bound to engender tensions and controversy, and community-based ecosystem management has its share of critics. Community-based approaches have been criticized as a way to protect the status quo politically and as an "insincere rear-guard holding action," a Trojan horse to protect the privileges and subsidies industry and local resource extractive communities have traditionally enjoyed (Coggins 1998). Community forums are portrayed as more conducive to the objectives of indústry, because industry is perceived to be more successful in these than at the federal level or in court (McCloskey 1996). It is argued that lessons learned from efforts to decentralize policy to the community action agencies of the 1960s and 1970s "suggests that only a subset of such organizations are likely to achieve the envisioned levels of involvement and democracy, and that even fewer are likely to sustain them over time. Absent broad and committed participation, community-based institutions are easily captured by those motivated more by jobs and power than substantive reform. . . ." (Henig 1998 pp. 7-8). Criticisms such as these underscore the need for community-based ecosystem management to reconcile concerns of equity, authority, stability, and sustaining community capacity (Burns this volume). Progress in ecological stewardship is more likely when a meshing occurs among the various roles of central and local governments. For example, the Montezuma County Federal Lands Program, working in concert with other members of the Ponderosa Pine Partnership, focuses on the restoration of a ponderosa pine ecosystem based on indepth ecological analysis, adaptive management by thinning and fire reintroduction, and ongoing monitoring (Larmer 1996).

Among low-income landowners, community-based ecosystem management can be as politically important as the struggle for basic rights. Ecosystem management among the poor raises concerns of environmental health, sustainable food supply (fish, hunting, crop production), sustainable income source (forest management, nonforest product production, value-added production and marketing), land tenure, and the problems of dealing with a shrinking economic base. Making ecosystem management and sustainability an issue that includes poor people can raise the ecological imperative to the level of economic justice among the poor. Economics and ecology must work together for everyone. For ecosystem management to progress it must have some degree of economic sustainability or feasibility. For the economics of ecosystem management to be supportable, the end result must be an improved ecological condition. Because of this mutual interdependency, economics and ecological restoration are inseparable.

Politically, it must also be recognized that interests favor participation in venues in which they have the greatest power and can obtain the outcomes they ultimately prefer. They also congregate around those forums with veto power: "because veto powers are broadly distributed, interest groups converge on persons who can exercise vetoes" (Lindblom 1980, p. 93). At any one time, preferences for centralized or decentralized solutions are more often a function of pragmatic strategic calculation than pure philosophical preference for one model of federalism over another. Environmentalists, for example, who prefer the national arena for resolution of public lands disputes, also seek local control over situations when they can show how the local community suffers from lack of environmental protection. Thus, when public health and social justice are equated with environmental protection, environmentalists have argued on behalf of decentralization (Press 1994, p. 9). Likewise, industry has argued both for local flexibility and national standardization, depending on the effect on its competitive business advantage.

Participation in community-based forums occurs for a variety of process, outcome, and strategic reasons. Some participants are motivated more by process concerns (achieving equity, empowerment, participation, justice), while others see participation as means to another end. They are more concerned with outcome, i.e., the substantive and tangible policy results that can be observed and measured on the ground, e.g., their economic future or the protection or restoration of special areas (Press 1994, pp. 100, 121). In many cross-jurisdictional partnerships, for example, people come together because of need, not because of an inherent desire to work together. A catalyst organization, or an individual who can inspire others, is a critical element in successful partnerships (Sample et al. 1995, p. v).

To succeed in a federal system, community-based approaches must be attuned to the national interest and national values. The reality is that it may be very difficult to negotiate issues locally while those issues remain unresolved nationally (Daniels and Walker 1995). "Existing policy regimes, embedded in national environmental laws, bureaucracies, and political institutions, are incredibly durable and resistant to change. Local partnerships therefore must do a better job explaining how their work can be integrated into these national policy regimes" (Brick 1998, p. 35).

Community-based ecosystem management is enhanced by increased participatory democracy, which tends to manifest itself in a redistribution of power. Redistribution of power, however, does not necessarily mean granting citizens or community groups decision-making authority. Power is broader than the legal concept of authority. More "power" for community groups will rarely mean more authority. Many community-based practitioners "do not want . . . to call the shots" (Gray and Kusel 1997 pp. 27-30). How then does increased participatory democracy lead to more power? While it may often be difficult to fathom the many facets of power, understanding the nature of power is nonetheless important.

The predominant theory of power focuses on overt resolution of observable conflict, concentrating on transparent decisions made within formal processes (Lukes 1974; see also Dahl 1961). Just by becoming participants in open, observable processes that indirectly influence decision-making authorities, community-based groups build a power base. Most debates will take place in this realm.

This theory of power, however, does not take into account informal actions occurring outside formal decision-making structures (such as the informal group meetings or informal networks described by Priester and Kent 1997). The mechanisms of power can be indirect or covert as well as direct; indirect power, for example, can come from money, persuasion, or exchange (as in trading favors). Thus, a second, more comprehensive, theory of power suggests another dimension of power, but one that has significant negative as well as positive effects. On the negative side, indirect power can be contingent on a "mobilization of bias," in which "a set of predominant values, beliefs, rituals, and institutional procedures ('rules of the game') . . . operate systematically and consistently to the benefit of certain persons and groups at the expense of others" (Bachrach and Baratz 1970 cited in Gaventa 1980, p. 14). The mobilization of bias theory also describes how indirect power leads to group exclusion as well as "non-decision making," and helps to explain the difficulties of political participation for southern low-income landowners. This facet of indirect power has serious implications for community-based ecosystem management, for participant exclusion is antithetical to the quest for inclusiveness. Special efforts to increase inclusiveness can somewhat counteract the bias.

Lukes (1974, p. 23) points out a third dimension of power, in which those exercising power influence, shape, or determine citizen wants through indoctrination. Commitment to building critical thinking skills can do much to distinguish education from indoctrination. Both the second and third dimensions of power generate feelings of powerlessness (Gaventa 1980). It is hoped that those elements of community-based ecosystem management that emphasize inclusiveness, collaboration, education, and civility can overcome these more nefarious elements of the exercise of power.

Whatever the local or regional situation, increasing participatory democracy and collaborative stewardship is essential to the sustainability of community-based ecosystem management. While these decision making and stewardship efforts will include both formal arrangement and informal partnerships, they will have to develop approaches that are shared, are equitable, are inclusive, and provide mutual benefits for lands, watersheds, communities, and the broader public interests.

> *Recommendations*–Proponents of community-based ecosystem management should seek to increase the political feasibility of this approach by

understanding the nature and distribution of political power and the implications of proposed changes in this distribution.

All institutions should enhance community-based ecosystem management by increasing the use of participatory democratic processes.

CONCLUSIONS

Implementing effective and equitable community-based ecosystem management will require that citizens, policymakers, and scientists pay considerable attention to institutions and governance issues. This involves examining the alternative institutional models that might best be suited for a community group's needs, and the challenges as well as opportunities presented by institutional change.

Community-based ecosystem management is about restructuring relationships and creating new interdependencies. Its goals are to foster local community participation, inclusiveness, and long-term sustainability of human communities and natural resources. Its full potential can be achieved when laws and policies, policy tools, coordination mechanisms, organizational cultures, and distributions of political power build community capacity, serve traditionally underrepresented populations, and embody democratic principles of open public deliberation, inclusiveness, and social justice.

Experience with community-based ecosystem management has already revealed that there is no one best institutional model. Innovations vary in purpose, organizational history, networking patterns, and available resources. Yet, while a significant number of success stories exist, considerable challenges remain to ensure the longevity of these experiments and realize the visions of community-based ecosystem management. Legal structures and processes will need to promote a citizen-based model of public deliberation rather than an adversarial, interest-based model of participation. Low-income and minority citizens can no longer be excluded from program design and implementation. Attention will need to be directed toward enabling communities to exploit the full range of policy tools, including economic incentives, educational and technical assistance programs, and regulatory regimes that can be applied to their particular problems. Effective cross-jurisdiction coordination will require mandated commitment and resources, not just rhetoric. Agency cultures will need to foster a collaborative spirit and a willingness to share resources and power rather than remaining protective of turf and resistant to change. All-too-often aloof scientific institutions will need to create a civic science that provides more avenues for direct citizen participation. Citizens, policymakers, and scientists can, and should, all be involved in the design and conduct of research programs that examine the governing needs and processes of different community groups, the adequacy of laws, and policies, the performance of various policy tools, incentives for changing organizational cultures, and new approaches to scientific inquiry and information transfer (Cortner et al. 1996). Rather than overlooking the ramifications of changes in the distribution of political power, issues surrounding the use and abuse of power, national and local interests, and political equality will need to be put squarely on the table and confronted. The institutional issues that challenge community-based ecosystem management go directly to the structural core of the American political system.

The intense political debates and the ultimate compromises of the Constitutional Convention created a federal republic and a representative democracy. Yet, the legacy of Jefferson's political thought ever-increasingly pushes toward more direct democracy and more inclusive participation. While community-based ecosystem management reflects the Jeffersonian spirit of direct democracy, it cannot be separated from the tensions that constantly define a dynamic and complex system of political pull and tug and the checks and balances built into our federal system of representative democracy. These forces ensure that neither centralization nor decentralization, national nor community interest, governmental nor individual interest consistently prevail. The issues of power, the national interest, and political feasibility that characterize all walks of political life will continue to characterize community-based management.

Moreover, one cannot reasonably contemplate adjustment in our institutions and governance at this time without considering the importance of civility. Kemmis suggests:

> No amount of reforming institutions that are widely and rightly perceived to be beyond human scale will heal our political culture until we begin to pay attention once again to democracy as a human enterprise. Without healing the human base of politics, we will not restore democracy itself. One thing alone will give us the capacity to heal our politics and to confront the problems and opportunities that politics must address. That one thing is a deeply renewed human experience of citizenship. (Kemmis 1995, pp. 6-7)

The capacity of citizens to learn and solve their problems must be acknowledged. The importance of every person's needs must be acknowledged and collaboratively addressed. Individually and collectively, citizens must teach and model the concepts of respect and dignity, which are different from surrender; trust, which is different from blind faith; goodwill, which is different from self-absorption; and exploration, which is different from blind consent. The opportunity and challenge for community-based ecosystem management is to explicitly recognize its political dimensions and to continually experiment and innovate with ways to improve American democracy as well as achieve long-term ecological sustainability.

REFERENCES

Bellah, R.N. et al. 1991. *The Good Society.* New York: Alfred A. Knopf.

Bliss, J.C., T.L. Walkingstick, and C. Bailey. 1998. Development or dependency? Sustaining Alabama's forest communities. *Journal of Forestry* 96(3): 24-30.

Boyle, B.J. et al. 1994. *Policies and Mythologies of the US Forest Service.* Seattle, WA: University of Washington College of Forest Resources Institute for Resources in Society.

Brendler, T. and S. Crosman. n.d. The Federal Advisory Committee Act: implications for public involvement on the national forests. The Forest Trust, Santa Fe, NM.

Brick, P. 1998. Of imposters, optimists, and kings. *Chronicle of Community* 2(2):34-38.

Carey, H.H. and M.L. Loeks. 1995. Restructuring the timber economy. In J.D. Echeverria and R.B. Eby (eds.), *Let the People Judge: Wise Use and the Private Rights Movement*, pp. 253-260 Washington, DC: Island Press.

Coggins, G.C. 1998. Of Californicators, quislings and crazies. *Chronicle of Community* 2(2):27-33.

Cortner, H.J. and M.A. Moote. 1999. *The Politics of Ecosystem Management.* Washington, DC: Island Press.

Cortner, H.J. et al. 1996. *Institutional Barriers and Incentives for Ecosystem Management: A Problem Analysis.* Gen. Tech. Report PNW-GTR-354. Portland, OR: USDA Forest Service Forest Experiment Station.

Dahl, R.A. 1961. *Who Governs? Democracy and Power in an American City.* New Haven, CT: Yale University Press.

Daniels, S.E. and G.B. Walker. 1995. Resolving local disputes amidst national controversy: a public forest management case study. *International Journal of Conflict Management* 6:290-311.

Davis, S.K. 1997. Fighting over public lands: interest groups, states, and the federal government. In C. Davis (ed.), *Western Public Lands and Environmental Politics*, pp. 11-31. Boulder, CO: Westview.

Duane, T.P. 1997. Community participation in ecosystem management. *Ecology Law Quarterly* 24(4): 771-797.

Frentz, I., D. Voth, S. Burns, and C. Sperry. 1999. Linking community strategic planning and forest planning. Fayetteville, AR: University of Arkansas School of Human and Environmental Sciences.

Frentz, I.C., D.E. Voth, S. Anderson, and W.F. Pell. 1997. A talk in the woods: an impartial advisory committee eases conflict on the Ouachita. *Journal of Forestry* 95(1): 26-31.

Gaventa, J. 1980. *Power and Powerlessness: Quiescence and Rebellion in an Appalachian Valley.* Urbana, IL: University of Illinois Press.

GNRO (Governor's Natural Resource Office). 1997. Oregon Coastal Salmon Restoration Initiative executive summary and overview. Salem, OR.

Gray, G. and J. Kusel. 1997. Changing the rules. *American Forests* 103(4): 27-30.

Henig, J.R. 1998. Market and community-based proposals for school reform. *Policy Currents (American Political Science Association)* 8(1):7-8.

John, D. 1994. *Civic Environmentalism: Alternatives to Regulation in States and Communities*. Washington, DC: CQ Press.

Keiter, R.B., L. Milkman, and T. Boling. 1995. Legal perspectives on ecosystem management: legitimizing a new federal land management policy. Paper presented at the Ecological Stewardship Workshop, Tucson.

Kemmis, D. 1995. *The Good City and the Good Life*. Boston: Houghton Mifflin.

Knight, R.L. and P.B. Landres (eds). 1998. *Stewardship Across Boundaries*. Washington, DC: Island Press.

Larmer, P. 1996. A Colorado county tries a novel approach: work the system. *High Country News* 28(9):11-13.

Lindblom, C. 1980. *The Policy-Making Process*. 2nd ed. Englewood Cliff, NJ: Prentice-Hall.

Lukes, S. 1974. *Power: A Radical View*. London, UK: Macmillan.

McCloskey, M. 1996. The skeptic: collaboration has its limits. *High Country News* 28 (9):7.

National Commission on Small Farms. 1998. *A time to act: A report of the USDA National Commission on Small Farms*. Washington, DC: US Department of Agriculture.

National Research Council. 1998. *Forested Landscapes in Perspective: Prospects and Opportunities for Sustainable Management of America's Nonfederal Forests*. Washington, DC: National Academy Press.

Natural Resources Law Center. n.d. *The Watershed Source Book: Watershed-Based Solutions to Natural Resource Problems*. Boulder: University of Colorado Natural Resources Law Center.

Nelson, R.H. 1995. *Public Lands and Private Rights: The Failure of Scientific Management*. Lanham, MD: Rowman and Littlefield.

Ostrom, E. 1986. An agenda for the study of institutions. *Public Choice* 48:3-25

Padover, S.K. (ed). 1939. *Thomas Jefferson on Democracy*. New York, NY: Mentor Books.

Priester, K. and J.A. Kent. 1997. Social ecology: a new pathway to watershed restoration. In J.E. Williams, C.A. Wood, and M.P. Dombeck (eds.), *Watershed Restoration: Principles and Practices*. Bethesda, MD: American Fisheries Society.

President's Council on Sustainable Development. 1997. *Sustainable Communities Task Force Report*. Washington, DC.

Press, D. 1994. *Democratic Dilemmas in the Age of Ecology: Trees and Toxics in the American West*. Durham, NC: Duke University Press.

Richard. T. 1995. Timber on a tightrope: a demonstration project to reshape the future. *Inner Voice* August: 16-17

Richard. T. and S. Burns. 1998. Beyond 'scoping': citizens and San Juan National Forest managers, learning together. *Journal of Forestry* 96(4):39-43.

Rieke, E.A. 1997. The Federal Advisory Committee Act, rules and executive orders: judicial interpretations and suggested revisions. Boulder, CO: University of Colorado Natural Resources Law Center.

Ringgold, P. 1998. *Land Stewardship Contracting in National Forests: A Community Guide to Existing Authorities*. Washington, DC: Pinchot Institute for Conservation.

Sample, V.A. et al. 1995. *Building Partnerships for Ecosystem Management on Mixed Ownership Landscapes: Regional Perspectives*. Washington, DC: American Forests Forest Policy Center.

Snow, D. 1998. Finding the dollars to do good things. *Chronicle of Community* 2(2):53-55.

Thoreau Institute. 1998. The Forest Options Group proposal. *Different Drummer* 14:6-7.

Tuchmann, E.T. et al. 1996. The Northwest Forest Plan: A report to the President and Congress. Portland, OR: Office of Forestry and Economic Development.

USDA Forest Service. Collaborative Stewardship Team. 1997. Report of the Collaborative Stewardship Team. Washington, DC: USDA Forest Service.

USDA Forest Service. 1998. Forest Service Action Strategy for State and Private Forestry Services. Washington, DC: USDA Forest Service.

Voth, D. and B. Pell. 1998. Ozark-Ouachita Highlands Assessment–National Forest Communities Project. Briefing paper presented at the Showcasing Community-Based Elements of Forest Planning and Assessment workshop for the USDA Forest Service staff and partners. Washington, DC, USDA Forest Service.

Yaffee, S.L. and J.M. Wondolleck. 1997. Building bridges across agency boundaries. In K.A. Kohm and J.F. Franklin (eds), *Creating a Forestry for the 21st Century*, pp. 381-396. Washington, DC: Island Press.

Process: Redefining Relationships

Margaret A. Moote
Beverly A. Brown
Eric Kingsley
Steven X. Lee
Steve Marshall
Donald E. Voth
Gregg B. Walker

SUMMARY. Process in community-based ecosystem management (CBEM) involves a fundamental reframing of human relationships with each other and with other ecosystem components. Inherent to this reframing is a respect for diversity and for different ways of knowing, a

Margaret A. Moote is affiliated with the Udall Center for Studies in Public Policy, University of Arizona, 803/811 East 1st Street, Tucson, AZ 85721.

Beverly A. Brown is affiliated with the Jefferson Center for Education and Research, P.O. Box 27, Wolf Creek, OR 97497.

Eric Kingsley is affiliated with the New Hampshire Timberland Owners Association, 54 Portsmouth Street, Concord, NH 03301.

Steven X. Lee is affiliated with The Heritage Museum and the Benjamin Banneker Historical Park & Museum, 4509 Prospect Circle, Baltimore, MD 21216.

Steve Marshall is affiliated with the USDA Forest Service, Cooperative Forestry, 201 14th Street SW, Washington, DC 20250.

Donald E. Voth is affiliated with the Department of Rural Sociology, University of Arkansas, A 227, Fayetteville, AR 72701.

Gregg B. Walker is affiliated with the Departments of Speech Communication and Forest Resources, Oregon State University, 104 Shepard Hall, Corvallis, OR 97331-6199.

[Haworth co-indexing entry note]: "Process: Redefining Relationships." Moote, Margaret A. et al. Co-published simultaneously in *Journal of Sustainable Forestry* (Food Products Press, an imprint of The Haworth Press, Inc.) Vol. 12, No. 3/4, 2001, pp. 97-116; and: *Understanding Community-Based Forest Ecosystem Management* (ed: Gerald J. Gray, Maia J. Enzer, and Jonathan Kusel) Food Products Press, an imprint of The Haworth Press, Inc., 2001, pp. 97-116. Single or multiple copies of this article are available for a fee from The Haworth Document Delivery Service [1-800-342-9678, 9:00 a.m. - 5:00 p.m. (EST). E-mail address: getinfo@haworthpressinc.com].

willingness to meet others on their own terms, and a willingness to respect others' knowledge even while openly sharing one's own. CBEM process is characterized by an embracing of diverse cultural and social groups within a community, equal access to and open exchange of information, mutual learning, flexibility, and collective vision. CBEM process also is impeded by long-standing power imbalances inherent in our culture and in our governance systems. Overcoming these power imbalances will not be easy, and CBEM faces a number of process challenges. These include: lack of incentives for some parties to participate, traditionally disempowered groups, administrative bureaucracy, and legitimacy. To be sustainable, CBEM process ultimately will have to integrate the key dimensions of inclusiveness, accessibility, transparency, mutual learning, adaptability, and collective vision within existing social and governance processes.

KEYWORDS. Community-based, forest(ry), ecosystem management, inclusiveness, diversity, community capacity

Effective community-based ecosystem management process involves a fundamental reframing of relationships. It involves new patterns of thinking, including a recognition of interdependence among humans and other ecosystem components and among different human communities. It involves reformed patterns of recognition, standing, and legitimacy. This reframing must occur in communities of place, in interest-based communities, and among scientists, professionals, politicians, and others.

Inherent to this reframing of relationships is a respect for diversity and different ways of knowing. Effective CBEM processes explicitly seek out and provide access to diverse groups, particularly those who might not typically come to the table. Information and decision-making procedures are clear and open to all interested parties. There is a strong emphasis on mutual learning, which often occurs through collaborative group activities. CBEM process is often guided by a collective vision, developed through an exploration of the full range of participants' interests and values. The process is flexible, changing as participants or other conditions change. The first part of this chapter discusses these dimensions of CBEM process. We do not provide a "process guide," however; readers seeking more direction in process design should consult the guides listed at the end of the chapter.

We would be remiss to discuss key dimensions of good CBEM process without also discussing the frustrations of many CBEM practitioners. The most significant of these stem from power imbalances inherent in our culture and in our governance systems. In the second part of this chapter we discuss the challenges posed by and possible approaches to working with power differentials and bureaucratic administrative processes.

KEY DIMENSIONS OF PROCESS

Inclusiveness and Diversity

The foundation of community-based ecosystem management is community, which is defined as:

> the people with common interests living in a particular area; . . . an interacting population of various kinds of individuals (as species) in a common location; a group of people with a common interest living together within a larger society; . . . a group linked by a common policy . . . (Webster's New Collegiate Dictionary 1993)

In community-based ecosystem management we are referring to all the above; there are many communities, both ecological and human, involved in and affected by community-based ecosystem management. The sense of community further must be understood not only by the letter of its definition, but the spirit of it as well. The diversity of both human and ecological communities requires that community-based ecosystem management processes be explicitly inclusive.

The Human Community: The Stewards–Effective community-based ecosystem management process proactively engages all community groups, providing a forum that allows everyone equal voice and participation, regardless of their relative size or influence within the larger community. The legitimacy of the CBEM process depends primarily on a high percentage of participation from those diverse interests most affected by a particular CBEM project and with a proximate stake in the outcome. The measure is not so much in the numbers or legitimacy of individual participants, but in the consistency of broad representation in the overall process (Kusel et al. 1996). With a high proportion of directly affected groups involved, the process is far more likely to be grounded in the details of the specific area under consideration and result in an accurate and legitimate response. An effective CBEM process further ensures genuine representation not only in the physical bodies of diverse peoples (i.e., the percentage of minorities among participants around a given table), but equally in the traditions, cultural concepts, and self-determination of the respective peoples.

In CBEM process, the principle of inclusiveness is, from inception, applied at all levels. It is crucial that community members from diverse backgrounds actively participate from the beginning in planning, design, decision making, implementation, and monitoring. CBEM processes therefore are open, encouraging broad participation in the generation and analysis of information; they recognize and respect nontechnical forms of knowledge (Kusel et al. 1996). CBEM processes further must accommodate new forms of knowledge and multiple sources of information, balancing both expert and lay input.

Most relevantly, there is a need to recognize the value in diverse knowledge systems (i.e., the empirical knowledge of indigenous cultures) and to have equal and direct funding for diverse organizations and their programs (particularly ethnic and economic minorities who have traditionally received little or no direct support). Only with this level of recognition and support can minority groups genuinely function as equal players. Herein lies the key to true empowerment and self-determination for diverse groups within a community. Here too is a key to their participation and sustained interest in CBEM.

From Egocentric to Ecocentric–A particular challenge to CBEM is our ability to recognize the values and roles of each species in the system. The term "ecosystem," like "multi-culturalism," does not imply the gathering of diverse instruments to mimic the sound of one. To the contrary, both terms denote an arena in which each component adds its own sound to the symphony. Inclusiveness in CBEM includes the ability to recognize and respect the values and roles of other species. Under the banner of ecosystem management, programs are often planned and implemented around one or a few aspects of an ecosystem, such as only the trees in a forest or a single endangered species. The potential damage effected to the greater ecosystem (including other human groups) is seldom considered. The foremost interest and measure of success begins to be the benefit for the one, over the good for the whole.

Human communities are both stewards of, and dynamic and dominant subsystems within, the greater ecological community. Full recognition and inclusion of people of diverse cultural backgrounds is essential and indicative of a broader sense of inclusiveness. If we cannot responsibly address the range of diversity within our own species, it is likely we cannot responsibly value and manage the needs of the multiple species that delicately balance an ecosystem. Ultimately, the human community is responsible to itself to the degree that it embraces and represents all its human constituents equitably. It is effective and responsible to the ecosystem to the degree that it recognizes, represents, and values all members of the ecological system.

Accessibility

A broadly accessible process is critical to ensure diversity and inclusiveness. Access is often narrowly interpreted to mean receiving an invitation and having the physical and financial ability to attend meetings. Although both

CASE EXAMPLE: A GREENWAY PROJECT

Phase I (Planning & Design)

In the early 1990s, a municipal government, in conjunction with the state government and a local parks agency, decided to build a greenway that would pass through a corridor of several inner-city neighborhoods and a historic wilderness park. Although the city's population was predominantly African American, and the project impacted mostly African American neighborhoods, African American representation in the inception of the project's planning, design, and development was minimal. African American representation was greatly disproportionate–to nonexistent–on the city's project task force, project design team, and the committee that defined the project's request for proposals and made the contractors selection.

The project's community information and planning sessions were presented by the city as the forum and mechanism for community input, and the African American presence at them was substantial. However, here the community had no real power or decision-making authority. When the community arrived at the sessions, the agenda (like the decision to develop the project through their neighborhoods) had already been set.

Though attendance at the information and planning sessions alone could be interpreted to indicate a significant number of African Americans participated, in reality the first phase of this greenway project did not exemplify genuine inclusiveness or diversity. For the project to have been truly inclusive in its process, African American representation should have been at least equal in number to that of other groups. From the outset, African Americans should have been able to participate equally in decision-making at the project's conception, planning, and layout stages; in determining the process, designers, and contractors for the project's development; and in the agenda and schedule for its progression. Just as significantly, with the many dollars involved in the project, African American companies should have also been engaged in some of the lead contractor roles. Instead, they were allocated very few subcontracted positions (as is typical for minorities). Beneath the surface, phase one of this project did not equally empower and economically further the local African American community.

While this project was presented as an effort to enhance the urban ecosystem, much of the greenway's route was planned to pass through a very large wilderness park. This historic park, with old growth forest and uncommonly rich biodiversity, is home to many rare and threatened native plant and migratory bird species, as well as more common wildlife (deer, fox, turtles, etc.). The impact on these species' habitat was not comprehensively considered in the greenway plan. No surveys, studies, or inventories were conducted to document significant wildlife habitat areas to be conserved or how the communities of plant and animal species would be impacted, even though this park is the last vestige of traditional forest left to them in the region. Rather than ecological considerations, the scenery, construction-friendly topography, and neighborhood and public service accessibility were the primary factors used to determine the greenway's course. This too does not exemplify inclusiveness or diversity in ecosystem planning in this project. Again, the project empowered and benefitted a single groups of the ecosystem, the human community.

CASE EXAMPLE: A GREENWAY PROJECT

Phase II (Project Implementation and Development)

Over the course of six years of planning and development, the local government and private park agency were flexible in their process and open to learning. As the project ensued, a more genuinely community-based ecosystem management effort began to emerge. Largely through the work of the private park agency and an African-American community relations representative, ongoing community information sessions helped stimulate a highly motivated and interested citizenry. The community organized, generated more inclusive representation, and began to influence the project in meaningful ways. The project planners began to collaborate with other environmental programs being conducted in the greenway project area and incorporated more diverse environmental concerns.

As a plan and organization were developed to manage the trail, the project came to be much more community-based. The trail management organization was developed by building a coalition among the array of neighborhood and park groups whose neighborhoods and interests were within the greenway's course. In partnership with the city, this trail management organization, which is made up of diverse individuals and community group members, will oversee greenway construction and coordinate its management with contiguous human and ecological communities. From an imperfect beginning, this project has progressed to be an exemplary model of many of the qualities of good community-based ecosystem management.

these conditions are crucial, access in CBEM involves tailoring the process to match the diverse needs and communication styles of affected groups.

Access considerations include the cultural manner in which a constituency is accustomed to meeting and making decisions, the language (professional or not, English or not, based on advanced literacy or not) in which affairs are typically conducted, and the history of groups' experiences of inclusion or exclusion. The latter is especially important if they suffered retaliation in some substantial manner for their participation in the past. If a group has a history of inclusion, were they ever able to achieve any success in their sphere of concern? Special consideration should be given to groups that have long been unable to access civic deliberations. Criticizing the absence or blocking the participation of a directly affected group because it "is not up to speed" is a failure of the process.

Symbolically, design choices communicate messages to parties about their standing and legitimacy in the process. For example, the decision to hold a workshop on a weekday may communicate to working citizens that their input is unimportant. Conducting a field trip only in English may discourage non-English and non-native-English speakers from participating and send a message that their ideas are not valued as much as those of native English speakers.

Providing access to CBEM means using innovative and situation-specific participation methods. Some groups may be willing to come to a multi-party gathering, for example, only after the conveners or decisionmakers have met with them individually, at a place and time determined by the individual group. The process may need to be multi-faceted, in terms of both method and place. Some people might prefer to participate privately, via a letter, phone call, or Internet reply. Some parties might attend only meetings in their own town and be reluctant to travel to a community 20 miles away. Non-majority cultures or working class people from any background may be skeptical about participation processes, particularly if they have a history of exclusion, oppression, or being used as pawns. Many may not see a meeting announcement published in the public notices section of the local newspaper but may learn about it from a community organization, informal cultural network, bulletin from a religious organization, broadcast public service announcement, newsletter, or a posting on the local grocery store bulletin board. Person-to-person recruitment by known and trusted family members or friends is often the most effective method.

Transparency

An effective CBEM process is clearly presented and understandable to all interested parties. It is clear who the participants are, what the schedule for decisionmaking is, who has ultimate authority for making those decisions, and how input to the process will be used.

Traditional natural resource management and community development processes most often are funded or undertaken by government agencies. The very complexity of agencies (and other institutions, such as universities), make them confusing and inaccessible to many community members. Exacerbating the situation, scientific disciplines and government agencies have developed specialized information and jargon that increase their remoteness. Internal organizational cultures in agencies, universities, and professional societies may further discourage information sharing by fostering the attitude, "we are the experts–you couldn't possibly understand." Administrative bureaucracy and agency cultures have led communities to a deep mistrust of agencies and "outside experts."

Part of the CBEM process involves restoring trust between the communities and the institutions that generate and archive information. This in turn requires a major restructuring of information and how it is communicated. An effective CBEM process makes information and data analysis tools, such as computers and modeling programs, widely available to all. Administrators and technical experts give up some discretion and freely share information and information analysis with communities: Transparency implies equal and open access to both information and planning processes. In particular, communities require some assurance that their input will be considered when decisions are being made.

Mutual Learning

For community-based ecosystem management policies to be made and carried out effectively at the community level, parties must be open to learning from one another and acknowledge that learning is ongoing. Such learning needs to be inclusive and interactive, emphasizing activities that encourage input from the full variety of cultures, income groups, and perspectives. No single party–including any individual, group, agency, organization, or discipline–holds the key to understanding a particular CBEM situation. For any one party to assume that it knows best, understands fully, or has all the answers, is presumptuous and inconsistent with the interrelatedness inherent in community-based ecosystem management. The best scientific information, for example, means little if it is not considered in tandem with social values, needs, and knowledge.

Both theory and experience have shown us that mutual learning is best achieved through collective action. Participation in activities such as field trips, field work (e.g., monitoring, training, rehabilitation, planting seedlings), and workshop tasks such as planning, problem-solving, and analytical and information-sharing discussions foster learning and understanding better than passive speeches, hearings, and videos. This type of active learning requires those people traditionally considered "experts" to shift from primary presenters or speakers to join with other participants as resource people, facilitators of learning processes, and coordinators of interaction. The active learning process respects community-based knowledge by emphasizing opportunities for people to draw on their experience and expertise–and learn from one another. CBEM process helps participants actively learn as much as possible about the important issues at hand, rather than implicitly assuming that they know as much about the situation as they might need and that all their perceptions are accurate. An active learning focus offers a considerable opportunity for new ideas to emerge in both process (how we might come together to make this decision) and outcome (what are the choices before us and what are their relative strengths).

In CBEM, learning can happen in the context of particular places, involving people who live in those places and are familiar with many different facets of the human and ecological communities. Because it is place-based and may include communities of interest, CBEM offers an ideal opportunity for active, mutual learning.

Collective Vision

CBEM processes function best when guided by a clear vision–a well-specified statement of process goals that includes all participants' interests, which are often based in values. Values are deeply held constructs; we often see them as intrinsic parts of our being. They serve as inner reference points for our outward behavior. Values and interests are more significant to us than particular positions and thus do not change as rapidly as positions. Therefore, if our actions are based on values, we are less likely to shrink before obstacles.

Exploring interests and values not only defines vision and provides a basis for trust, it also provides a foundation for commitment. If a community is able to find broadly defined common values–such as protecting the environment for future generations–it will have a stronger sense of itself and a basis for collective action. The extent to which common interests or common values can be unearthed or created is thus an indicator of the community's ability to both create and sustain a common vision.

Adaptability

The complex and dynamic nature of ecological and social systems means there can be no explicit guidelines or management prescriptions for CBEM; uncertainty will always be inherent to it. In other words, organizations, laws, policies, and management practices need to be flexible, so they may rapidly adapt to changes in social values, community composition, ecological conditions, political pressures, available data, and knowledge (McLain 1994).

Adaptive management has been offered as one way to address the staggering information requirements of community-based ecosystem management while allowing that management to move forward in the face of uncertainty. In essence, adaptive management involves repeatedly monitoring the outcomes of alternative management practices, either in the field or through simulation modeling, and making adjustments based on what is learned (McLain 1994, Lee 1993). Adaptive management uses scientists, resource managers, policymakers, interest groups, and citizens to collectively identify management problems and set initial goals. Proposed means of achieving the goals, such as management prescriptions, are treated as working hypotheses. By treating these prescriptions as research hypotheses, resource managers can continuously monitor and modify their decisions and practices as needed (McLain and Lee 1996; Lee 1993). As Kai Lee notes, "experiments often bring surprises, but if resource management is recognized as inherently uncertain, then surprises become opportunities to learn rather than failure to predict" (Lee 1993, p. 56).

The concept of all-party monitoring illustrates many characteristics of a good CBEM process, including adaptive management. If a community has committed itself to a fundamental concept of inclusiveness, then the process of monitoring will incorporate different ways of knowing and acting. By having representatives from each stakeholder group, including national interests, collectively monitor the implementation of a land management decision, all parties are assured that their interests are being met. Effective all-party monitoring will include not only mainstream concepts of academic, technical, and "scientific" monitoring, but also monitoring that is crucial to non-mainstream groups, such as that of rights to speak without retaliation or employer provision of dignified working conditions.

PROCESS CHALLENGES

There are many process pitfalls and obstacles that can derail efforts to involve citizens and stakeholders in the management of ecosystems. The questions of who gets to the table, what energy sustains participation, how decisions are made, and how the interests of unrepresented parties are safeguarded are huge hurdles faced by communities. Layer that with a multitude of federal laws, concern for property rights, and the vested interests of those who benefit from the status quo, and the challenges to CBEM can appear insurmountable.

Many of these challenges are due to the fact that most CBEM efforts currently operate outside the existing governance structure. As such, they lack recognition, power, a means of reaching out to the larger community, and long-term funding sources. They must begin from scratch to design their own processes for outreach, communication, decisionmaking, and fundraising. This is not all bad: operating outside the existing governance structure also makes these efforts, by necessity, inventive and adaptable. Because they usually are not efforts by the status quo, they may represent community sectors and perspectives traditionally excluded from community and ecosystem management decisionmaking. Creating a new process is exhausting, however. It involves frequent meetings and many hours of volunteer effort to design a "shadow" structure for planning and decisionmaking that may or may not influence established governance processes.

The following sections discuss the challenges of power imbalances and bureaucratic administrative processes in more detail and suggest ways of empowering communities while addressing the concerns of established interests.

Working with Government

Community-based processes need assistance from, or at least acquiescence of, government, especially the agencies responsible for managing or regulating public natural resources. Sharing administrative decision making and forging new community-government collaborations are part of the "reframing of relationships" inherent in CBEM. Yet integrating CBEM process into existing bureaucracies raises challenges for both agencies and communities.

Aside from electoral politics, the administrative process is the major existing avenue for public participation. Agencies differ widely in the manner in which they are allowed or required to engage their publics. At the federal level, they are usually encouraged to achieve "maximum feasible participation," even though the mode of participation is limited by such things as the Federal Advisory Committee Act (Cortner et al., 2001). Traditionally, individuals and interest groups have participated in government through voting, lobbying, appealing, and litigating. These processes are inadequate for CBEM, which requires flexibility, site-specific decision making, and ongoing interaction between local communities and government decisionmakers.

The most important administrative barrier to agency involvement in CBEM process is ambiguity of authority. The public administrator must adhere to mandates coming down from the top, while incorporating and respecting those coming up from the community. Although the decision-making process may be shared, when decisions involve public resources, final authority still rests with the public agency. The extent to which a process is collaborative thus depends on the willingness of the public manager–and the agency itself–to share authority.

A second and closely related problem concerns the concepts of standing and representation. Unless they are legislated–and in natural resource policy generally they are not, excepting, of course, the limitations imposed by FACA–the representation of "affected publics" in participatory administration is left to the responsible officials to define. When it comes to citizens' rights to have input into a proposed action of the U.S. Forest Service, for instance, the status of that input is quite ambiguous. The agency is typically required only to "consider" public comments. Public input in the administrative process is never equivalent to a vote in the electoral or "normal" political process. Indeed, FACA would appear to prevent public agencies from using such votes. Federal regulations in place since the Reagan administration further limit federal abilities to perform citizen surveys in order to gauge public opinion.

Finally, the dynamics of bureaucracy in virtually all large organizations result in a separation of functions–and even in the isolation of one function from another. Of particular importance for agencies charged with natural resource management is a radical separation of planning functions, research functions, actual land management functions, and those functions that traditionally have dealt with communities of place. The Forest Service's community assistance programs, for instance, are not considered to have anything to do with forest management and/or forest planning. Thus a community where the Forest Service's Rural Community Assistance program has stimulated very effective and inclusive countywide planning and development processes may not even be contacted when the local national forest initiates its forest plan revision. Were land management agencies to bridge their own functional divides, they would find many opportunities to link long-term strategic planning, ecosystem management efforts, and community development projects.

Opportunities for Innovation–The public officials, then, find themselves in a situation of considerable ambiguity but one that allows for innovation and creativity. Public officials can carry out their responsibilities in a manner that is sensitive to the site-specific and community-specific situation, needs, and capabilities. In so doing, they can build collaborative partnerships on a highly specific basis with those most involved and affected in each ecosystem. They can utilize local and interest-group knowledge in their decision making and local human resources in implementing and monitoring those decisions.

Public land managers may find a number of reasons to participate in CBEM, including sharing with others the burden of working through a complex decision, access to new ideas and tools that partners bring to the table, and resiliency (appeal-proofing) of agency decisions. If the agency can participate in a way that lessens the likelihood of costly appeals and court battles, utilizing CBEM may prove to be a cost-effective decision. Add to this the intangible benefits of working closely with communities, increased staff morale when confrontation is replaced by cooperation, and the promise of an agency adapting to the times and it is clear that reasons exist for land management agencies to embrace a new way of doing business.

In public lands management, the work of a CBEM group can give a decision maker confidence, support, and (quite importantly) political cover in moving forward with decisions. Support from a CBEM group, while holding no statutory weight, can be leveraged to show a regional forester or judge that broad support does exist for this action. As long as this support does not seek to undermine the rule of law, a CBEM group can be a positive and influential force.

The role of local government in CBEM activities, and especially in the management of public lands, allows for further innovation and creativity. Local governments are not subject to the FACA restrictions, and so can serve as intermediaries between communities and the federal government. Local governments have their roots in the community and may be more responsive to citizens' concerns. Local leadership thus can communicate directly with federal agencies through their local governments or through an organization sponsored by local governments, and federal agencies can act on such advice without being liable to legal challenge under FACA. It is important to note that this may not be an option in many communities as local leadership structure does not always reflect the cultural and social complexity of local communities. In those cases, allowing local government to speak for the community may give undue legitimacy to viewpoints held by the locality or region's more powerful economic and social interests. In other cases, however, people are turning to local governments almost in desperation, finding no other effective mechanism for affecting decisions about public lands management.

Roles for Communities–Participatory process requires not only willing government officials but a populace willing and informed enough to work with government. It is important for CBEM proponents to fully understand the administrative situation, so as to know what the land manager actually can do. They must also be prepared to give key support to land managers who are willing to innovate to the advantage of CBEM. There is great potential for communities to assist natural resource agencies in redefining the relationships among their own functions to embrace a more participatory form of governance.

Working in Baltimore in the context of the government's War on Poverty, and sensitive to the fact that nondecisions are frequently more important than decisions, Bachrach and Baratz (1970) devised a community decision-making sequence highlighting the key points on which those seeking change should focus. This procedure, while focusing on local-level decision making, can be applied to other levels of government. Bachrach and Baratz' sequence includes three phases: getting an issue on the agenda, working through the decision arena, and implementation. At each phase, the community faces obstacles.

To get an issue on the agenda, the community must tackle the dominant power structure, which is likely to stick to the status quo and block issues not in its interest. Many important issues are never raised in the decision-making arena; they are simply "nondecisions." These nondecisions may, in fact, be as important as decisions. Changing the existing bias to allow the consideration of issues never before brought up involves sensitivity both to the link between the local power structure and its current "mobilization of bias," and to the deeply held values within the community.

To make it through the decision arena, the community must work within a formal decision-making agenda. This involves learning and dealing with protocol, writing resolutions, and identifying appropriate decisionmaking institutions. Achieving a positive policy decision may require political influence.

Even when a positive policy decision has been made, those resisting change may hamper its implementation by ensuring that there is no funding for an essential function (e.g., code enforcement). CBEM efforts, therefore, must monitor not only outcome conditions, but the policy process itself.

Different skills, techniques, and processes are required to overcome different policy barriers. Fortunately, within a community there typically exist individuals with a wide variety of skills and very different relationships to community power structures. The process remains daunting, however, especially when it is built exclusively on volunteer citizen time and funding sources that are, at best, short-term and uncertain. The irony of social change is that those most likely to undertake it–the poor, the disenfranchised–are those who carry the heaviest burden and take the greatest risks.

Advocacy and conflict are complicated, delicate processes. It is always essential, when engaging in advocacy, to keep an eye toward that point, perhaps far in the future, when advocacy gives way to collaboration and cooperation. Importantly, CBEM efforts frequently seek collaboration and cooperation across lines where this has not occurred. Achieving communication across these lines may initially require an advocacy stance or other means of leveraging power.

Leveraging Marginalized Groups

Even if a group believes that realistic conditions of access and participation exist, without some leverage to help it attain its objectives, participation is an empty affair. Leverage has typically been accomplished in natural resource affairs through an established ability to effect powerful alliances (or in the case of agencies, directives and mandates), efficiently pursue mainstream organizing techniques, and have or be linked to substantial sources of money.

Among groups that do not have ready access to resources, the road to creating leverage is a long one and requires extra measures of persistence and creativity. Low-income groups are consistently frustrated by their lack of leverage with agencies, policymakers, and corporations–and by the ability of powerful interests, including national interest groups, to simply veto their efforts through judicial and legislative strategies. Groups reaching across large cultural differences and varying levels of capacity are aware that they will probably fail in efforts to challenge nationally and locally powerful interests. Without credible opportunities to make a difference, these groups may avoid participation–even if other conditions are met–in order to save their energies for issues in which they have a greater possibility of success.

Among the many ethnic cultures and low-income groups marginalized by mainstream middle-class activism–a list that includes low-income European Americans–willingness to participate hinges on their ability to create leverage. Of course the simple matters of fairness and justice should be leverage enough, but in our system of power leverage, this is rarely sufficient.

Fortunately, important leverages are available to typically marginalized constituencies. One is resistance. For instance, if low-income forest workers do not consider the decision-making and implementation processes legitimate, they will gladly "take"–that is, poach–what they consider to be legitimately theirs. A second leverage is strategic creative organizing and the power to assemble information. In some cases, critical information on which the desired outcomes of CBEM are premised may be available only through the marginalized groups' efforts or knowledge. For instance, forest worker-gatherer groups may witness degradation of fish and wildlife resources, destruction of habitats essential for certain economically important under-the-canopy species, or unjust working conditions that impact the wages and power of the majority of people in a locality or occupation. With this information people can organize for creative and strategic leverage–for instance, assembling information on unjust working conditions and, with allies, submitting it to the media and other venues for public scrutiny.

Building Linkages–To create local leverage is very costly to typically marginalized groups and requires the willingness of allies who support community-generated efforts. The emphasis here is on lateral support–not on top-down "education" or the taking over or directing of community-based efforts. If outside groups preempt a previously marginalized group's process of building capacity or strategizing, or decide it just needs to be "educated," the outside group is creating greater leverage for itself but removing it from those making an effort to become strong. All parties must participate in CBEM on their own terms and with their independent sources of leverage. Paradoxically, this independence creates the greatest opportunity for diverse groups to work together to bring about mutually supported goals.

There are other reasons to build linkages. The first concerns non-human ecology. CBEM frequently requires mobilization at watershed or landscape scale–well beyond the level of one human community. Thus, several nested human communities must be involved in the process. Second, the reality of the political dimension requires communities to build networks among communities of place and communities of interest. Rather than be intimidated by the potential of interest communities to "throw a monkey wrench" into a local process, local communities should endeavor to reach a point where their broader community networks can be allies in their empowerment efforts. It should also be their objective–to the extent feasible–to use the leverage they have at multiple government levels to effect change on the ground.

Engaging the Status Quo

Getting buy-in from the status quo is a major challenge for CBEM. Those individuals who now possess something are reluctant, if not hostile, to enter into a conversation on how they can most gracefully relinquish what they have. "No person, group, or industry . . . will give up its historical preferences, subsidies, and special treatments without a to-the-death fight in Congress and courts, no matter what" (Coggins 1998 pp. 27-33).

While engaging the powers that be is important for all CBEM efforts, it is critical for efforts on public lands. As The Wilderness Society has pointed out, "there is no statutory suggestion that propinquity to the lands in question conveys a greater than usual right" (Blumberg and Knuffke 1998 pp. 41-44). Addressing local communities' lack of power in a proactive manner is essential; national organizations of every stripe all too often regard what happens in one small area as a precedent that will spread across the landscape like wildfire, and oppose it on principle. However, national organizations themselves are criticized for similar reasons: They are spread across the nation, frequently don't have the staff to devote to numerous community-based efforts, and may be seen by some local interests as "big city environmentalists," "greedy multinationals," or simply "people from away." National groups thus both threaten and are threatened by local process. Efforts must be made to engage these interests in CBEM dialogue, provide them with input in ways other than the traditional meeting/work session, and seek their understanding (and hopefully consent or support) for the community effort. Given the influence national organizations currently wield, to do any less is akin to rearranging deck chairs on the Titanic.

There are some incentives for national interests to participate in CBEM, even established groups–for instance, agencies, large environmental interests, or industry–that can easily leverage or even determine implementation of projects or outcomes. One incentive lies in their ability to build relationships that will support ecosystem-enhancing activities while promoting their political, conservation, or profit-generating goals. Participation in CBEM process could lend a legitimacy to activities, insulating powerful actors from constant challenge. If considerable transparency is built in, the process encourages trust as well as legitimacy.

On private land, CBEM faces other challenges. Landowners have property rights and can make individual decisions within the limits of the law. Private landowners must be convinced that their rights will not be relinquished and that by engaging in CBEM they will achieve a net gain. Reasons landowners give for participating in landscape-level planning include: becoming better stewards of their property by viewing it in a larger context, being able to have an impact on an area larger than just their ownership, having an alternative to undesired regulation, or being able to seek a unified voice when regulations do come (Williams and Ellefson 1997). Landowners may also work in a larger block to achieve economically efficient management costs or to get better pricing for products. In some cases, landowners may participate in a community-based process out of a sense of civic stewardship or responsibility to the community. Whatever the reason, private landowners must have some incentive–whether altruistic, regulatory, or monetary–to enter into CBEM.

Legitimizing the Process

The founders of the United States integrated a concept of checks and balances into our formal governance structure and legitimized it in popular concepts of fairness and justice. Ideally, an effective structure of checks and balances prevents the establishment of unequal power vested in one or a few entities. In the current situation, the equivalent of a checks and balances systems for environmental and natural resource affairs operates mostly at the regional and national level. To participate, organizations have to muster a sophisticated mixture of time, money, legal expertise, and professional follow-through to pursue challenges to resource management decisions in the courts, legislatures, or corporate board rooms. The results of these long administrative and formal court decisions typically end up in "winner takes all" decisions that extend far beyond site-specific ecosystems and communities. This results in prescriptions or effects that have indiscriminate, often unintended, effects at the level of local ecosystems and in networks of directly affected people.

Corporate, political, and large environmental interests have used this "winner takes all" system of court challenges and legislative tinkering effectively enough to create considerable frustration at the ecosystem level. For instance, the idea that "salvage is industrial logging in disguise" may be on target in one ecosystem, but it may create a barrier to "salvage is a critical fuels-reduction strategy," which may be entirely accurate in a densely thicketed forest where natural fire has been suppressed for many years. Those who do not or cannot accept a management decision can delay, and often derail, proposed actions by using the National Forest Management Act, Endangered Species Act, National Environmental Policy Act, and many other federal laws. Local communities may work together to achieve broad buy-in, have decisions firmly rooted in science and public values, and propose actions that reflect a broad consensus. However, under existing law a lone outlier, even one with no discernable interest in a decision, can use the process to promote (and sometimes enforce) his or her own vision of what should occur. The current systems, whether exhaustive administrative review within federal land agencies or direct action in the federal courts, frequently create gridlock, forcing all management actions into abeyance while positions are argued. The rise of CBEM has been, in large part, a response to the current system of power brokerage in the formal courts and legislatures, as communities have sought to create leverage and power of their own.

With the current lack of a checks and balances system that functions at the local and regional levels, those with a healthy suspicion of CBEM ask a legitimate question: "What happens if a CBEM process chooses a course of action that causes major degradation within the affected ecosystem?" For CBEM to acquire legitimacy, the creation of a process of review and challenge may be necessary at several levels, beginning at the local community, or site-specific, level.

An accessible and scale-appropriate system of checks and balances would provide local entities a clear, legitimate right to participate as well as local leverage to establish "standing" and the power to be effective. It would ensure that all parties recognize nonparticipation has consequences. In order to be accountable, the system of checks and balances would have to be transparent. It should include a review process prior to implementation, a challenge process that involves all parties, and a final body vested with the authority to ratify and implement the decision.

Instead of mimicking the "winner takes all" strategy, and proceeding through a sequential line of appeals, we need a new way of thinking about checks and balances that is appropriate to the openness, diversity, and inclusiveness characteristic of CBEM process. This might include a step akin to the mediation and arbitration required in high-level labor disputes. All parties to a local-level dispute could be expected to participate–much as all parties are optimally at the table in CBEM and all-party monitoring processes. Thus, instead of having one party pitted against another on a formal point of law, the local level of checks and balances could involve a transparent, public, all-party mediation process. The advantage of an intermediate system of checks and balances is to create greater accountability to challengers in light of the CBEM-specific situation at hand. The goal would be to create a greater number of mutually acceptable decisions before challenging parties opt for monkey wrenching or legislative/judicial decisions that block any action at the local level across the entire nation.

Sustaining the Process

As noted above, the demands of CBEM are heavy on all participants–government, industry, interest groups, and individual citizens. CBEM process is messy, time consuming, and nonlinear. It is not entirely surprising, therefore, that many early CBEM efforts have been ephemeral, lasting two or three years and then folding in the face of participant burnout, lack of funding, lack of recognition (and even active money wrenching) by powers that be–and with few tangible successes. Those that continue face leadership turnover and even loss of all the original participants, which can result in a revision of the goals and/or process to something that is not CBEM. Even with the reforms discussed above, demands on individuals, communities, and institutions will remain considerable.

How can the integrated needs of communities and the ecosystems of which they are a part be addressed in the long term? Is this even desirable? If the goal of CBEM is sustained community and ecosystem well-being, then the sustainability of CBEM efforts clearly is desirable. CBEM "activities are largely irrelevant if they are not accompanied by a concerted effort to strengthen institutions and to build an indigenous capacity to sustain resource management and development actions" (White et al., 1994, p. 114). The former question is less easily answered. Challenges for continued CBEM include: participant burnout, lack of funding, membership and leadership turnover, changing goals, and the risk of exclusivity (see Daly, 2001).

Relying on a handful of individuals operating without external support can only last for so long. "Without funding, they (collaborative groups) will have no professional staff, and thus will lack an institutional structure. Without a permanent institutional structure they are destined to wither away as the perceived crisis passes" (Coggins 1998 pp. 27-33). While foundations and the like may fund such efforts in the short-term, this is inherently unsustainable. CBEM groups must strive to build a successful track record and a constituency that will recognize their work with contracts, financial support, and something that resembles membership. Only then will a CBEM process have the institutional infrastructure to sustain efforts over any ecologically meaningful length of time.

Some CBEM efforts have been institutionalized in quasi-governmental organizations that give them some standing in the local community and some government support, including funding. Examples include watershed councils (see Clark, 2001), landowner cooperatives, and coordinated resource management groups that have memoranda of agreement with government agencies (see Cortner et al., 2001). The process of institutionalization, however, brings other challenges to CBEM efforts, including co-optation by the status quo, exclusivity, and rigidity. As CBEM efforts are formalized, they risk losing the flexibility and adaptability that make them accessible to and representative of diverse interests. The cost of recognition may be organizing CBEM in a way that feels "safe" to established interests, especially governments. The risk of formalizing process guidelines for representation, decisionmaking, and accountability is creating a process so formal it mimics traditional processes.

The challenge, then, is to remain sensitive to the key dimensions of CBEM–inclusiveness; diversity; access; mutual, active learning; common vision; and adaptability–while formalizing the process enough to ensure its continuance. Ideally, the process of institutionalizing CBEM would help fold these dimensions into existing governance structures, making them more inclusive, flexible, and open–in essence, infusing the existing system of bureaucratic representative democracy with participatory processes.

REFERENCES

Blumberg, L. and D. Knuffke. 1998. Count us out: Why The Wilderness Society opposed the Quincy Library Group legislation. *Chronicle of Community.* 2(2):41-44.

Bachrach, P. and M.S. Baratz. 1970. *Power and Poverty: Theory and Practice.* New York: Oxford Press.

Clark, L.R. 2001. The Evolution of Watershed Councils and the Oregon Plan: Tempering the Populist Tradition. *Journal of Sustainable Forestry* 13(1/2):205-221.

Coggins, G.C. 1998. Of Californicators, quislings, and crazies. *Chronicle of Community* 2(2):27-33.

Cortner, H.J. et al. 2001. Governance and Institutions: Opportunities and Challenges. *Journal of Sustainable Forestry* 12(3/4):65-96.

Daly, C. 2001. Surviving the Transition: Issues for Maturing Community Forestry Groups. *Journal of Sustainable Forestry* 13(1/2):321-334.

Kusel, J. et al. 1996. The role of the public in adaptive ecosystem management. Pages 611-622 in: *Sierra Nevada Ecosystem Project: Final Report to Congress, Volume II, Assessment and Scientific Basis for Management Options*. Davis: University of California, Centers for Water and Wildland Resources.

Lee, K.N. 1993. *Compass and Gyroscope: Integrating Science and Politics for the Environment*. Washington, DC: Island Press.

McLain, R. 1993. Toward more effective ecological learning: procedural guidelines for the USMAB landscape sustainability project (draft manuscript). Seattle, WA: University of Washington College of Forest Resources.

McLain, R. and R.G. Lee. 1996. Adaptive management: Promises and pitfalls. *Environmental Management* 20(4): 437-448.

Merriam-Webster's Collegiate Dictionary. Tenth Edition. 1993. Springfield, Massachusetts: Merriam-Websters, Incorporated.

White, A.T. et al. 1994. *Collaboration and Community-Based Management of Coral Reefs: Lessons from Experience*. West Hartford, CT: Kumarian Press.

Williams, E.M. and P.V. Ellefson. 1997. Going into partnership to manage a landscape. *Journal of Forestry* 95(3):29-33.

Process Guides

The following list of recommended readings is provided for those looking for more specific guidelines to CBEM process.

Bock, L., K. Fedor, D. Gangloff, and K. Lipkis. 1996. Building Effective Partnerships for City Trees. Washington, DC: American

Conservation Technology Information Center. Watershed Partnership Guides (Building Local Partnerships, Leading and Communicating, Managing Conflict, Getting to Know Your Local Watershed, Putting Together a Watershed Management Plan). West Lafayette, IN: Conservation Technology Information Center.

Kusel, J., S.C. Doak, S. Carpenter, and V.E. Sturtevant. 1996. The role of the public in adaptive ecosystem management. Davis, CA: University of California, Centers for Water and Wildland Resources.

Moote, A. 1997. Partnership Handbook: A Resource and Guidebook for Community-Based Partnership Groups Addressing Environmental, Natural Resource, or Land Use Issues. Tucson, AZ: University of Arizona, Water Resources Research Center. Available on the Web (ag.arizona.edu/partners/).

The Practice of Stewardship: Caring for and Healing Ecosystems and Communities

Bradley L. Ack
Carol Daly
Yvonne Everett
Juan Mendoza
Mary Mitsos
Ron Ochs

SUMMARY. Community-based ecosystem management efforts are proliferating around the United States. While these projects differ greatly in style, substance, players, and places, they share common elements; the philosophy of stewardship appears to be one of the most powerful. The effort to promote stewardship of local ecosystems has shown remarkable power to engage broad and diverse interests. The authors

---

Bradley L. Ack is affiliated with the Grand Canyon Trust, 2601 North Fort Valley Road, Flagstaff, AZ 86001.

Carol Daly is affiliated with the Flathead Economic Policy Center, 919 Elk Park Road, Columbia Falls, MT 59912.

Yvonne Everett is affiliated with the Department of Natural Resources Planning and Interpretation, Humboldt State University, Harpst Street, Arcata, CA 95521.

Juan Mendoza is affiliated with the Willamette Valley Reforestation, Inc., P.O. Box 137, Molalla, OR 97038.

Mary Mitsos is affiliated with the Pinchot Institute for Conservation, 1616 P Street NW, Washington, DC 20036.

Ron Ochs is affiliated with the USDA Forest Service, Cooperative Programs, 333 SW 1st Avenue, Portland, OR 97204.

[Haworth co-indexing entry note]: "The Practice of Stewardship: Caring for and Healing Ecosystems and Communities." Ack, Bradley L. et al. Co-published simultaneously in *Journal of Sustainable Forestry* (Food Products Press, an imprint of The Haworth Press, Inc.) Vol. 12, No. 3/4, 2001, pp. 117-141; and: *Understanding Community-Based Forest Ecosystem Management* (ed: Gerald J. Gray, Maia J. Enzer, and Jonathan Kusel) Food Products Press, an imprint of The Haworth Press, Inc., 2001, pp. 117-141. Single or multiple copies of this article are available for a fee from The Haworth Document Delivery Service [1-800-342-9678, 9:00 a.m. - 5:00 p.m. (EST). E-mail address: getinfo@haworthpressinc.com].

hypothesize that the concept of stewardship is central to many community-based ecosystem management efforts, acting as the 'glue' that holds these efforts together. This chapter explores the emerging philosophy of stewardship, its defining characteristics, and how it differs for community-based ecosystem management efforts. It examines a framework for stewardship at the practitioner's level, with an emphasis on key concerns local practitioners need to address in their projects. The chapter closes with an examination of some of the largest challenges ahead–consistency and accountability, the economics of stewardship, and the need to build a favorable institutional environment–in the continued development of stewardship as an organizing principle for community-based ecosystem management. *[Article copies available for a fee from The Haworth Document Delivery Service: 1-800-342-9678. E-mail address: <getinfo@haworthpressinc.com> Website: <http://www.HaworthPress.com>]*

KEYWORDS. Community-based, forest(ry), ecosystem management, stewardship, land ethic, ecosystem workforce

> A thing is right when it tends to preserve the integrity, stability, and beauty of the biotic community. It is wrong when it tends otherwise. (Leopold 1949)

> . . . we face the urgent need to re-create a civil fabric within which people can at least agree to accept the possible legitimacy of other's views, to stem the hemorrhage of trust in others and faith in the future that any forms of support and investment require . . . (Romm 1993)

> Without personal and collective commitment, without an ethic based on the acceptance of responsibility, efforts to sustain natural resources protection and environmental quality cannot succeed. (Presidents Council on Sustainable Development 1996)

Community-based ecosystem management (CBEM) efforts are springing up across the landscape like mushrooms after a soaking rain. Across the country, people are coming together to get more involved in its complex issues, bringing diverse backgrounds, perspectives, and interests to the discussion.

While these community-based efforts differ greatly in style and substance, and take in an enormous variety of people and places, they seem to share a number of common elements. The concept of *stewardship*–people caring for and healing the land they live in over the long term–appears to be one of the most pervasive and powerful of those elements, demonstrating a remarkable ability to engage broad and diverse interests together in an effort to do right by the land and the community.

The authors hypothesize that this concept of stewardship is at the very core of community-based ecosystem management. The pull of stewardship's principles draws people in, guides their actions, and holds their efforts together.

However, stewardship and CBEM are not synonymous, nor are they necessarily always linked. Community-based ecosystem management is a social and institutional arrangement for resource management; its goals and processes are specifically determined in each case and may vary widely. Stewardship, on the other hand, is a philosophy and approach to the care of land and ecosystems, a philosophy laden with the values of long-term health for ecosystems and the communities that depend on them. Stewardship may be practiced by individuals, communities, companies, organizations, and governments–singly or together.

This paper looks at the idea of stewardship primarily within the context of community-based ecosystem management efforts. Why does stewardship have such a remarkable power to engage broad and diverse interests in a common effort to do right by the land and the community? What are stewardship's defining characteristics and what seems different about it in community-based ecosystem management efforts? The paper also looks at how stewardship is most effectively practiced at the community level–if you are going to do it, what are the key things you have to think about and do well? Finally, we examine some of the daunting challenges that lie in the path of stewardship's continued development and maturation as a guiding philosophy of community-based ecosystem management.

WHAT IS STEWARDSHIP?

Stewardship is a philosophy of care for and long term commitment to the land. It is about the exercise of moral and civic responsibility to protect, restore, conserve, and prudently use the earth's ecosystems and all that they sustain. Stewardship is an activity to be undertaken with humility, respect for other people and life forms, and a willingness to learn and adapt. It requires an awareness that our individual and collective actions on the land can have far-reaching and often unknowable consequences. Good stewards heed the physicians' oath: "First, do no harm."

Stewardship is both an art and a science. While stewardship depends on knowledge and skills from the natural sciences, the crux of its challenge is human behavior and our ability to organize ourselves for concerted action. Stewardship means recognizing that many of our actions have myriad consequences, and that we have a responsibility to understand and manage our behavior accordingly.

While stewardship in community-based ecosystem management begins with sustaining ecosystems and natural communities, it is also about human needs; at the root our need for survival and the quality of that survival. The demands and pressures upon all the earth's ecosystems intensify as human populations expand, triggering greater consumption of resources and creation of waste byproducts. Individuals and communities look to their natural environment to provide various forms of "income"–raw materials, jobs, recreation, nutrient and waste recycling, cultural and aesthetic experiences, and more. Balancing both these needs and the health of the land gets increasingly complex.

Stewardship is really enlightened self interest, the understanding that we must take care of the natural world if it is to take care of us and provide us with the services and goods we need and (within reason) desire.

Ecosystem stewardship can be compared to the management of a financial trust, created as an asset for its beneficiaries in perpetuity. The first duty of the trustee is to protect the principal, the corpus of the trust. Income from the trust may then be used in ways that are in the best interest of the beneficiaries.

In making management decisions, the trustee is required to exercise reasonable care and skill and to avoid undue risk to the corpus. Before taking any action, s/he must consider the effect of that action on the condition and performance of the total investment portfolio, not just particular portions of it. A conscientious, best effort is demanded, but there is a recognition that there are limits to knowledge, and that perfection will not always result. Trusteeship, like stewardship, is both an art and a science (Box 1).

As a trustee, the aspiring steward(s) requires an intimate knowledge and understanding of ecosystems. Stewardship requires that attention be paid to the interconnectedness of all ecosystem components, processes, and flows. Prospective stewards must learn about the component elements of a given ecosystem (its composition), how they relate to each other (its processes), and what the pieces do individually and collectively (its functions). They must also study and try to understand how the ecosystem varies over time and reacts to different kinds of human and natural disturbances.

BOX 1. The Many Faces of Stewardship

Stewardship has many faces. Stewardship is residents cleaning up a stream that flows through the local park, suburbanites developing backyard wildlife areas, and ranchers placing conservation easements on their lands. Stewardship is diverse entities working together to restore a degraded forest, agencies restoring water flows to dammed rivers, families recycling and consuming less. Stewardship is volunteers working to eradicate exotic plants from wildlands, private landowners pooling resources for integrated land management planning, and fishermen adopting lobster conservation methods. Stewardship is when communities act as responsible crew members on spaceship earth.

Stewards place the long-term health of the land before short-term gain. Sustainability of the corpus, in our case the ecosystem, is the overriding objective.

Stewardship has reemerged as a philosophy and practice as people seek to recreate lost connections with the land, take back responsibility for caring for that land, heal the wounds, and ensure the future for coming generations. Aldo Leopold, one of the first Americans to discuss the need to renew these connections, put it in terms of a new land ethic, or "bringing the land back into the community" (Leopold 1949). Stewardship is an attempt by individuals and communities to practice this ethic.

SO WHAT'S NEW? STEWARDSHIP AS APPLIED IN CBEM

While the idea of stewardship may be as old as the hills, it has taken on some new dimensions within the context of CBEM.

Community-based ecosystem management is characterized by the creation of multiparty, multi-interest collaborative processes and groups. Although they may go by a variety of names–watershed councils, forest partnerships, and advisory committees, to name a few–they share many characteristics. At their best, CBEM efforts operate openly and transparently, are diverse and inclusive in their membership, encourage active and adaptive learning, facilitate collaborative decisionmaking, build trust among participants, and ultimately create a shared vision of ecological sustainability that is honored and pursued by the group collectively and by members individually.

The Feather River Coordinated Resource Management Project, for example, is a long standing coalition of private land owners, community members, representatives from federal and state agencies, Indian tribes, nongovernmental organizations, and resource conservation districts. Their cooperation is restoring natural communities and processes to four stream reaches of the Feather River in Plumas County, California.

Many CBEM efforts seem to share some conceptual starting points. One of these is an awareness that communities of place and interest must take responsibility for the well-being of the ecosystem(s) in which they belong. The growing popularity of the philosophy of stewardship within community-based ecosystem management efforts seems to parallel an increased understanding of the connections between healthy communities and healthy ecosystems (U.S. EPA 1997).

Tillamook County, located in the furthermost portion of northwest Oregon, known as the "land of cheese, trees, and ocean breeze," is extremely reliant on the health and productivity of local ecosystems. Timber, dairy, fisheries, and tourism define its culture, history, and economy. When timber outputs, salmon numbers, and water quality levels all began to decline, the community realized its future depended on its collective actions to protect and restore the ecosystem. These issues could not and would not be solved by the federal or state government alone. The result, the Tillamook Performance Partnership, is a voluntary commitment by dozens of local entities and interests to work together to improve water quality, fish populations and habitat, flood mitigation, and local per capita income levels.

The philosophy of stewardship in the context of CBEM has proved able to bring diverse and apparently conflicting interests (forest workers, miners, recreationists, conservationists, farmers and ranchers, scientists, business owners, and people of varying cultures, ages, educational backgrounds, and economic capacities) together in a common focus on the needs and capabilities of their landscapes (Box 2).

On the eastern seaboard, federal and state agencies, nonprofit groups, stakeholders, and industry groups are collaborating to restore the ecosystem health of the Chesapeake Bay, a 64,000-square-mile watershed that spans six states' boundaries. Projects are designed within subwatersheds using an educational approach that assists local communities in understanding how improvements in the health of the bay will directly affect them. Through this process the federal government has also become more aware of the importance of the Bay's health to surrounding communities and has begun to invest in its restoration.

However, stewardship within CBEM projects does not, a priori, represent

BOX 2. Putting the Land First: Stewardship Within Community-Based Ecosystem Management

Stewardship represents important shifts in philosophy toward:

- Recognizing human dependence upon healthy ecosystems and the interconnectedness of ecosystems
- Incorporating multiple interests, multiple parties and multiple values in ecosystem management
- Seeking collaboration before confrontation
- Emphasizing restoration of ecosystems and of impaired processes or flows
- Applying adaptive models of learning and managing and valuing indigenous, traditional and experiential knowledge
- Focusing scientific research on understanding ecosystem dynamics and involving communities in research
- Recognizing the need for skilled labor and knowledge-intensive management to restore and sustain ecosystems
- Focusing on investment and outcomes rather than on revenues and outputs from ecosystems
- Being accountable to future generations

some new consensus about resource management and how it should occur. Instead, it represents a starting point, a central concept that people wrestle with in terms of what to do on the land. Although the participants in CBEM processes may–and probably often do–disagree about how exactly stewardship should be practiced, they begin from the idea that it should be practiced. With that agreement for a starting point, working out the "how" is possible.

Perhaps more interesting is the scale at which this "wrestling" is done–it is primarily local, a level at which people can actually know and have experienced the ecosystem that they are discussing. At its best, this wrestling is also done the way other community business is done, with some attention to preserving broader relationships. Difficult issues are engaged in a way that civility and civil discourse are maintained, leaving the fabric of the community enhanced, rather than torn asunder.

In Flagstaff, Arizona, increased understanding of the relationship between open space, wildlife habitat, growth management, and local quality of life drove a three-year, multi-interest, multi-agency planning process to protect critical open space around the community. The resulting Open Spaces and Greenways Plan was unanimously approved by all the region's major jurisdictions. This positive experience created new social capital and led to other major area-wide efforts, including a broader visioning and action-planning project to build a healthy and sustainable community.

Stewardship in CBEM projects does represent a shift in the locus of some elements of ecosystem planning and management. This shift is from centralized government or corporate centers to the places where the results and consequences of plans and decisions will be played out. While this shift may cause alarm among some people beyond the local area, this is not about local or community control. It is, however, about local communities and individuals contributing more of their knowledge, information, experience, and energy to the management process.

Nevertheless, communities acting alone simply cannot, even if they so desired, mobilize the resources required to do all the necessary stewardship work. The collective challenges transcend the community level, and there is a need for the active involvement of nonlocal interests (regional, national, and global) in assuring proper functioning and management of ecosystems.

On the other hand, good stewardship can certainly be advanced by having those on the ground responsible for working out ecosystem-specific details. This is where CBEM serves its most valuable function, linking, supporting, informing, enhancing, and building upon the stewardship efforts of local individual, corporate, and government landowners and managers.

In the southwest corner of Oregon, the Applegate Partnership's objectives are to maintain ecosystem health, community well-being, and economic diversification while fostering research, monitoring, and community outreach.

Formed in 1992, the partnership has engaged a diverse group of residents, federal agencies, and industry and environmental representatives to develop and implement a long-term vision and plan for the Applegate watershed.

HOW YOU DO IT: BUILDING BLOCKS FOR THE PRACTICE OF STEWARDSHIP

In this section we look at some factors key to achieving effective stewardship within community-based ecosystem management efforts. First we examine the processes through which communities initiate stewardship efforts and develop the community capacity to maintain them over time, then we delve into challenges in the technical substance and on-the-land practice of stewardship in CBEM. Many of the ideas presented draw upon the authors' experience with forest ecosystems and CBEM projects.

Building the Foundation: Trust, Partnership, and Common Ground

While there is extensive literature on group process, and the subject is explored in greater depth elsewhere in this volume (see Moote et al. this volume), several process issues are particularly relevant to the practice of stewardship in CBEM.

Participation–Bringing players to the table is an essential first step. If CBEM is to achieve its full potential, its initiators must make the process open, accessible, and collaborative, and they must seek out and include as diverse and broad a range of interests and opinions as is possible. In doing so, it may be necessary to overcome a variety of obstacles, including (but certainly not limited to) social and cultural differences among participants, past adversarial relationships, differing levels of knowledge and experience, reluctance to share authority, concerns about property rights, and fear of being pressured to compromise values.

It is absolutely essential to find some common ground, some shared concern that is strong enough to overcome people's doubts and reluctance and draw them into the process. The philosophy of stewardship and ecosystem health defines that common ground in many CBEM efforts. It focuses participants' attention on the "why" of their involvement, rather than on the more contentious issues of who, when, how, and how much. In communities where tensions over natural resource management are high, or where the results of a lack of past stewardship are visible and sorely felt, the opportunity to come together to assume responsibility for the future rather than to assign blame for the past can be compelling.

Trust–Once people are at the table, the processes of understanding, learning, and building trust continue. How and whether participants' mutual trust is created, strengthened, tested, and built upon plays out differently in every case, and its evolution needs to be accepted as a slow, incremental process. Trust has to be earned where it cannot be freely given. Being able to fall back on an acknowledged, shared commitment to being good stewards, on why they are together at the same table, can be enormously helpful in seeing diverse participants through difficult times in collaborative CBEM work.

Group Process–Finally, there is the issue of process itself. Once CBEM processes have begun, groups often find it difficult to maintain momentum and keep stakeholders involved and motivated over the long haul. Moving from a mutual belief in the philosophy of stewardship to the more difficult decisions about where, when, how, and how much that philosophy should be played out on the ground can be trying for even the most dedicated participants. Because stewardship is inherently risk-averse, and must be practiced for posterity as well as for more immediate outcomes, its implementation can be frustrating for people who "just want to get the job done." For many practitioners, working through the process of mutual education and collaborative decisionmaking may seem like a slow-motion root canal, but process is the foundation for product. Patience and then more patience is the golden rule.

Community Capacity for Ecosystem Management and Stewardship

Community capacity as used here refers to a community's collective ability to take on the challenges and requirements of stewardship efforts. Practicing good stewardship within a community-based ecosystem management project requires many skills, including community organizing, project development and management, science and research, and communications.

In many cases, those involved in CBEM efforts struggle with these wide-ranging needs. Many communities may have little access to services, training, and funding opportunities, or if opportunities for assistance are available, they may not know how to take advantage of them. CBEM efforts must be able to marshal the diverse resources needed to organize and begin stewardship efforts. Outside support may be needed to help foster necessary skills. Some groups involved in CBEM are reaching out to one another to provide peer-to-peer training and are also seeking to learn from international examples.

Leadership–Leadership is critical to the success of CBEM projects. In many successful cases, a volunteer group, a nongovernmental organization, or one or more dedicated individuals provides the needed leadership. Often the most successful at providing this leadership is a group or an individual seen as being "above the fray" or not having a specific agenda.

Another starting point for some CBEM efforts has occurred when an individual or group from each "side" jointly reach out to each other and together provide the leadership and direction needed to begin a community-based project. When previous adversaries are able to come together to seek common ground, it provides a powerful draw for others.

Regardless of their genesis, leaders in CBEM processes must have a gentle touch. The best leaders acknowledge that no one person or group has all the answers and that a spirit of experimentation and adaptation must prevail. Successful leaders must also be able to synthesize and incorporate changing information and modify their efforts accordingly.

Financial Resources–Money is critical, although the amounts are not necessarily large. While funds are sometimes available from inside the community, help often must be sought from outside investors, foundations or government programs.

When low-income and minority groups organize they sometimes find funders will not support them directly, but insist on channeling any support through other "mainstream" groups. Other groups find funders are more interested in regional or national efforts, believing that local projects have limited impact.

Even when the willingness is there, some foundations have begun to realize they are not well structured to support community-based ecosystem management efforts. One large foundation recently created a special fund specifically to support CBEM efforts. The new fund features an outside panel of advisors to review and approve funding, a much shorter response time, and greatly reduced paperwork requirements. This is the kind of thinking that is needed from the philanthropic community in supporting CBEM.

Some groups are working toward making stewardship activities self-supporting. The Watershed Research and Training Center in Hayfork, CA, for example, is working with its local Forest Service ranger district to develop markets for byproducts from fuels reduction and forest restoration activities. Small diameter logs that only two years ago would have ended up as waste are being processed into flooring and wood furniture. This new added value from "trash" wood now nearly covers the cost of the restoration treatment (Watershed Research and Training Center 1998).

Knowledge–If one of the requisites of stewardship is humility, it is because we recognize even the wisest of us actually knows little about ecosystem components, processes, and flows. The vastness and complexity of natural systems are still only minimally understood, even with all the resources at our disposal. While the desire to be a good steward may be intuitive, the practice of good stewardship will need to be learned over time.

For example, newer residents in a rural area are often fiercely protective of its finer environmental qualities, at the same time as they are unwittingly compromising them. Drawing down water tables for residential and recreational irrigation in desert environments, dividing prime agricultural lands into five-acre "ranchettes," cutting the best trees on a property to make the mortgage payments, carving out an estate in a wildlife travel corridor and then placing an insurmountable fence around it–all may occur as people seek to "get closer to nature" without realizing their activities are harming what they seek to embrace.

Yet newcomers are not the only ones who must learn stewardship. Farmers and ranchers doing things "the way Granddad did" may believe they are perpetuating responsible, tried-and-true management techniques while, in fact, changing circumstances and evolving scientific understanding have led instead to a need for change.

Successful stewardship requires considerable experience with ecology and ecological knowledge. There are many ways to know and understand a landscape–through rigorous scientific examination and simulations, personal practical experience and observation, the shared knowledge and remembrances of others, cultural and historic tradition, classroom and on-the-land instruction by experts, trial and error, and spiritual/aesthetic experiences and relationships. All sources of knowledge have a place in the practice of stewardship in CBEM.

Many community stewardship projects can benefit from ties to universities and government research stations through participatory appraisal and research approaches and new incarnations of traditional university extension services. Under the Northwest Forest Plan, 10 Adaptive Management Areas (AMAs) have been specifically designated for testing standards and guidelines and learning to implement ecosystem management (Box 3). In these AMAs, BLM and USDA Forest Service managers, community members, and researchers are working together on a range of adaptive ecosystem management projects identified by the community (USDI and USDA 1997).

Outreach–In addition to educating themselves about stewardship, participants in CBEM projects must also actively educate the broader public. This includes public policymakers, land owners and managers, users and consumers of environmental goods and services, business owners and workers whose activities impact natural systems, and interested members of the general public (whether they live deep in the Ozark Mountains or high in a Chicago apartment tower).

Outreach programs can be vitally important in helping residents in an area learn about and practice stewardship. Some of the most frequently used methods are drawing interested residents into a local CBEM collaborative, inviting them on field tours of stewardship projects, providing educational materials and workshops (on designing defensible fire spaces around homes in the urban-wildlands interface, for instance, or on preparing stewardship plans for their small forestlands), making available a trained and capable work force to provide land planning and management services, and engaging residents in hands-on ecosystem monitoring and restoration projects (Box 4).

BOX 3. Teaching Stewardship–The Next Generation

Many communities are investing in the future with programs to teach stewardship to their children. *Adopt-A-Watershed* partners with 400 community organizations and some 200 schools to tie K-12 science curricula to local watersheds. In each school, students conduct field studies, and restoration and community action projects using their local watersheds as living laboratories. Children in San Diego, California and Cibeque, Arizona, plant trees, count frogs, and measure soil erosion around their schools, then go back to see what has changed each year. The children hear from people working the land, including farmers, fisheries biologists, and foresters–to give them a range of perspectives on communities and ecosystems. The program involves school children in meaningful work with their communities and helps them develop a sense of environmental stewardship (Adopt-A-Watershed 1997).

The Mattole Salmon Group (MSG) is made up of residents of the Mattole watershed along the northern coast of California. This mountainous watershed, originally cloaked in redwoods, was heavily logged during the 1950s and 60s; by the 1980s, 91 percent of the old-growth redwood was gone. Logging left behind exposed land and roads that washed into fish-bearing creeks and the Mattole River, waterways that once were home to fabled runs of king and chinook salmon. With fish totals numbering more than 30,000 per year in the 1960s, these runs dwindled to just a few hundred fish as sediment choked the river.

In 1981 the MSG began to trap, survey, spawn, raise and release native salmon in the Mattole. In the 14 years since the program began, half a million king salmon have been released and salmon numbers are on the rise. From the beginning, children have been involved in raising and releasing the native salmon yearly. The Mattole restoration effort has been built through hands-on work that provides people, especially children, with common experiential information on their watershed (excerpted from Bernard and Young 1997).

BOX 4. Local Knowledge Can Enhance Scientific Data

Using local knowledge to supplement federal and industrial landowners' scientific data is a major effort of the Swan Ecosystem Center in Montana. The Center not only provides an accessible repository for data on wildlife sightings and movements and other local ecosystem information, it also is collecting and making available to citizens the results of years of study of the local ecosystem by university, environmental organizations, and federal agency researchers. Taking the opportunity a step further, SEC has created a learning center for tourists and local residents alike, complete with appropriate displays, explanatory materials, and a planned interpretative trail through a nearby ponderosa pine restoration site.

Workforce–Ultimately, most community-based ecosystem management efforts will rely on workers to make their particular vision and program a reality on-the-ground. Stewardship is often quite labor intensive and requires a well-trained and motivated work force.

In the best cases, workers are involved in the stewardship process from the beginning, not brought in only when the project moves to field work. Workers, including both those who live in close-by communities and those who migrate in to work, will end up with great responsibility for stewardship. To be successful, they must have the capacity and the ability to earn a fair income.

Unfortunately, at least in the case of forests, workers are often underpaid and undervalued. Industrial labor productivity studies have long indicated that worker health and welfare are key ingredients in workplace performance. Industries that successfully maintain skilled workforces also generally provide health benefits, bonuses, and child care for their workers. However, even though their efforts to restore healthy ecosystems, clean air, and clean water benefit all of society, workers face high rates of poverty, minimum-wage jobs, and unemployment, with all the associated stress upon their families. These conditions are not conducive to stewardship.

Attempts to complete work on public lands by bringing in crews of unskilled labor (e.g., prison labor, federal welfare-to-work participants) may exacerbate the problem. Not only do these crews usually lack necessary stewardship skills and orientations, they may also be in direct competition with local businesses–displaced timber workers who have retrained to become restoration contractors.

Jobs-in-the-Woods is a prominent example of how some communities have developed ecosystem management skills and created a trained local stewardship workforce (Box 5).

On the Land: Stewardship Substance and Practice

The actual on-the-ground practice and implementation of stewardship is, in many ways, a "messy" proposition, made even more so by the diversity of players within community-based ecosystem management efforts. There are no neat and easy answers, no guidebooks to follow. There are, however, some key points that practitioners must keep in mind.

Adaptability and Scale–Practitioners across the nation are experimenting with different approaches to stewardship and developing and testing specific practices that fit local ecosystems, cultures, and capacities. Given the experimental nature of the field, one of the most important principles is the concept of adaptive management. Projects should be designed to be flexible, with feedback loops that channel lessons learned back into ongoing implementation.

In the northwest corner of Montana, for example, the Flathead Forestry Project (FFP) has been implementing small stewardship projects for a number of years and learning a great deal about how to expand its efforts. Undertaking stewardship projects on federal lands can be extremely challenging; as contractors are often unfamiliar with this type of work and the traditional

BOX 5. Jobs-In-The-Woods: Development of an Ecosystem Management Workforce

In the Pacific Northwest, the transition in public lands policy toward ecosystem management has required adjustments on many levels. Forest communities housing skilled workforces for the logging industry are struggling to shift both their economies and their suddenly unemployed residents to the new realities of ecosystem management work in the woods. The federal Jobs-In-The-Woods program was created to retrain forest workers for ecosystem management.

A range of new training programs have emerged in Washington, Oregon, and California. In most cases, coalitions of nongovernmental organizations, community colleges, public land management agencies, and community services providers have formed to create these programs. They have developed curricula that cover a range of skills from applied ecosystem ecology and forest inventory to restoration and management activities, safety and interpersonal skills, and contracting and business management.

A preliminary analysis of the impacts of five Jobs-In-The-Woods programs finds workers who have successfully completed these programs are now using those new skills and are demonstrating independence and flexibility in undertaking the work. Land management agencies value these skills and are developing trust in the workers, which is reducing the cost of on-site supervision and monitoring. Initial evidence also indicated that the project crews compare favorably in cost and quality to the likely alternative workforces (Public Knowledge, Inc. and Labor Education Research Center 1998).

The cost of the program is still high in funding and human terms. Many of the newly trained workers remain unemployed, still ahead of the curve as the expected new ecosystem jobs are slow to emerge. However, the overall evaluation suggests that such costs will fall over time as ecosystem management is implemented. While it is premature to assess the impacts on ecosystems, having an ecologically sophisticated, skilled workforce can only help (Public Knowledge, Inc. and Labor Education Research Center 1998).

contracting mechanisms do not work well. Through small-scale experimentation, the FFP has been working on changing contractor incentives from the volume of logs removed to a focus on the condition of the land at the contract's completion. One of the goals: a new stewardship contract that will provide better results on the land while costing the Forest Service less.

In order to minimize potentially negative, unforeseen impacts of experimentation on ecosystems, small-area stewardship implementation projects may be preferable during initial learning phases. "Scaling up" to expand areas of impact and numbers of people involved in is a challenge in itself.

Adaptability is very much related to appropriate scales for stewardship. There is no fixed area or geographic scale that is correct, as scale is dependent upon the intent of the project as well as local capacity. Landscapes, river basins, or watersheds may be appropriate scales for stewardship activities, but stewardship scales can also extend down to just a few acres. When stewardship processes cover large or densely populated areas, they usually involve complex mixes of jurisdictions and diverse stakeholders. Working with different land owners means stewardship practitioners must deal with multiple governmental, business, and institutional structures that are often cumbersome at best.

The Grand Canyon Forests Partnership in northern Arizona was created to restore forest ecosystem health across 130,000 acres of forested lands. Land owners in the project area include the U.S. Forest Service, the Arizona State Lands Department, Coconino County, the city of Flagstaff, and a host of private owners. All these agencies, along with others with direct management responsibilities such as U.S. Fish and Wildlife Service and Arizona Game and Fish, are members of the Partnership and participate in the design and implementation of its work. Nevertheless, even with this high degree of cooperation and agency willingness, it remains challenging to get the work done across diverse boundaries.

Technology–The question of technology is very important to the on-the-land practice of stewardship, whether you are talking about forest health, stream restoration, or control of noxious weeds in prairie ecosystems. There are a number of specific issues relating to the appropriateness of the technology: (1) Does the technology actually do the job that you want to do? Technologies created for industrial-scale resource extraction often must be adapted to achieve stewardship objectives. Other times, new technologies need to be developed. (2) Is the technology affordable? Does it rely on resources that the community has in abundance or those that are scarce? For example, technologies that cost a fortune and use almost no labor, are inappropriate for poorer areas with high unemployment. (3) Does the scale of the technology match up with the scale of the project? (4) Is the equipment easy to maintain and repair when broken? It is important that the people who will use the chosen machine can repair and maintain it easily; having to send for parts from Switzerland can kill a project.

These are just some of the questions that address the appropriateness of a specific technology. What works will vary with every project's circumstances, but the concept of appropriateness remains important. For example, in forestry, new technologies are being designed to be light on the land, highly mobile, and to work with much smaller diameter trees than were targeted in the past. Rather than being capital intensive, the new technology is often more labor intensive–it puts workers back to work. Rather than investing in one or two large machines, equipment operators now may own several more-diversified pieces of equipment that allow them to take advantage of a wider range of work opportunities locally.

Management–Making stewardship happen on the ground requires management systems that stimulate and reward outstanding work, rather than punish poor performance. In many cases, existing institutional management regimes hinder rather than promote stewardship.

There is often an incongruity between project timeline planning and the actual process of working on the land, particularly when stewardship is undertaken in cooperation with large institutions. To do good stewardship, workers need flexibility and must be able to adjust work orders to account for changes in weather, actual field conditions, and other vagaries.

The ecosystem and the changing seasons must drive activities when planning stewardship on-the-ground. Nature does not conform to administrative and fiscal cycles or to 8-hour work days. Where workers are forced to follow schedules inconsistent with on-the-ground reality, the ability to do time for stewardship may suffer. Often there is not enough time to even complete projects at all.

The need for adaptive management described above means that management systems have to be iterative, with much dialogue between workers and managers and a significant monitoring component. Traditional contracts may not be very effective in this type of environment and other arrangements will need to be developed.

Efforts in Sweet Home, Oregon, have demonstrated the benefits of integrating social, economic, and ecological concerns in project management. To better use local workers' skills and abilities, the Sweet Home Partnership has packaged projects from federal, state, and local partners in part to promote quality jobs for local citizens. Sweet Home has worked to develop contracts that provide more predictable, stable, and higher quality employment. By packaging diverse projects together, Sweet Home has seen increased quality of work, local community stability, and increased utilization and reliance on local workforce capacity and capabilities.

Accountability–While flexibility and new arrangements are important, somebody still must be accountable. In many areas, nongovernmental organizations (NGOs) are taking on an important role, stepping in to assume risk, help raise and/or track needed funds, provide training for workers, and perform many other necessary tasks of stewardship implementation. These organizations can also help add a measure of additional accountability for both internal and external audiences.

The Redwood Community Action Agency (RCAA) from Eureka, California, has been undertaking ecosystem restoration work for more than 20 years. While functioning as the accountable entity, it has demonstrated an ability to collaborate with numerous partners to achieve difficult goals. For example, in restoring the Hammond Trail, originally a coastal railroad right-of-way, RCAA has worked with state, county, and private land owners to provide public walking and bicycle access along miles of the beautiful north coast for residents and visitors to the Eureka-Arcata area. RCAA is also pioneering effective and economical road maintenance techniques designed to reduce siltation in fish-bearing creeks.

People involved in CBEM efforts need to think through how to create checks and balances that allow for adaptability yet provide strong accountability. Accountability measures must be created with the input of all stakeholders, including (when possible) the process' critics. It is essential that the monitoring and evaluation of CBEM projects be open to all interested parties to ensure adaptive learning, give the larger community a sense of ownership in stewardship efforts, and provide deserved recognition to exemplary stewardship (for more on monitoring and accountability, see Bliss et al. this volume).

WHAT ARE THE BIG CHALLENGES AHEAD? INTRODUCING THE "700 POUND GORILLAS"

While stewardship-oriented CBEM efforts have emerged in many different places, shapes, and forms, overall they are still relatively small in their impact on the American landscape. Interest in replicating and expanding these community-based ecosystem management efforts has begun to build as the philosophy and practice of stewardship draw more adherents and enthusiasts.

But a number of challenges, common to these diverse efforts, must be addressed if stewardship-oriented CBEM is to grow. The following three key challenges demand serious attention in the near term. This is not an exhaustive list but one of priorities. While the previous section examined challenges at the project level, this section explores the challenges that transcend the community level to require attention on a regional and national scale.

Building Confidence: Promoting Consistency, Common Standards and Accountability in Stewardship

Community-based ecosystem management projects are springing up in diverse ecosystems, and they are testing and utilizing a broad range of approaches. While this is one of CBEM's most interesting characteristics, this strength could become a major weakness.

One of the biggest concerns is that the stewardship ethic will not always be at the front and center of CBEM efforts. Some are concerned the concept of stewardship will become a Trojan horse to cover for continued resource exploitation. Others are concerned that even with good intentions, the capacity to do stewardship simply doesn't exist at the local level in most places.

Representatives of the national environmental community have been up-front about their concerns–how can they be confident that what happens in these myriad CBEM efforts is really about stewardship of the ecosystem? While there are local environmental organizations and environmentalists active in CBEM efforts, communication and understanding between them and the national organizations is not always as good as it should be. Without being everywhere, the national organizations wonder how to assure that the environmental interests of their large constituencies are honored. This concern cannot be ignored if the field is to develop and mature.

Part of their concern stems from the lack of consistency and recognizable, agreed-upon "standards" of stewardship across diverse CBEM efforts. If stewardship is to mature into a broadly accepted management approach, this issue of consistency, and the related question of standardization of terms, goals and outcomes, must be rigorously addressed. Without clarification and agreement, and some way to ensure consistency and accountability, the practice of stewardship is unlikely to gain the wider legitimacy that it needs to prosper.

The authors recognize that any attempt to standardize or formalize stewardship is an idea fraught with peril. Given the diversity of ecosystems, people, and levels of community capacity in CBEM efforts, the variety of management approaches being tested, and the inherent need for flexibility and adaptive management, standardization could stifle stewardship efforts. Nevertheless, the issue won't simply go away.

The authors believe that a broad community of interest must be engaged to think through how best to deal with these issues of consistency and accountability. While we do not agree ourselves whether standards and guidelines are the best approach to consistency, we do agree that the issue must be fully and openly discussed. This is an area in which the national environmental groups need to participate fully, rather than resist. Given the significant potential for good conservation within CBEM and stewardship efforts, there is ample reason for these organizations to engage.

One starting point may be to examine the processes undertaken to develop guidelines for sustainable forestry certification (Viana et al. 1996) (see Box 6).

Any attempt to ensure consistency and accountability must deal with the actual performance of stewardship activities. While it is not easy to generalize across ecosystems, it may be possible to define broad performance standards in terms of outcomes. Existing federal and state law provides the foundation upon which performance standards then build.

For example, we might be able to define and agree upon a set of principles such as the following: Stewardship activities must: maintain or improve conditions for viable populations of native plant and animal species; maintain or improve water quality; maintain or improve a viable network of native biological communities; recover and conserve native species; maintain struc-

BOX 6. Green Certification for Forest Products

The Forest Stewardship Council (FSC) is currently developing guidelines for sustainable management of private forestlands. FSC is an international umbrella organization including environmental groups, community and industry representatives, and certification organizations. Landowners who agree to adhere to strict standards of forest management, including both ecological and social criteria, may have their forest products certified by FSC as "sustainably harvested." FSC tracks the products through harvest, processing, and the marketplace. There, consumers seeking to buy responsibly can purchase FSC-certified products with the certainty that they came from forests managed to the best standards available.

tural diversity, genetic diversity, function, resilience, and evolutionary processes of ecosystems; and protect and conserve resources needed by people.

All-party monitoring is another approach that can be explored to assure accountability. The Lead Partnership Group in northern California and southern Oregon is exploring all-party monitoring to involve skeptics and to enhance the legitimacy of its community-based efforts with others at the regional and national levels (Kusel, Gray, and Enzer 1996). All-party monitoring can bring outside interests together with members of CBEM efforts to define and measure the objectives and performance of stewardship activities.

Balancing the Books: Navigating the Economics of Stewardship

Another major challenge facing stewardship practice within community-based ecosystem management efforts is the myopic vision of the marketplace. If stewardship activities on the land do not make a profit, or at the very least pay for themselves, then they are not economic and shouldn't occur–or so goes the logic of the free market. If we allow this logic to be our measure, stewardship practices will be at a sore disadvantage, and the future of our ecosystems will continue to be at risk.

From a purely financial standpoint, there are many aspects of stewardship that simply do not provide cash flow, at least not in the near term. Instead they are investments back into the land, required to correct excesses of the past and repair damaged ecosystems. They are the rebuilding of the corpus of the trust, to extend the earlier analogy of trusteeship. Many of the current costs of stewardship are a direct result of past exploitation and overextraction from these ecosystems.

Over time, ecosystems that are managed according to stewardship principles will become more and more productive and certainly will generate a variety of goods and services with significant values. How long this process will take depends on the ecosystem and its state of health at the start of the stewardship efforts.

Yet even though stewardship activities will and do produce tangible and measurable benefits, many of these goods and services do not have a market value–that is, they are not currently bought and sold. Therefore, we have no easy mechanism to capture and quantify their real value, at least not at this stage in our economic evolution. How we begin to measure and account for these so-called nonmarket values is an extremely important issue for stewardship and its future.

Many communities are beginning to realize the nonmarket value of ecosystem services and the significant monetary cost associated with *not* maintaining the ecosystem. In the early 1990s, New York City residents were faced with a choice. Either build a water filtration system for which construction costs alone would run $5 to $8 billion, or begin working with upstate residents in the Catskills to restore and maintain the watershed that supplies the city's clean drinking water. Working with rural residents to care for the upland watershed proved much less expensive. It also helped those rural communities survive economically and helped stem the conversion of agricultural lands to suburban development.

So while stewardship is clearly about better practices and caring for the land, its success will be in part dependent on better accounting and valuation systems and an understanding of the benefits, goods, and services a healthy ecosystem provides. This is not a new concept, but the discussions so far have stayed largely at the theoretical level (Daly and Cobb 1989).

CBEM efforts provide an excellent forum to begin bringing these discussions down to the community level and raising people's awareness of the blind spots in our current accounting system. These efforts also put information in real terms: What is the actual value of wildlife, clean water, and recreational access? What are the real costs in increased fire risk, loss of habitat, and simplification of once-complex ecosystems?

Meanwhile, many CBEM efforts will continue to face difficulty paying for the good works that need to be done. These financial challenges can lead to unproductive and even dangerous tensions between environmental imperatives and economic issues. For stewardship to meet the standards and aspirations that we have described, environmental goals cannot be dependent on their short-term financial viability.

This begs the question, if stewardship cannot fully "pay its own way" in the short-term, then who should pay for it? How much national will is there to pay for the stewardship of our collective resources? Are people willing to pay for the care and healing of our environment, and how do we turn that willingness into actual money? How deep is the public's awareness of the need for stewardship, its benefits, and its costs? What kinds of creative finance mechanisms might be developed, such as the use of tax credits or abatements and other economic incentives? These questions need urgent attention from economists, social scientists, finance specialists, policymakers and others (see Luzadis et al. this volume).

If we cannot solve the economic challenge, stewardship will never develop its potential to be widely practiced.

Building a Favorable Institutional Environment: State, Regional, and National Support for Stewardship

Good stewardship of ecosystems requires a web of partnerships across and throughout agencies and organizations and up and down the vertical layers of government and civil society. Even where community-based efforts develop a high degree of capacity for stewardship activities, they will still depend on institutions beyond their control to determine the wider environment within which stewardship activities will be practiced. These may be state or federal government agencies, larger industrial landowners or national interest groups.

In recognition of this, many community-based efforts have worked hard to build relationships with government agencies and nongovernmental organizations at the local, regional, national, and even international levels (Yaffee et al. 1996). In the Henry's Fork watershed in Idaho, a community-based watershed council has found this task easier than most because of the watershed's blue ribbon trout streams. National and even international groups have been kept informed and provide input into the community-based restoration process. However, if the various agencies are to respond effectively to these CBEM efforts, changes in how they do business are required. This is especially true within federal agencies.

The following are some of the key institutional challenges and barriers we see to better support of community-based ecosystem management efforts.

Balancing Accountability and Flexibility–Governmental efforts to support stewardship and CBEM efforts are hampered by the need to both promote standardization and accountability within the agency and allow the flexibility required by community-based ecosystem management efforts. Flexibility is needed to address differing local ecological, social and economic conditions. Agency approaches to stewardship will need to place more emphasis on adaptive management than on control.

There are a number of ways that agencies might improve their flexibility without a loss of accountability. In each case, the people in the various agencies need to think this through with their nonagency partners. Some suggestions include:

1. Develop quantifiable and/or qualitative, end-results objectives for stewardship projects and manage and evaluate performance based on end results rather than on means.
2. Create intra- and interagency teams to help community-based efforts navigate through the institutional maze.
3. Foster interagency coordination of information and consolidate relevant information into usable and standard format.
4. Create mechanisms for interagency funding, oversight, and adaptive management.

Reinvestment into Ecosystems–The incentives and funding structures that govern many agencies often run counter to good stewardship. For example, within the Forest Service, efforts to encourage forest ecosystem restoration have been financially disadvantaged because any revenues generated from utilization of the by-products of ecosystem restoration have not been allowed to go toward other restoration costs. This means the agency has to pay both for the removal of the materials, and all other restoration costs, such as road obliteration, prescribed fire, and noxious weed control. In this case, restoration becomes almost impossible without large infusions of appropriated dollars.

This same problem applies to other stewardship challenges. For example, on the Gifford Pinchot National Forest in the state of Washington, considerable effort has been taken to improve the stewardship of special forest products, a resource under rapidly increasing pressure. For each additional dollar invested in management, at least $2 in revenue is being generated. Yet these new revenues are not available to reinvest in the ongoing management costs, leaving the ability to manage curtailed.

This dilemma–revenues that are often not available for reinvestment in the local landscape–is all too common. Instead, stewardship costs must compete for limited appropriated funds–a difficult proposition in the face of existing commitments–or they must be borne by the private sector.

Ability to Enter into Stewardship Arrangements–Systems and processes developed by the federal government to accomplish land management objectives on federal lands were designed independently of each other. For example, resource extraction activities use certain tools and processes (e.g., timber sale contracts), while wildlife management services use a completely different set of tools and processes. Stewardship requires a more integrated approach, therefore new mechanisms must be created.

In general, agencies need to develop new tools that respond to new needs on the land. For example, the bundled contract may allow for increased flexibility while increasing effectiveness and efficiency. Currently, agencies are restricted from consolidating ecosystem stewardship activities into one contract or agreement if the activities are a mix of revenue generation and provision of services. In ecosystem restoration projects, many tasks are necessary in the same place, and combining them into one contract may greatly increase effectiveness and efficiency. Such bundled contracts can also have a positive impact on workers by giving them a wider variety of work across a longer time period (for a discussion of tools available to the U.S. Forest Service, see Ringgold 1998).

Partnerships: Utilizing the Skills and Abilities of Non-Agency Collaborators–The size of government continues to decline while the demands of ecosystem stewardship continue to escalate. It is ever more of an imperative that the skills and abilities of nongovernmental partners be recognized and used to achieve mutually shared goals. The reliance in the past exclusively on federal employees to achieve desired results must evolve toward more shared responsibility.

Increased emphasis on partnership often requires the modification of existing policies. The Federal Advisory Committees Act, for example, has been used in a number of cases to stifle real partnership. Other laws limit the ability of federal agencies to use nonfederal employees to get work done on the ground. Further, where partnerships are emerging, it is important that the burden of financial risk not fall disproportionately on NGO shoulders.

The state of Oregon has realized that ecosystems cannot be restored and maintained by mandates or solely through government efforts but rather requires collaborative partnerships across political boundaries. This realization manifested in the passage of state legislation in 1995 that "encourages local government groups to form voluntary local watershed councils" to organize and respond to the needs of ecosystems. The state, along with other partners, is providing funding to help these community-based efforts obtain the leadership needed and infrastructure required to accomplish their work. Since the legislation passed, more than 80 councils have formed and are operational in Oregon, completing watershed-wide plans, implementing some of the highest priority restoration work, and monitoring performance.

Working with Private Landowners–Community-based ecosystem management efforts often span many land ownership patterns. While many private landowners may be committed to stewardship, they are not always financially able to implement what is needed on their lands. Since ecosystems are interdependent and contiguous, their inability to accomplish this work can negitively affect larger CBEM efforts accomplishing their objectives.

Agencies can play a useful role here. One example is a new authority granted to the Bureau of Land Management and the Forest Service. Each agency was authorized to utilize its resources anywhere within a watershed to accomplish its or the management objectives (commonly known as the Wyden Amendment). This new authority has enabled these agencies to work with landowners to achieve shared goals. This kind of integrated thinking will be essential if agencies are to become more effective partners in CBEM.

CONCLUSIONS

The field of community-based ecosystem management is at once fascinating, terrifying, hopeful, and exceedingly complex. It is still in its infancy, slowly emerging as a response to the transgressions of the past and dictates of the future. These early years will be defining, and the choices we make and actions we take over the near term will determine if community-based ecosystem management is just a passing fancy or the beginning of a fundamental shift in how we as a society care for and heal our damaged ecosystems. The powerful philosophy of stewardship is at the center of this grand experiment.

Stewardship is both an art and a science. While it has the natural sciences at its core, the crux of the issue and the challenge is about human behavior and our ability to organize for concerted action. As with all ambitious human endeavors, stewardship and community-based ecosystem management are often messy, confusing, and characterized by fits and starts.

Despite the myriad challenges described in this paper, there is still every reason to push forward with local ecosystem stewardship projects. A sea-change of thoughts and actions always takes the time and energy of numerous people. Much can be accomplished with patience, tenacity, an understanding of current situations, and a large dose of humility for what we do not know.

While it is quite easy for folks to sit back on the sidelines and criticize stewardship-oriented CBEM efforts, in the end it is far more useful and satisfying to roll up one's sleeves and get into the game. The great thing about stewardship is that anyone can get involved. We need to engage ever-larger numbers of people and we especially need problem solvers.

For those who are already engaged in the practice of stewardship in community-based projects, we must recognize the experimental nature of the field, the keen interest from many levels of society, and the need to be creative, open, and willing to listen to people's concerns. If we become defensive and dogmatic at this early stage in our collective development, survival to maturity is unlikely.

In the final analysis, stewardship is an ethic whose time may have arrived. It transcends ecosystems, agencies, interest groups, and community-based efforts. As a re-emerging environmental and community ethic, it should be given much care and feeding.

The opportunities that stewardship offers to care for and heal our damaged ecosystems and to reengage our society in questions of the land justify this nurturing. The ability to pass on a healthy and vibrant natural environment to future generations may be our reward.

REFERENCES

Adopt-A-Watershed. 1997. *Prospectus Fiscal Year 1997-1998*. San Fransisco, CA.

Bernard, Ted and Jora Young. 1997. *The Ecology of Hope: Communities Collaborate for Sustainability*. Gabriola Island, British Columbia: New Society Publishers.

Daly, H.E. and J.B. Cobb Jr. 1989. *For the Common Good*. Boston: Beacon Press

Kusel, J., Gray, G.J. and M.J. Enzer (eds.) 1996. *Proceedings of the Lead Partnership Group Northern California/Southern Oregon Roundtable on Communities of Place, Partnerships and Forest Health*. Washington, DC: American Forests/Forest Community Research.

Leopold, A. 1949. *A Sand County Almanac–with Essays on Conservation from Round River*. New York: Ballantine Books.

President's Council on Sustainable Development. 1996. *Sustainable America: A New Consensus for Prosperity, Opportunity and a Healthy Environment*. Washington DC: President's Council on Sustainable Development.

Public Knowledge, Inc. and Labor Education and Research Center. 1998. *The High Skill Approach to Ecosystem Management: Combining Economic, Ecological and Social Objectives*. Eugene, Oregon: University of Oregon.

Ringgold, P. 1998. *Land Stewardship Contracting in the National Forests: A Community Guide to Existing Authorities*. Washington, DC: Pinchot Institute For Conservation.

Romm, J. 1993. Sustainable Forestry, An Adaptive Social Process. *In* Aplet, G.H., Johnson, N. Olson, J.T. and V.A. Sample (eds) *Defining Sustainable Forestry*. Washington, DC: Island Press.

U.S. Environmental Protection Agency. 1997. *Community-Based Environmental Protection: A Resource Book for Protecting Ecosystems and Communities,* U.S. EPA (EPA 230-B-96-003), Washington, DC.

U.S. Department of the Interior and U.S. Department of Agriculture. 1997. Adaptive Management Area 1997 Success Stories: Partnerships . . . Working, Learning and Managing Together. http://www.teleport.com/~amanet

Viana, V.M., J. Ervin, R.Z. Donavan, C. Elliott and H. Gholz. 1996. *Certification of Forest Products: Issues and Perspectives*. Washington, DC: Island Press

Watershed Research and Training Center. 1998. Chopsticks Small Diameter Demonstration Project. Report to USDA Forest Service Grant #r5-14-97-30. Hayfork, California.

Yaffee, S.L. et al. 1996. *Ecosystem Management in the United States*. Washington DC: Island Press.

Community-Based Ecosystem Monitoring

John Bliss
Greg Aplet
Cate Hartzell
Peggy Harwood
Paul Jahnige
David Kittredge
Stephan Lewandowski
Mary Lou Soscia

SUMMARY. Community-based ecosystem monitoring refers to a range of observation and measurement activities involving participation by community members and designed to learn about ecological and

---

John Bliss is affiliated with the Oregon State University, Department of Forest Resources, 203 Peavy Hall, Corvallis, OR 97331-5703.

Greg Aplet is affiliated with The Wilderness Society, 7475 Dakin Street, Suite 410, Denver, CO 80221.

Cate Hartzell is affiliated with the Collaborative Learning Circle, P.O. Box 1139, Ashland, OR 97520.

Peggy Harwood is affiliated with the Bureau of Land Management, 1849 C Street NW, Washington, DC 20240.

Paul Jahnige is affiliated with the Community Resources, P.O. Box 226, Williamsburg, MA 01096.

David Kittredge is affiliated with the Department of Forestry & Wildlife Management, University of Massachusetts, Holdsworth Hall, Amherst, MA 01003.

Stephan Lewandowski is affiliated with the Canandaigua Lake Watershed Task Force, 8 Weld Street, Naples, NY 14512.

Mary Lou Soscia is affiliated with the US EPA, Office of Ecosystems and Communities, 811 South West Sixth Avenue, Portland, OR 97204.

[Haworth co-indexing entry note]: "Community-Based Ecosystem Monitoring." Bliss, John et al. Co-published simultaneously in *Journal of Sustainable Forestry* (Food Products Press, an imprint of The Haworth Press, Inc.) Vol. 12, No. 3/4, 2001, pp. 143-167; and: *Understanding Community-Based Forest Ecosystem Management* (ed: Gerald J. Gray, Maia J. Enzer, and Jonathan Kusel) Food Products Press, an imprint of The Haworth Press, Inc., 2001, pp. 143-167. Single or multiple copies of this article are available for a fee from The Haworth Document Delivery Service [1-800-342-9678, 9:00 a.m. - 5:00 p.m. (EST). E-mail address: getinfo@haworthpressinc.com].

social factors affecting a community. This chapter presents observations from community-based ecosystem monitoring activities throughout the United States. It discusses factors leading to the emergence of community-based ecosystem monitoring, multiparty monitoring and its role in building social capital, the monitoring process, the integration of social and ecological factors, and ongoing challenges in community-based monitoring. Sidebars illustrate the chapter's central themes and lessons with examples from particular community monitoring initiatives. Community-based monitoring has been motivated by concern for special places, in response to perceived environmental threats, and as part of an effort to overcome longstanding conflict between diverse stakeholder groups. Multiparty monitoring has the potential to address each of these needs. Significant challenges include achieving effective, diverse community participation, integrating social indicators into ecosystem monitoring and analysis, identifying an appropriate level of rigor for specific monitoring objectives, and effectively integrating monitoring into an adaptive decisionmaking process. *[Article copies available for a fee from The Haworth Document Delivery Service: 1-800-342-9678. E-mail address: <getinfo@haworthpressinc.com> Website: <http://www.HaworthPress.com>]*

KEYWORDS. Community-based, forest(ry), ecosystem management, monitoring, adaptive, accountability

> If the biota, in the course of aeons, has built something we like but do not understand, then who but a fool would discard seemingly useless parts? To keep every cog and wheel is the first precaution of intelligent tinkering.
>
> –Aldo Leopold, *Round River Essay, 1949*

Intelligent tinkering–be it puttering in one's garden, managing a family forest, or restoring an ecosystem–requires knowing what the component parts are and understanding how they fit together. Our principle source of knowledge and understanding is observation. Identifying and observing the social and ecological components of a community in order to understand how they fit together and function as a system is the central objective of community-based ecosystem monitoring.

WHAT IS COMMUNITY-BASED ECOSYSTEM MONITORING?

"To monitor," from the Latin *monere* ("to warn"), is defined in *Webster's New Collegiate Dictionary* as "to watch, observe, or check, especially for a special purpose; to keep track of, regulate, or control the operation of." Community-based ecosystem monitoring is a relatively recent, emerging innovation. We use the term to refer to a range of activities that are motivated by a perceived need for information on ecological and social factors affecting a community and that involve participation by community members. The actual motivations behind individual monitoring efforts, the complexity of the projects, and the breadth of community participation vary tremendously. As discussed in this chapter, monitoring is an integral component of adaptive management–a process within a process, induced by the need for reliable, relevant data to inform decisionmaking. Monitoring is a first step toward taking responsibility for the condition of one's home, community, or ecosystem. This theme of accountability–of individuals, communities, and governments–pervades discussions of monitoring and ecosystem management.

In this chapter we share observations from monitoring activities underway throughout the United States. It is not intended as a comprehensive treatment of monitoring, nor a monitoring handbook. Rather, it presents a snapshot of where community-based monitoring for ecosystem management is today. It grew primarily out of the personal experiences of individuals practicing some aspect of community-based ecosystem management, balanced by the perspectives of public agency personnel and academics. The chapter reflects a range of experiences and contexts–eastern and western U.S., urban and rural, public land and private land, high-tech and low-tech–but it maintains a community focus throughout (Box 1).

We emphasize multiparty monitoring because it reflects the principle of inclusiveness embraced by many community-based efforts. The discussion reflects our concerns with social as well as ecological conditions, equity as well as efficiency in monitoring activities, and an overall commitment to pursuing sustainable strategies for community development. In our use of the term, "ecosystem" includes human, social, and cultural elements as well as nonhuman biological and physical ones.

In the following sections we discuss some of the factors behind the emergence of community-based ecosystem monitoring. Next, we address the central question of who participates, and provide a rationale for multiparty monitoring. We then review the steps in the monitoring process; identifying goals and participants, selecting social and environmental indicators, and setting up a monitoring protocol. We also consider monitoring types, scale, and funding. We close by reviewing several lessons and challenges borne of these experiences. Throughout the chapter are sidebars that illustrate points made in the text through the experiences of various community-based monitoring efforts.

BOX 1. One Practitioner's Philosophy of Monitoring

Theories and questions about perception are as old as philosophy itself. Can we trust our senses? Is there more or less than we perceive? For some purposes, monitoring need be nothing more than close perception of our surroundings or ourselves. We have many tools to help us get a closer or clearer view of what we perceive–eyeglasses, microscopes, telescopes, cameras and spectrometers–but the process of attention and observation is the same throughout. We have developed even more powerful tools to help us observe and store, analyze, and interpret our observations.

We make a mistake, however, if we allow our tools to dictate the process of observation or mistake the new perspectives they add for something other than the fruits of observation. Observation alone can be very powerful. For example, refraining from using most available technology in favor of long-term, close, and caring observation of "the other," pioneering geneticist Barbara McClintock used her powers of observation to unravel the genetic mystery of "jumping genes" in corn plants (Keller 1983). Initially dismissed by the scientific community because of her unorthodox methods, McClintock was eventually awarded a Nobel Prize for her discovery.

McClintock's example teaches us to check our means and our perceptions, then set goals and select tools appropriate to those goals. Second, if we are sufficiently humble, perceptive, and lucky, we may hope that "the other"–watershed, forest, mountain, or creature–will come forward to teach us. McClintock suggests that we cannot compel the other to approach us, but we can cultivate our readiness and receptivity. Further, she affirms that the relationship between observer and observed will translate into feelings of concern and attachment. Her testimony is a powerful answer to the question "why monitor?" We pay attention in order to learn and to form these attachments, which then define a place as our home.

Monitoring allows us to take a fresh look at our home, the earth. Tools change our perspective, freshen our view, and invite our reconsideration of what is known, but it is critical to stay in touch with the basic elements of observation even as we employ an array of tools. Without the important goal of understanding and protecting our home our technology may alienate us further from that teaching presence. But through careful observation we learn, and by learning we form the attachments that engender stewardship. By bringing heads, hands, and hearts to the task, we may arrive at the combination of knowledge, wisdom, and humility required to protect and restore our home.

WHY MONITOR?

For some communities, concern for special places motivates ecosystem monitoring. Elsewhere, community-based monitoring activities have emerged where individuals perceived an environmental or community problem or threat, organized to address the problem, and began collecting data about it. In other situations, existing groups have initiated monitoring to provide baseline data on neighborhood conditions, enhance citizen awareness of local conditions, or contribute to local planning. In some communities, monitoring is part of a broader community development or ecosystem management strategy.

These efforts often arise where trust in public or private land management decisions has broken down. In the Pacific Northwest, for example, many communities are still reeling from the profound changes that have occurred in the region's timber-based economy. After years of acrimonious debate and conflict, some communities are turning to collaborative monitoring to build a shared understanding of how the physical, social, political, and economic elements of their environment are connected. Through working together, citizens hope to reestablish mutual trust, generate an ethic of community stewardship, and develop shared goals and the framework of accountability required to meet those goals.

Despite all these good reasons, monitoring remains one of the least appreciated and most neglected aspects of land management. In general monitoring is undertaken only if there is both the will and the means for it to be accomplished. The means, of course, is a function of funding, which is not always available. Many land management agencies insist they would monitor if only the money were made available in their budgets.

But a lack of means is not the only barrier; it also requires the will. All the money in the world will not make monitoring happen if there is no incentive–or worse, if there are disincentives. The disincentives for monitoring are many and help to explain the lack of financial support. First, monitoring creates new demands on personnel and resources that are perennially stretched thin. Second, it raises concerns about privacy and autonomy, particularly on private lands. Third, monitoring results in the creation of new information which, to the extent that information is power, can lead to shifts in the locus of power. It is a cardinal law of political science that those in power will resist change in the locus of power.

Perhaps the most powerful disincentive for monitoring can be traced to the word's root: to warn. Monitoring warns if something bad is happening. It provides the mechanism to judge the soundness of management actions. In short, it holds landowners, managers, and policymakers accountable for their decisions. This asymmetry ("Monitoring only tells me if I was wrong") can be a powerful disincentive if information has only negative consequences. Clearly, the good that can come from monitoring will only be realized when new approaches overcome traditional barriers and instill the will and the means for its accomplishment.

MULTIPARTY MONITORING

There are numerous approaches to planning and implementing monitoring systems. They can involve a single regulatory agency or complex webs of agencies, organizations, community groups, and citizens. One approach that has received significant attention from community groups is multiparty monitoring, in which an attempt is made to identify and involve both local and distant stakeholders in the system under scrutiny. Stakeholders may include private landowners, individual citizens, representatives of nongovernmental organizations (NGOs), businesses, public agencies, church and school groups, labor organizations, and others who have a stake in or commitment to the community.

Multiparty monitoring is, in part, a response to a perceived lack of existing environmental and social monitoring arrangements. It reflects a national trend toward broader participation in environmental policy and management, especially on public lands. It is also attractive for its potential to forge new relationships between former adversaries and engage distant stakeholders and local citizens to address shared concerns (Box 2).

Building Social Capital Through Multiparty Monitoring

Building constructive social relationships can be a significant benefit of community-based monitoring. By engaging citizens to work together on shared objectives, monitoring can help build "social capital," cooperative, interdependent relationships that comprise the foundation of community. Social capital that comes through an increasing community awareness of problems and promises strengthens existing institutions and builds new social networks. Other examples of building social capital include identifying and

> **BOX 2. All Party Monitoring in the Pacific Northwest**
>
> In the Pacific Northwest, efforts are underway to diffuse tensions between local community residents and national environmental groups who also consider themselves stakeholders in public land resources. All party monitoring seeks participation by all potential stakeholders in order to increase the likelihood that diverse issues are addressed, and to reduce resistance to management decisions.
>
> Representatives of six place-based multi-stakeholder groups in southern Oregon and northern California have met for five years as the Lead Partnership Group (LPG) to explore common goals and challenges. Since 1995, LPG has hosted tours and discussions between local, regional and national forest-related interest groups to consider collaborative projects, areas of mutual concern, and how distant stakeholders can participate in monitoring land management activities. In response to concerns raised during these discussions, LPG secured support for three separate pilot projects that explore the concept of all party monitoring.
>
> The projects, in California's Plumas and Trinity National Forests and the Rogue River National Forest in Oregon, will develop models for involving regional and national interest groups, agency representatives and community members in the development of monitoring plans that integrate multiple interests and concerns. In addition to exploring roles varied stakeholders can play, the pilot projects will focus on how information and communication can be structured to respond to both local and distant stakeholders' interests.

engaging volunteers, creating new agency connections, building leadership skills and responsibilities, brokering solutions to persistent conflicts, and identifying overlooked or unrecognized resource values.

Monitoring can be the catalyst for strengthening emerging social institutions that foster ecosystem management across political or land management boundaries. Many watershed councils, natural resource councils, water watch groups, stream teams, and neighborhood associations have formalized their partnerships after conducting baseline monitoring. Some groups develop informal compacts to find and express agreement on ecosystem issues and concerns (Canandaigua Lake Watershed Task Force 1995, Kohn [n.d.]). The process of developing a compact clarifies stakeholder concerns and engenders compromise and cooperation. Once formed, such groups have the potential to embrace diverse interests across political boundaries to provide education, technical expertise, and data management (Ticknor 1992) (Box 3).

Who Participates?

A direct link exists between those who participate in monitoring and that which is monitored. Different constituent groups within a community see the landscape through different eyes, have access or are denied access to different resources, and derive value from different places and uses. Broadening participation in monitoring, then, is important both as a means to build social capital and as a means of improving the inventory and evaluation of the ecological, economic, and social elements of significance to the community and the linkages between them. Moreover, by making the process of monitoring more visible and participatory, participants increase the likelihood that what they learn will inform future resource management decisions.

Many community residents are uniquely qualified to participate in monitoring. Bird-watchers in many communities regularly conduct systematic bird counts; tree planters, loggers, and nontimber forest products workers are reservoirs of knowledge and observational skills. Long-time community residents can contribute greatly to monitoring efforts if properly engaged and supported. Increasing the capacity for systematic local observation over time and structuring mechanisms for integrating those observations into monitoring programs hold great promise.

For example, it is estimated that tens of thousands of European American, Native American, Latino, and Southeast Asian workers comb the forests of the Pacific Northwest gathering nontimber forest products (Brown 1998). They collect orange chanterelle and matsutake mushrooms, evergreen boughs, salal, huckleberry greens, and medicinal herbs for sale in domestic and international markets. Others plant trees and engage in ecosystem restoration activities. Their livelihood depends upon keen observation skills and knowledge of the forest floor–skills that are essential to ecosystem monitoring. In

BOX 3. Community Education Through Monitoring: The Canandaigua Lake Watershed Task Force

Formed in 1989 in a 174-square-mile watershed in the western Finger Lakes region of New York state, the Canandaigua Lake Watershed Task Force functions as a catalyst for local efforts to protect and enhance water quality through public policy education. The Task Force incorporates mostly local efforts of existing agencies, organizations, and individuals in a four-county area comprised of 11 towns, two villages, and one city. The Task Force identifies pertinent information gaps and collects and disseminates information to fill them through issue-based committees, local news media, public forums, surveys, volunteer monitors, workshops, publications, displays and special events.

In 1994 the Task Force assembled a portrait and technical assessment of the watershed emphasizing social and environmental factors. The data collected were arranged by watershed, pollution source, and subwatershed. The resulting 600-page document, "State of the Canandaigua Lake Watershed-1994," included an analysis of subwatersheds using the Generalized Watershed Loading Function computer model. It also contained an appraisal of watershed values including the provision of drinking water for 50,000 residents, recreation, and real property value.

An important outcome of the project was the formation of a watershed council made up of the chief elected officers of watershed municipalities. Bound by a Watershed Compact, the group met monthly for three years to review, revise, and accept more than 100 remedial actions (in 16 categorical sources of pollution) as components of a watershed management plan. A second outcome was the creation of an enhanced watershed sampling and testing program. The program includes in-lake and stream components, baseline and event sampling regimes, protocols for volunteer monitoring, and contractual work by several local colleges. Funding for the $50,000 monitoring program has been provided by private contributors, local governments, and in-kind contributions from agencies and organizations.

Due to limited resources for monitoring, the Canandaigua Lake Watershed Task Force has been extremely opportunistic and aggressive in responding to community concerns about water quality. Multiple, coordinated actions at various levels have been common. A good example is the response to road salt contamination. Chloride monitoring in streams identified several "hot spots" while an annual survey of highway officials has encouraged reduced use of salt and improved salt storage practices. Signs reading "Sensible Salting Requires Sensible Driving" have been erected in critical areas.

The monitoring and evaluation of water quality have been continuous concerns. Ongoing efforts in the community have integrated and linked with many other social processes. Early on, the community was convinced that sufficient background information existed to enable practical solutions to move ahead. Eventually, through community-building exercises, a need for more and better information was realized, and local agencies, municipalities, funders, and individuals have all responded.

forums hosted by the Jefferson Center, a nonprofit organization in southern Oregon, representatives of these workers listed participation in ecosystem monitoring among their top priorities. Their work has a significant impact on the forest, and their lives are profoundly affected by the structure of the nontimber forest products industry. These workers (and, forest workers of related industries) are most able to identify ecological and social factors and relationships that have significance in their lives. Involving nontimber forest products workers in the design and implementation of community-based ecosystem monitoring will result in more comprehensive identification of relevant social and ecological indicators.

Another example of local citizens' unique qualifications to contribute to community monitoring comes from East Baltimore's Madison East End community. There, a community leader organized local youth to conduct a neighborhood assessment using the "Knowing Your Community, Showing Your Community" handbook developed by a local nongovernmental organization, Community Resources. The outcome not only provided valuable information, it encouraged young people to become involved in community improvement activities. The youth focused on vacant lots, a major concern in this neighborhood. They mapped locations, photographed conditions (particularly noting trash, appearance, and maintenance), and interviewed neighbors about the condition of the lots and solicited their ideas for improvement. They also observed how people interacted with the sites: Did they walk near them? Play in them? Throw trash on them? This assessment then became the youth group's foundation for deciding what steps to take to improve lots in their neighborhood. Several objectives were simultaneously achieved: kids were engaged in constructive activity, valuable information was obtained, and the monitoring activity led to positive short- and long-term steps toward improving the community.

Multiparty monitoring may not be appropriate for every situation. Involving a wide array of diverse interests adds organizational complexity to an already complicated task. It requires large amounts of time and energy devoted to process, particularly to developing communication, understanding, and trust. It requires sensitivity and responsiveness to the range of values, emotions, and capabilities of those involved and entails risks for participants. Because it builds upon consensus, individual interests can disrupt it, with negative consequences for relationships between participants. Finally, those in positions of authority or control may actively resist multiparty monitoring, perceiving it as a threat to their status.

Many communities have decided that the potential rewards of multiparty monitoring outweigh these formidable difficulties and are pursuing the strategy as a means toward advancing community-based ecosystem management goals.

THE MONITORING PROCESS

Monitoring and the Adaptive Decisionmaking Cycle

Monitoring is part of a larger adaptive decisionmaking cycle (sometimes called adaptive management [FEMAT 1993]) designed to facilitate action in the face of ecological, economic, or social uncertainty. Under adaptive decisionmaking, continuous data collection and reevaluation inform and trigger adjustments to management decisions (Figure 1). The term "adaptive" is critical to the approach because it implies that action is taken in response to new information. Monitoring is one process through which new information is developed.

Within the adaptive decisionmaking cycle, monitoring involves a cycle of issue and participant identification, planning, implementation, evaluation, and adaptation (Figure 1). Monitoring may be initiated in response to a perceived threat to a community resource, as in the case of numerous water watch groups across the country. It may grow naturally out of long-term conservation work, in which case long-term ecosystem goals may already have been defined. In some cases it results from more formal processes initiated by state or federal agencies, as in the case of the Oregon Salmon Plan.

Identifying Monitoring Goals and Participants

Whatever the motivation for instituting a monitoring plan, the first steps must include clarifying goals and identifying stakeholders. These two critical tasks are intertwined, reflecting the connections between ecological and social systems. How issues are framed, what questions are asked, and what indicators are monitored is a function not only of social and ecological

FIGURE 1. The monitoring cycle meshes with the adaptive decision-making cycle.

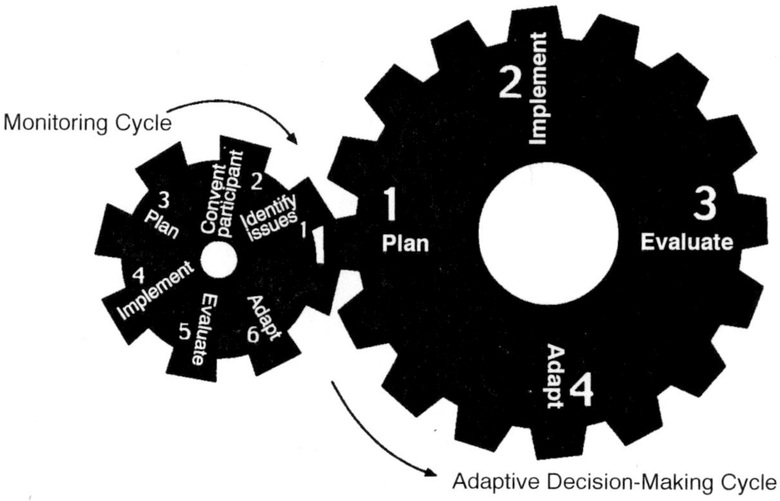

conditions but, just as importantly, of who is involved in the process. Who should serve on the monitoring team deserves as much thought and care as go into determining what to monitor. When the initial participants convene, they share relevant information, research, assessments, and plans to build a common understanding of current conditions and constraints. This body of knowledge might include indigenous, local, ethnic, cultural, and anecdotal information as well as relevant published scientific literature.

Developing this common base of understanding is followed by reviewing long-term goals and interim measurable objectives. Interim objectives are needed to help track project progress because long-term goals (e.g., recovery of an endangered species or development of a sustainable forest-based economy) will likely involve time frames beyond the scope of most monitoring programs. Developing interim objectives entails a thorough discussion of participants' project-specific needs, questions, and concerns. In multiparty monitoring, participants must synthesize a plan that responds to diverse values and perspectives. While the planning may be laborious, the result will be more useful to and receive greater acceptance from stakeholders than a monitoring plan developed by more efficient but less participatory means.

Selecting Social and Ecological Indicators

The next step is to identify indicators relevant to interim targets and long-term goals, measurable with available technical expertise and funding, and well distributed across both space and time. Indicators should be chosen with a clear understanding of how they will be used to evaluate progress toward goals. Monitoring goals and associated indicators vary widely with individual community characteristics and priorities. The sidebars in this chapter provide examples from specific monitoring efforts (see Box 4 and Box 5).

While most existing community-based monitoring involves only environmental variables, there is a growing recognition of the interdependence of environmental and social factors. Policies designed to remedy a particular environmental problem usually affect community members differentially. Linkages between timber harvest levels on public lands, employment, and household incomes in timber-dependent communities are a clear example. The distribution of the costs and benefits of endangered species protection on private land is another. Monitoring environmental change alone has little utility unless the links to social change are understood. Effective community-based ecosystem monitoring requires attention to social as well as ecological variables (Box 6).

The sustainability indicators developed by the President's Council on Sustainable Development (Table 1) reflect the linkages between social, econom-

BOX 4. Deciding What to Monitor: Philadelphia Recreation's Earth YES! Program

In 1995, the Philadelphia Department of Recreation developed an environmental education program for inner-city middle school youth that was designed to achieve a combination of educational, environmental and institutional goals. Monitoring and evaluation became a central part of the Earth YES! program to (1) support fundraising efforts, (2) identify program strengths and areas for improvement, (3) encourage Earth YES! to become a "learning program," and (4) document the program and its lessons for others.

The program brought together a Technical Advisory Committee comprised of Department of Recreation staff, Earth YES! instructors, and representatives from partner organizations to articulate and agree upon program goals, desired results, and monitoring indicators. Earth YES! participants then monitored their progress toward these goals and results through a set of simple methods, including daily attendance sheets, a pre- and post-"Environmental Jeopardy" game, and structured observations by outside observers. As a team, Earth YES! participants evaluate their progress through weekly meetings, year-end focus groups, and an annual retreat. They use ongoing monitoring and evaluation together to continually build and improve this exciting program.

A central goal was to educate Philadelphia's young people about urban environmental issues and concepts using hands-on activities in their own neighborhoods. One desired result was that over 200 inner-city youth, ages 10-14, complete the program each year. Indicators of success for this result included data on student age, race, family income, and daily attendance. A second desired result, an increase in participant knowledge of environmental concepts and issues, was measured through pre- and post-program tests, observation of students by instructors, and observation of instructors by the program director.

ic, and environmental health. The Council's broad long-term goals are economic prosperity, environmental protection, and social equity. While the goals are national in scope and the indicators are stated quite generally, they illustrate the relationship between indicators and goals.

Monitoring Protocol

The next step in the monitoring cycle is developing a monitoring protocol–deciding who will measure the indicators and how, when, and where they will be measured. Clearly, the monitoring protocol is directly affected by the availability of funding, expertise, and labor. In some cases, funding and expertise may be readily available, with agencies, scientists and volunteers eager to participate. In other cases, developing these resources may take years.

Among the important issues to take into account when developing monitoring protocols are data quality and data availability. High quality data consist of complete records that include both qualitative and quantitative

BOX 5. Oregon's Salmon Plan: An Experiment in Public-Private Collaboration

To residents of the Pacific Northwest, salmon symbolize the region's ideal of living in touch with nature. The dramatic decline in salmon populations is seen as a threat not only to the species but also to the region's very identity. The 1995 proposal by the National Marine Fisheries Service to list coho salmon as threatened under the Endangered Species Act triggered a major, innovative experiment to restore salmon habitat and avoid an ESA listing. In 1997 the state legislature approved expenditure of $30 million–$13.6 million from a tax on forest products volunteered by forest landowners and industry–to implement what has become known as The Oregon Plan. In essence, the Plan aims to supplement existing regulations (such as the state's Forest Practices Act) with voluntary measures designed not only to prevent "takings" of salmon but also to restore its habitat and increase its populations. Funding was made available to some 60 existing or newly created watershed councils–voluntary associations of landowners, community residents, environmental organizations, and public agencies–to collaborate on restoration efforts (Christensen 1998). In the first year, some 1,200 restoration activity reports were filed on projects including steam bank stabilization, fencing, riparian habitat restoration, culvert replacement, and "putting to bed" unused logging roads (Oregon Forest Resources Institute, n.d.).

Monitoring efforts have been a major component of watershed council work. Volunteers have inventoried and mapped culverts, dams, stream crossings, forest roads, and other obstructions affecting fish passage. They have conducted salmon spawning and stream temperature surveys. Instructions for conducting and reporting these surveys are available on the World Wide Web (www.oregon-plan.org), thereby encouraging broad participation. An Independent Multi-Disciplinary Science Team monitors achievements of watershed councils through an annual audit.

This experiment in bottom-up, voluntary collaboration between state government and local stakeholders is not without controversy and risk. In 1998 a federal magistrate ruled that the NMFS had violated the ESA by not listing the coho, forcing its listing. By prior agreement, that action nullified the dedicated forest products tax, thereby putting Plan funding in jeopardy. Governor John Kitzhaber subsequently reaffirmed salmon recovery as a top priority of his administration, and joined with other Pacific Northwest Governors to successfully lobby for federal funding to expand restoration efforts. Conservation groups around the country are watching to see if watershed councils will achieve the potential this experiment in collaboration promises.

observations (data) and information about the data needed to assess its quality (metadata), such as who collected the data, dates and locations of observations, methods used, and other information. Monitoring does not always have to be highly technical, but it will be most useful when conducted in a consistent, repeatable manner.

Making information accessible throughout a project is essential to the success of multiparty monitoring initiatives. Monitoring programs should specify where observation records will be housed and who will be responsible for maintaining these records. Planners might arrange for storing and providing access to monitoring records at local libraries, public schools, government offices, or institutions of higher education.

BOX 6. Community-Based Social and Economic Monitoring in Trinity County

In the early 1990s, residents of Hayfork, in northern California's Trinity County, watched logging activity in the surrounding forests drop sharply as a result of major changes in the management of federal forest land. To counter the impact this had on their community, residents rolled up their sleeves and formed the Watershed Research and Training Center (WRTC). As they began to rebuild the economic foundation of their community, they decided to check out their assumptions about the relationship between timber industry activity and their community's socioeconomic health (Danks and Jungwirth 1998).

To do this, the WRTC collected existing data from sources such as school districts, the U.S. Census, the U.S. Forest Service, and state agencies. They analyzed data on income, poverty rates, employment rates, public assistance, labor force participation, education, timber harvests, and crime. Of special interest were the relationships between timber volume and value, poverty, and employment. Community-level indicators were used where possible since each community has a different relationship to forestry activities. However, since Trinity County has little economic diversity, no urban centers and a strong countywide identity, county-level census and other secondary data were also very useful.

The WRTC continues to track data. In fact, community agencies, institutions, and residents now collect information of mutual interest and take it to the Center to be recorded. Graphs of the correlations hang like wallpaper at the WRTC and are the first stop for anyone who comes to tour the facility. They are a symbol of what motivates the Center, and they give the community a way to track its progress. The information and analyses have been presented in speeches, articles, newsletters, reports, and proposals; and even in a display at the county fair. The information monitored has been useful in other local, county, and federal agency assessment and planning efforts.

Monitoring results provide Hayfork residents with evidence of the need to incubate value-added businesses using local resources and to concentrate on training and policy changes that create local jobs in ecosystem management. The WRTC relies on ecological and socioeconomic monitoring to gauge its success in linking forest stewardship with the community's well-being.

Evaluation and Adaptation

Finally, monitoring plans should include an evaluation step to assess the composition of the team, the appropriateness of the goals and indicators, and the effectiveness of the monitoring protocol. If the evaluation uncovers problems, the team can make appropriate adjustments. In this way monitoring resembles a cycle within the adaptive decision-making cycle.

Land management agencies historically have emphasized planning and implementation and paid insufficient attention to monitoring and evaluation. Even when monitoring has occurred, it has not always led to a modification of plans based on the new information produced. Multiparty monitoring does not ensure that this loop will be closed, but participation by committed stakeholders increases the likelihood that it will.

TABLE 1. Summary of Goals and Indicators, President's Council on Sustainable Development (1996)

GOALS	INDICATORS
1. Health and the environment	Clean air, drinking water, toxic exposures, diseases and mortality
2. Economic prosperity	Economic performance, employment, poverty, savings and investment rates, natural resources and environmental accounting, productivity
3. Equity	Income trends, environmental equity, social equity
4. Conservation of nature	Ecosystems, habitat loss, threatened and endangered species, nutrients and toxins, exotic species, global environmental change
5. Stewardship	Materials consumption, waste reduction, energy efficiency, renewable resource use
6. Sustainable communities	Community economic viability, safe neighborhoods, public parks, investment in future generations, transportation patterns, community access to information, shelter, metropolitan income patterns, infant mortality
7. Civic engagement	Public participation, social capital, citizen participation, collaboration
8. Population	Population growth, status of women, unintended pregnancies, teen pregnancies, immigration
9. International responsibility	International assistance, environmental assistance, assessment of progresss, environmental technology exports, research leadership
10. Education	Information access, curriculum development, national standards, community participation, national achievement, graduation rates

Types of Monitoring

Several different types of monitoring exist to serve various monitoring goals (Table 2). Baseline monitoring provides a snapshot of current conditions that exist in an ecosystem or community. Similarly, inventory monitoring provides baseline information but is designed to be replicated to allow for trend analysis. Implementation monitoring seeks to determine whether an action has been executed. For example, were trees planted at 2×2-meter^2 spacing? Were hay bales and silt fence placed in riparian areas? Implementation monitoring represents more-or-less immediate follow-up to ensure actions were executed according to goals. Effectiveness monitoring determines whether goals were actually achieved. For example, have silt levels been reduced in streams based on our actions? Have inner-city ambient temperatures been reduced due to the planting of street trees? How effective have our measures been? Finally, validation monitoring provides for rigorous scientif-

TABLE 2. Different Monitoring Objectives Call for Differing Degrees of Complexity and Various Mixes of Participants

TYPE OF MONITORING	TYPICAL PARTICIPANTS[1]	LEVEL OF COMPLEXITY, SOPHISTICATION[2]
Baseline: What are current conditions?	Students, teachers, landowners, natural resource professionals, residents, contractors & workers, interest groups, volunteers, certification organizations, public land managers, regulators, regulated community, academics	B, C
Inventory: How have conditions changed?	Natural resource professionals, volunteers, interest groups, public land managers, youth	A, B, C
Implementation: Did we do what we said we'd do?	Contractors and workers, interest groups, volunteers, landowners, certification organizations, public land managers, regulators, regulated community	B, C
Effectiveness: Did we achieve our goals?	Academic institutions, natural resource professionals, urban residents, contractors & workers, interest groups, volunteers, landowners, public land managers	B, C
Validation: Were these science-based observations?	Academic faculty, graduate students, natural resource professionals, certification organizations, public land managers	C

[1] These "typical participants" are not an exclusive list. They represent the people *most likely* to be involved with monitoring for a given monitoring type.
[2] Levels of complexity include:
A–one-time monitoring, little or no planning, low budget, little or no expertise or specialized equipment required, few variables monitored.
B–monitoring repeated occasionally, simple planning, moderate budget, some expertise required for planning and evaluation but not necessarily for acquisition, several variables.
C–monitoring with regular periodicity, sophisticated planning and execution, requires considerable funding, commitment, and expertise among participants in all phases, specialized equipment required, many variables monitored.

ic testing of hypotheses, and thus requires sampling intensities and estimates of variation to permit the detection of significant differences over time and space.

Scale of Monitoring

We need and use information at many different spatial and temporal scales. The size of the geographic area to be monitored and the time frame over which monitoring is to occur should coincide with the overall objectives of monitoring, and should be commensurate with the resources available. While communities may tend to focus on their immediate geographic area, and on a relatively short time frame, understanding what is happening locally often depends on some general understanding of what is happening over a larger landscape and a longer period of time (Box 7).

BOX 7. Ecosystem Monitoring on Private Lands

Nationwide, nonindustrial private landowners and their families own 58 percent of all American forests. In the Northeast, as much as 85-90 percent of all forestland falls into this ownership category. Landowner participation is essential if ecosystem stewardship is to succeed here. Landowners are key both as caretakers of individual properties and as members of the larger community. One way to foster a sense of the broader landscape scale and increase awareness of the cumulative effects of individual management actions is to involve owners in monitoring ecosystem characteristics and dynamics. All the members of a community are affected by changes in the landscape, such as its conversion from forest to residential or by commercial development, which is currently occurring throughout much of the country.

Monitoring on private lands can be accomplished through several means. Regulating certain activities such as timber harvesting through forest practice acts generally requires monitoring potential impacts to wetlands, water resources, and rare species habitats (Kittredge 1996). Monitoring can take place at a coarser scale in a community by observing changes in land use, perhaps by an open-space committee or local land trust using windshield surveys. Similarly, building permit records can be monitored to detect changes in land use (Greenbaum and O'Donnell 1987). Volunteers organized by Massachusetts Audubon have completed three statewide monitoring projects, each focusing on a specific vertebrate type (i.e., breeding birds, butterflies, and herpetiles). Volunteers assigned to specific topographic quadrangles document species presence. Volunteers seek landowner permission to access posted lands but otherwise roam freely to investigate likely habitat for the presence of species. This monitoring approach documents distribution of species but not abundance (Jackson 1998).

Nonindustrial private lands are also monitored through participation in the USDA Forest Stewardship Program, administered through the offices of state foresters. Participating landowners receive cost-sharing grants to underwrite the development of stewardship plans. In Massachusetts, every plan must include a discussion of the landscape or ecosystem in which the individual property is located. It also must inventory timber, and include a map describing the property, cover types, and special features. Once the Plan is approved, landowners are eligible for further cost-sharing monies to help underwrite a wide variety of activities, including contracting with consulting biologists to document and advise on the presence and stewardship of rare species or communities.

In addition to temporal and spatial scales, monitoring efforts vary along a scale of complexity (number of variables measured) and technical sophistication (level of expertise and instrumentation) (see Table 2). Some objectives may require a monitoring design that measures a few variables over a limited geographic area for a short period of time and requires limited technology and expertise. Other objectives demand far more sophisticated approaches involving many variables, large geographic areas, lengthy time frames, and a high degree of scientific rigor. More complex monitoring programs require participants to make a greater commitment of time, money, and effort.

The complexity of the monitoring program should match the needs of the project, since each level of complexity reflects tradeoffs in expense, information quality, and utility. To illustrate, imagine designing a program to monitor an environmental issue of concern, such as an endangered species, that will likely require future judicial decision-making. Monitoring at level A (see Table 2), with minimal rigorous scientific input would provide inadequate information. In contrast, level A monitoring might prove just right for a community experiencing a creeping shift in land use, resulting in fragmented forest habitat. Level A monitoring might help marshal community support for addressing the issue by providing just the right level of attention and sufficient information to identify trends and stimulate public interest. Level C monitoring, on the other hand, would be very costly, appear intimidating, and might fail to sufficiently engage the public in the issue.

Funding

Adequate long-term funding is essential to establish and maintain monitoring programs. Lack of funding has been a direct cause of data fragmentation and incompatibility and inadequate, inconsistent standards. Funding for monitoring and evaluation may be viewed as an investment in the community and can be built into ongoing program budgets for public works, schools, and environmental protection. It is critical to ensure project and program monitoring to build monitoring and evaluation costs into budget development of all community-based ecosystem projects and programs.

Public sources of funding for monitoring and evaluation include legislatively appropriated funds (such as government assistance programs), special levies, trust funds, ratepayer funds, taxes, fines and penalties, lottery funds, casino revenues, tax incentives, and stamps. Private sources of funding include corporations, foundations, and non-governmental organizations such as "green banks" (Refer to Luzadis et al. this volume). The diversity of individuals and entities involved in community-based ecosystem efforts may provide an opportunity to pool and leverage resources from diverse sources. This diversity may also encourage development of creative new approaches to funding (Box 8).

LESSONS AND CHALLENGES

In contrast to the dry, technical image connoted by the term, monitoring involves contentious issues, tensions, and difficult political implications. We close by considering some of these challenges and summarizing lessons learned from practical experience in dealing with them.

Integrating Social and Ecological Factors in Monitoring

A major challenge facing community-based ecosystem management is recognizing the complex links between human and nonhuman elements of

BOX 8. A Market-Based Approach to Monitoring: Forest Products Certification

Green certification–third-party evaluation of sustainable forest management and forest products–is a rapidly growing form of monitoring for ecosystem management. Community-based ecosystem management practitioners have begun to use the process of certification of land, products and workers as one way to combine a growing consumer demand for sustainably harvested products and their desire to encourage innovative, sustainable ecosystem management.

The Forest Stewardship Council (FSC), an international organization based in Mexico, used committees with multistakeholder representation to develop broad principles and criteria for green certification of forest management activities (FSC 1996). These include principles on compliance with existing laws; the rights of indigenous peoples, workers, and local communities; environmental impacts; conservation of environmental and culturally significant resources; and management planning, monitoring, and evaluation. Forest owners' management plans and lands are reviewed for their compliance with national and regional standards. If they comply, the timber products that are subsequently harvested off the land are certified in the marketplace as "green." Another avenue for certified timber products has been created through the certification of foresters who develop management plans and monitor compliance. Certified lands are monitored periodically for adherence to approved plans and conditions; lands can be decertified if conditions are not met.

There are several certifying organizations whose specific standards comply with those of FSC: Smartwood is a nonprofit certifier, Scientific Certification Systems (SCS) a for-profit. The FSC is developing standards and guidelines that could be used to certify the harvest of nontimber forest products as well. The benefit is not only the potential for a premium on the price obtained for the product but the growing assurance of a market that responds to changing consumer values. It remains to be seen whether green certified wood products will command a higher price in the marketplace. Nevertheless, some landowners and managers are choosing certification as a means to provide third-party verification of sustainability.

ecosystems. Our institutions of higher education produce social scientists (anthropologists, sociologists, economists) and physical scientists (chemists, ecologists, biologists), but few scientists with expertise that crosses these disciplinary lines. Governmental agencies follow similar lines.

In contrast, the problems with which communities grapple are seldom easily categorized as either "social" or "environmental," nor are they simply environmental problems with a human dimension. Rather, they are extraordinarily complex, with ecological, social, economic, political, and ethical dimensions. Those citizens working to improve their communities are often in a better position to comprehend the complexity of the issues than are specialists in any particular field. The community-based multiparty monitoring efforts we've described share a growing realization of the interconnectedness of ecosystems' social and ecological components. This is evidenced not only by their selection of indicators to be monitored, but also by the emphasis they place on achieving diverse participation in project planning and implementation. The recognition of the need for diverse, representative monitoring teams shows they realize different social groups are affected differently by environmental change. It may be that through persistently demanding research assistance, community-based ecosystem management efforts will stimulate more interdisciplinary collaboration between citizens, scientists, and land managers.

Fear of Change

The very idea of monitoring may cause concern among some community members since it suggests the status quo may need changing. Those with an interest in maintaining that status quo may find the prospect of monitoring discomforting or may actively resist it. Alternatively, prolonged periods of monitoring may be promoted as a means to delay change.

Rigor

A related tension arises over the level of scientific rigor required to achieve monitoring objectives. Whereas some monitoring purposes may require rigorous scientific methods, sophisticated technology, and highly skilled technicians, others may be better served by school projects conducted by children using paper cups. Both can be legitimate, appropriate types of monitoring, as long as the methods match the objectives. An underlying tension involves questions of the relative legitimacy of science-based knowledge, indigenous knowledge, and the mystique of academic professionalism. It seems reasonable to suggest that most community-based ecosystem management monitoring plans would benefit from a mix of high and low technology and professional and lay participation. The important point is to carefully determine the objectives for monitoring, then match the methods, participants, and approach appropriately.

Process

Monitoring programs should be designed for continuity but also for flexibility. Advances in understanding might require adjustments in the indicators monitored or the degree of complexity required, thus necessitating regular evaluations of the monitoring program. Moreover, the timing of monitoring is important. Tacking it onto other ecosystem management projects or programs as an afterthought is unlikely to be as effective as a monitoring plan that is carefully integrated with other project activities. Carefully planning for monitoring can help ensure that information becomes available in a timely manner.

Participation

The inclusive, democratic traditions embraced by the concept of multiparty monitoring are difficult to implement. Few communities are completely of one mind regarding significant social and environmental issues of local concern. It is challenging to achieve collaboration between competing agendas when only public land is involved. Community-based monitoring is even more challenging in landscapes of mixed ownership. In these cases monitoring must achieve a workable balance between the interests of the community and those of individual landowners. Without meaningful representation and participation by those who own the land, community-based monitoring cannot succeed.

Conducting collegial collaboration among local stakeholders is even more complex when it involves "distant stakeholders"–nonlocal individuals and organizations who claim stakes in local resources. Finding mechanisms to lessen tensions among local stakeholders, and between local and distant stakeholders, is one of the major challenges faced by communities involved in ecosystem management. However, bringing appropriate partners and stakeholders into the monitoring process early on leads to more comprehensive identification of relevant indicators and may help avoid later conflicts. Similarly, recognizing and collaborating with existing monitoring efforts avoids duplication and unnecessary competition, is more efficient, and builds beneficial relationships. Developing a shared understanding of the need for monitoring and then communicating that understanding throughout the community builds support for monitoring. Conversely, keeping monitoring activities hidden or inaccessible can doom the effort.

Differing Agendas

There may be tension between those who advocate monitoring primarily as a vehicle for community learning, those who see monitoring primarily as a first step toward fundamental shifts in environmental or social policy, and those who view monitoring as solely data collection for research purposes. Whereas the term "monitoring" suggests responsible stewardship to some, it may connote draconian images of "Big Brother" to others. Such different, conflicting, real, or perceived objectives accompany many multiparty monitoring efforts and suggest the wisdom of pursuing shared understanding and clearly articulated goals early in the process.

Accountability

Accountability–having to answer for one's actions–is intimately associated with stewardship and with monitoring. Accountability is a measure of promises made and promises kept. In financial matters, it involves ledgers and balances. In matters of law, it is the subject of contraets. In personal matters it is the stuff of ethics. All these notions of accountability–the fiduciary, legal, and ethical–are relevant to community-based monitoring. Accountability is related to all the steps in the monitoring cycle, from the inclusive participatory design of the process and the choice of variables monitored to the selection of methods and the achievement and sharing of results.

Accountability begins with the individual: the citizen whose sense of place and stewardship leads her to start a local monitoring effort, the forest worker who volunteers to participate in data collection, the civil servant charged with supporting a monitoring program with tax dollars, the landowner who sees his property as part of a larger landscape and takes the initiative to learn how his activities affect his neighbors.

As communities take on the responsibility of monitoring and evaluation programs, they assume accountability for the financial and human resources entrusted to them. The diverse funding strategies and sources employed by community-based efforts demand a sharp focus on accountability to ensure that funds are efficiently used in achieving the desired social and environmental objectives. Government agencies (federal, state, tribal, local, and regional) must ensure both project and program accountability where government funds are used. Ensuring fair and equitable management of the human resources dedicated to the monitoring effort is equally challenging and essential (Box 9).

CONCLUSION

Community-based ecosystem monitoring is in its infancy. Monitoring programs appear where communities seek to respond to environmental crises, protect special places, or build constructive relationships between contending stakeholders. Monitoring displays many different forms, levels of sophistication, and degrees of participation. In this chapter, we have presented some of the dominant characteristics, motivations, and challenges of community-based monitoring.

Whereas the term "monitoring" conjures up visions of water quality samples and species inventories, we've tried to emphasize it as a social enterprise with social as well as ecological variables, dynamics, and objectives. Identifying, measuring, and understanding the social variables relevant to ecosystem health are essential, since humans–individually and collectively–are the dominant influence in the environments we inhabit.

We contend that participation in community-based ecosystem monitoring by diverse stakeholder groups will lead to more comprehensive incorporation

BOX 9. In the Absence of Monitoring and Accountability

Aldo Leopold, writing in *A Sand County Almanac*, gives a compelling example of the consequences when an individual and the surrounding community fail to take responsibility–to be accountable–for the resources from which they both benefit. He tells the fate of a stand of pines near his farm in central Wisconsin, the only remaining stand of mature pines in the county, one with considerable ecological as well as historical value.

"My neighbor, who owns the grove, has treated it rather decently through the years. When his son got married, the grove furnished lumber for the new house, and it could spare such light cuttings. But when war prices of lumber soared skyward, the temptation to slash became too strong. Today the grove lies prostrate, and its long logs are feeding a hungry saw. By all the accepted rules of forestry, my neighbor was justified in slashing the grove. The stand was even-aged, mature, and invaded by heart-rot. Yet any schoolboy would know, in his heart, that there is something wrong about erasing the last remnant of pine timber from a county. When a farmer owns a rarity he should feel some obligation as its custodian, and a community should feel some obligation to help him carry the economic cost of custodianship. Yet our present land-use conscience is silent on such questions" (Leopold 1949, quoted in Leak et al. 1997).

Leopold knew, and claims that "any schoolboy" should have known, the value of conserving the county's last pine stand. The community failed to act either out of ignorance of the imperiled resource, or out of a lack of regard for it. The story illustrates the importance of monitoring, observing the local landscape, and its link to accountability, taking responsibility for what is observed. The stand was lost when neither the owner nor the community understood its significance to the county, and neither took responsibility for saving it.

of relevant variables. This is especially critical in landscapes where the land ownership pattern includes private lands. Achieving collaboration by diverse, often competing stakeholders is likely to be the greatest challenge of community-based monitoring. There are communities, however, where that challenge is being met, where community members have committed to seeking common long-range goals. Where this occurs, community bonds are strengthened even as environmental problems are tackled.

Ideally, community-based monitoring takes a broad view of ecosystem management that embraces human and nonhuman dimensions; is designed around an inclusive, multiparty process; is integrated into all aspects of project design; and continually informs management decisions. In a culture where individual choice, mobility, and the illusion of limitless resources are dominant themes, it is not surprising that the task of monitoring and integrating its lessons is neglected. Monitoring is often tedious, repetitive, and slow. It requires a long-term commitment to a place, an openness to learning, and a willingness to question our assumptions. If we proceed as observers, measurers, and learners, we will refine our ability to understand and appreciate the complexity of the systems of which we are a part.

REFERENCES

Brown, B. 1998. Voices from the woods. *American Forests* 103(4):35-37.

Canandaigua Lake Watershed Task Force. 1995. Canandaigua Lake Watershed Compact. Canandaigua, NY: Cananadaigua Lake Watershed Task Force.

Christensen, J. 1998. Bringing salmon back. *American Forests* 103(4):17-20.

Danks, C. and L. Jungwirth. 1998. Community-based socioeconomic assessment and monitoring of activities related to national forest management. A working paper of the Watershed Research and Training Center, Hayfork, California.

FEMAT. 1993. Forest ecosystem management: An ecological, economic, and social assessment. Report of the Forest Ecosystem Management Assessment Team. USDA Forest Service, USDC National Marine Fisheries Service, USDI Bureau of Land Management, USDI Fish and Wildlife Service, USDI National Park Service, and the Environmental Protection Agency, Washington, DC.

Force, J.E. and G.E. Machlis. 1997. The human ecosystem, Part II: Social indicators in ecosystem management. *Society and Natural Resources* 10:369-382.

Forest Stewardship Council. 1996. Forest Stewardship Council principles and criteria for forest management. Oxaca, Mexico: Forest Stewardship Council.

Greenbaum, D.S. and A. O'Donnell. 1987. Losing ground: The case for land conservation in Massachusetts. Lincoln, MA: Massachusetts Audubon Society.

Grigsby, K. 1997. Ecosystem monitoring handbook, Version 1.3, Rogue River National Forest integrated ecosystem monitoring framework. USDA Forest Service, PNW Region, 39 p.

Jackson, S. 1998. Personal communication to D. Kittredge. University of Massachusetts, Amherst, MA.

Keller, E.F. 1983. *A feeling for the organism: The life and work of Barbara McClintock*. NY: W.H. Freeman and Company.

Kittredge, D.B. 1996. Protection of habitat for rare wetland fauna during timber harvesting in Massachusetts. *Natural Areas Journal* 16:310-317.

Kohn, E. (n.d). The Central Street Compact: A neighborhood plan for shared preservation of landscape heritage. Plainfield, Massachusetts.

Leopold, A. 1949. Round River essay, in *A Sand County Almanac*. Balantine Books.

Leak, W.B., M. Yamasaki, D.B. Kittredge, Jr., N.I. Lamson, and M.L. Smith. 1997. Applied ecosystem management on nonindustrial forest land. Gen. Tech. Rep. NE-239. Radnor, PA: U.S.D.A. Forest Service, Northeastern Forest Experiment Station, 30 p.

Margoluis, R. and N. Salafsky. 1998. *Measures of success: Designing, monitoring, and managing conservation and development projects*. Covelo, CA: Island Press, 363 p.

Merriam-Webster Inc. 1987. Webster's Ninth New Collegiate Dictionary. Springfield, Mass: Merriam-Webster Inc.

Oregon Forest Resources Institute. (n.d). Special Report: Saving the salmon–Oregonians working together to manage environmental change. Portland, Oregon: Oregon Forest Resources Institute.

President's Council on Sustainable Development. 1996. *Sustainable America: A new consensus for prosperity, opportunity, and a healthy environment for the future.* Washington, DC: Government Printing Office.

Knight, R.L. and P.B. Landres, eds. 1998. *Stewardship across boundaries*. Washington, DC: Island Press.

Ticknor, W.D. 1992. New institutions for a new forestry. Unpublished paper presented to the Ben Roach Forum October 15, 1992. W.D. Ticknor Forestry Consultants, Orient, Ohio.

Investing in Ecosystems and Communities

Valerie A. Luzadis
Carolyn Alkire
Catherine M. Mater
Jeff Romm
William Stewart
Leah Wills
Duane R. Vaagen

SUMMARY. Our growing population demands more than ecosystems can provide without damage to them, and we lack the financial mechanisms to invest in their maintenance and restoration. Globalization of markets exacerbates this problem and creates another: it limits the connection between local value-added, secondary processing and local communities. Four strategies are suggested and several specific investment tools are described. Discussion includes how communities might employ these tools to strengthen ecosystem investment and the link

---

Valerie A. Luzadis is affiliated with the Faculty of Forestry, SUNY Environmental Science and Forestry, 320 Bray Hall, 1 Forestry Drive, Syracuse, NY 13210.

Carolyn Alkire is affiliated with The Wilderness Society, 900 17th Street NW, Washington, DC 20006-2596.

Catherine M. Mater is affiliated with the Mater Engineering, Ltd., 101 SW Boulevard, P.O. Box 0, Corvallis, OR 97339-0880.

Jeff Romm is affiliated with the College of Natural Resources, RIPM/ESPM, University of California-Barkeley, Barkeley, CA 94720-3114.

William Stewart is affiliated with the California Department of Forestry, Fire and Resource Assessment Program, P.O. Box 944246, Sacramento, CA 94244-2460.

Leah Wills is affiliated with the Plumas Corporation, P.O. Box 3880, Quincy, CA 95971.

Duane R. Vaagen is affiliated with the Vaagen Brothers Lumber, 585 West 5th, Colville, WA 99144.

[Haworth co-indexing entry note]: "Investing in Ecosystems and Communities." Luzadis, Valerie A. et al. Co-published simultaneously in *Journal of Sustainable Forestry* (Food Products Press, an imprint of The Haworth Press, Inc.) Vol. 12, No. 3/4, 2001, pp. 169-194; and: *Understanding Community-Based Forest Ecosystem Management* (ed: Gerald J. Gray, Maia J. Enzer, and Jonathan Kusel) Food Products Press, an imprint of The Haworth Press, Inc., 2001, pp. 169-194. Single or multiple copies of this article are available for a fee from The Haworth Document Delivery Service [1-800-342-9678, 9:00 a.m. - 5:00 p.m. (EST). E-mail address: getinfo@haworthpressinc.com].

between local value-added processing and community well-being.
[Article copies available for a fee from The Haworth Document Delivery Service: 1-800-342-9678. E-mail address: <getinfo@haworthpressinc.com> Website: <http://www.HaworthPress.com>]

KEYWORDS. Community-based, forest(ry), ecosystem management, investment, value-added

Ecosystems are the basis of life on earth. They are the source of diverse goods and services, typically at no direct cost to people. When healthy, ecosystems can store and regulate water quality and quantity; produce wood, food, feed, and fuel; sustain wildlife, plants, and other organisms; purify air; buffer human settlements against natural shocks; store carbon; and preserve an inventory of potential products and the scientific explanations of natural processes. Ecosystem services and resources such as these are natural forms of capital that produce valuable future benefits that depend upon the quality of the ecosystem today.

Human actions transform this natural capital into dollars which then move through financial markets to opportunities for even higher returns. Strong pressure from current markets discourages reinvestment in those ecosystems. Rates of natural regeneration rarely match financial rates of return, and the mechanisms do not currently exist to assess and charge for ecosystem uses or to invest in the original source of that value. Prices for the benefits from natural capital, the dollar value assigned, do not reflect their full cost. The result is a financial pressure that gradually reduces the capacity of ecosystems to adequately perform life-sustaining functions. The longer-term economic losses are widespread and extremely serious: increased costs to replace or repair ecosystem services (if possible) and a decline in jobs and social well-being of the human communities that depend upon them.

In short, our established structure of ecosystem finance increasingly does not satisfy social and political claims on ecosystem capacities, and market motives pull private investment elsewhere. Growing population, the resulting deficits in available funding, and conflicts over resource use and protection further destabilize the climate for ecosystem investment.

Framing the problems of and potential solutions to inadequate investment in ecosystems can help identify opportunities worth testing. When not enough money flows to an ecosystem to sustain the services it provides, the reasons are usually linked to location and the service itself. In addition, these reasons are nestled in contextual layers including local, regional, and federal institutions and markets; dynamic rural and urban cultures; and a growing global population. No one approach or technique–save a drastic reduction in human population–will answer all the questions or solve all our ecosystem problems. In fact, one of the challenges is the multitude of possibilities–how do we choose? And how do we adjust along the way to changes we cannot predict now? Flexibility is essential.

TRENDS INFLUENCING ECOSYSTEM INVESTMENT

Among the trends influencing ecosystem investment, first is a growing discrepancy between society's demands upon ecosystems and the financial support for activities that renew, sustain, and enhance ecosystem capacities. Our growing population demands more than ecosystems can provide without damage to them, and we lack the financial mechanisms to invest in their restoration and maintenance. Short-term and insufficient investment have exacerbated the discrepancy between societal demands and ecosystem health and sustainability. Furthermore, long-term ecological consequences of the current uses and activities are not fully understood.

One example of the discrepancies that exacerbate the problem is the incompatible economic and political demands the *ad valorem* taxation system places on ecosystems. This system, which many states use for revenue, taxes land on the basis of its "highest and best use," even if the land is not in that use. Other examples: monetary policies that promote borrowing; and increasing interest rates that motivate the selling of resources, resulting in lower product prices. This is characteristic of boom-bust cycles that gradually reduce stocks of resources. Some distribution policies subsidize certain ecosystem activities at the expense of other ecosystem services. This perpetuates the divergence between what we want ecosystems to provide socially and politically and the financial support to maintain healthy ecosystems.

A second influential trend is the globalization of financial markets and trade. This trend both expands possibilities to sell off natural assets reducing ecosystem capacity and improves prospects for investing in the restoration and maintenance of healthy and resilient ecosystems by increasing the market value. Global markets distance profits from resources reducing the likelihood of an investment back into the ecosystem (see Johnson et al., this volume, for further treatment of this subject). The lack of a financial feedback loop between economic activity and the ecosystems that provide services and raw materials is illustrated in Figure 1. This diagram shows the weak link between economic and ecological systems and is adapted from ecological economists' ideas of material and value flow (e.g., Daly 1977, 1996) and systems ecologists' notions of energy flow. The traditional notion of economic production as a closed system (e.g., Case and Fair 1996) has promoted the lack of connection between ecological and economic systems (Hall et al. 1986). Without adequate positive feedback from social and economic activities, labeled "investment in ecosystems" in Figure 1, ecosystems will continue to

FIGURE 1. Economic activity can be viewed as a subset of natural systems. The arrows between systems indicate system-level links and the arrows within the economic activity system indicate flows of materials (solid arrows) and monetary units (dashed line arrows). The weak link between economic and natural systems is indicated as the double, dotted line arrow labeled "investment in ecosystems." The weak link between communities and local secondary processes is also indicated by double, dotted line arrows. Adapted from Hall et al. 1986 and Daly 1977.

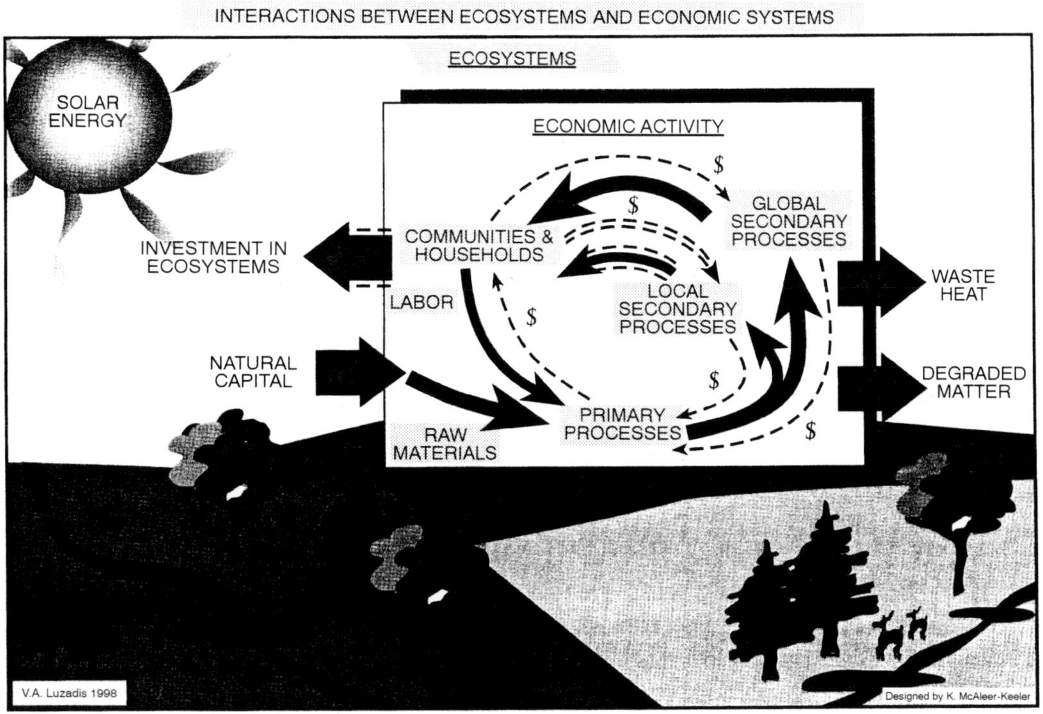

degrade, reducing their ability to provide life support functions into the future. New financial and social institutions are needed to enhance investment in ecosystems (see Cortner et al., this volume, for further discussion of institutions). Additionally, Figure 1 indicates a weak link between local secondary, value-added processing and local communities. The strength of the global market pulls resources away from local processors by offering higher rates of return without addressing ecosystem degradation. Local communities that depend on these ecosystems are hardest hit by this trend and struggle to attract the capital and economic development opportunities that will allow them to become or return to being healthy, viable human communities.

A third trend provides a potential counterbalance to the problem of ecosystem investment. Urban people continue to move into rural areas. The more complex, rural social structures that result increase public demands on ecosystems while adding to the skills and wealth that augment the potential investment capacity of ecosystem-dependent communities. These new ecosystem investment possibilities remain largely untapped but could provide a great opportunity to improve ecosystem viability. Communities first must overcome the disparity between established local social and economic traditions and the lack of new forms of social and financial organization that could engage new residents in ecosystem enhancement. Emerging social and economic networks can moderate grander political and economic trends in ways that reverse the drain on ecosystems.

In this paper we focus on the two weak links identified in Figure 1: lack of investment in ecosystems and the limited connection between local value-added secondary processing and local communities. We suggest four strategies to strengthen investment in ecosystems, describe several investment tools, and finish with a discussion of how communities might employ these tools to strengthen the link between local value-added processing and community well-being. Using an ecosystem approach to management is a long-term proposition that requires diligence, respect, trust, and hope. We focus primarily on innovative programs that can be implemented now and that will leverage existing skills and financing to maintain and improve ecosystem health. Longer-term strategies, including shifts in public expenditures, continue to be discussed nationwide and we address them briefly.

STRATEGIES

During the last few years the use of new financing tools and organizational efforts have demonstrated different ways of investing in ecosystem services. While having different denominators–money and human effort–many of these innovative investment tools are capturing unusual levels of investment and funding support while staying true to environmental and community values. And they are doing so without relying solely on regulatory intervention both domestically and internationally.

We present four strategies to stabilize ecosystem investment: innovations in private investment, jurisdictional boundaries, government funding, and community-based ecosystem management. These strategies reflect the direct experience and expertise of the authors of this paper and are expansive, not comprehensive.

Innovations in Private Investment

Business enterprises find ways to balance demand and capacity while creating competitive ecosystem investment opportunities that both benefit the public and profit those who make it possible. These innovative investment tools include:

1. Independent, third-party certification of forest sustainability
2. Carbon sequestration leverage packages
3. Resource banks
4. Partnership financing
5. Technology and new markets leveraging

Independent, third-party forest certification–By recognizing both the importance of community-based initiatives and the power of the marketplace, independent, third-party certification has gained credence as a viable market-based financial tool to fund ecosystem management throughout the world.

The international Forest Stewardship Council (FSC) was established in 1993 in response to concern about unverified environmental claims by wood products producers. The FSC gathered environmental, social, and industry representatives from all over the world and established principles and criteria to determine the sustainability of forest management. FSC-sanctioned certifiers assess forest sustainability and issue a seal of approval. To best implement these principles and criteria in ecosystems across the globe, FSC established regional community-based groups of social, environmental, and economic concerns. These groups then develop region-specific ecosystem standards and criteria. FSC required those regional standards to be technically defensible (scientifically based) with the understanding that to be successful, certified products would need to demonstrate clear performance in the marketplace (be recognized and preferred by retailers and consumers), making investment in certification financially viable to forestland owners and forest-product producers worldwide.

Since its inception, the investment worldwide in FSC-certification has been remarkable. Despite third-party certification being in its infancy, the volume of certified wood products has grown a great deal in just a few years

(Environmental Business News 1997). Further, demand for some certified wood products exceeds supply (Mater 1998a). As an example, Eric Bloomquist, president of Colonial Craft, the nation's largest producer of wood window grills, reports production of certified wood products in its Wisconsin and Minnesota operations could increase from approximately 10 percent of total annual product volume to more than 50 percent *if* more certified wood was available (Mater 1998a). The Menominee Indians in Wisconsin, with 281,000 acres of certified timber, are able to secure a free-and-clear (versus cost-recovery) premium of 10 percent on their certified hardwood veneer logs, with market demand exceeding four times their ability to supply while retaining their certification status (Mater 1998a).

The number of private forested acres that are certified is on the rise. In the United States, public forestland managers, from municipal watersheds to state-owned and -managed forestlands, have also recently chosen to invest in certification (Mater 1998b).

Carbon sequestration leverage packages–Carbon sequestration is the accumulation of carbon in terrestrial forms (as opposed to atmospheric)–by growing more trees, for example–and a strategy for limiting greenhouse gas buildup in the earth's atmosphere. Current efforts to improve investment packages for carbon sequestration projects throughout the world are slow to gain implementation but not new to the ecosystem front. What is new is the leverage packaging beginning to occur between carbon sequestration investment offerings and certification efforts on both private and public lands. This new collaboration, which has attracted attention internationally from private investors, is being considered for projects in the U.S. Certification efforts attract the attention of carbon sequestration project investors not only because of the attractive opportunity to convert land back to forests (e.g., sugar cane plantation conversions in Hawaii) but also because of the process, which actually reduces the carbon project's administrative costs. The assessment process a landowner undertakes to achieve certification provides the baseline technical information needed to set up carbon projects. More importantly, the annual monitoring and auditing required by certification contracts is the same as that required by investors to document the continuance of "carbon additions" needed to secure ongoing financial trade-offs for carbon projects. In fact, the annual auditing and monitoring process initiated by certification is so attractive to many carbon project investors that they design their contracts with landowners to match the five-year certification contracts. Landowners typically assume the annual auditing costs to maintain certification status. Since this is mutually beneficial to project management, carbon sequestration project investors are willing to pay the up-front costs to analyze certified forests for carbon sequestration (Sampson 1998).

Resource banks–The notion of banking natural assets has been discussed in recent years as a means of trading value for use, an attempt to balance the need for both natural assets and monetary value. This approach has been used by creative zoning boards in suburban areas (more on this later), but it is just now gaining a broader following as a useful conservation investment tool. In mid-1998, The Nature Conservancy (TNC) established the first forest bank, with the trade-marked name *The Clinch Valley Forest Bank*. This ecosystem investment tool is intended to provide private non-industrial forestland owners across the U.S. a third option in addition to conservation easements and relinquishing title. With the TNC Forest Bank, the landowner retains full title to the land but relinquishes title of the growing stock–trees. In exchange, the landowner receives a principal deposit in the Bank. Initial projections suggest that this deposit could provide an equal or better financial return than what the landowners might obtain by contracting to harvest the land themselves (Mater 1997a). Landowners can withdraw cash from the deposit over set periods and receive dividend payments similar to any other bank deposit. The TNC Bank will be a certified resource manager, thus managing and harvesting the land under certification criteria and guidelines adopted in the regions of interest. This means the Bank's harvested resource can be sold in the certified wood products market (Mater 1997a). Thus, the landowner retains aesthetic quality and a balanced ecosystem and has access to capital when needed but does not have to assume any of the time, energy, or responsibility for managing the forestland. The Bank gains access to important forest resources, especially in threatened ecosystem areas, and is able to harvest the lands in a certified fashion while protecting and improving sensitive ecosystems.

Partnership financing–Partnership financing was initiated by the international Caterpillar Corporation, manufacturers of logging equipment and other industry rolling stock and machinery. In 1997 Caterpillar undertook a major media campaign announcing the company's commitment to sustainable forestry practices and to certification. Caterpillar initiated the following investment package based on its recognition of certification as a growing force, coupled with the necessity to offer low-impact logging equipment that meets the alternative log harvesting practices often required for certification. The package offers:

1. Assistance in calculating the ecosystem benefits in gaining certification status through the use of low-impact logging equipment;
2. Bridge-financing to help potential equipment buyers pay for the low-impact logging equipment;
3. Information and FSC-approved certifier contacts to help forest operators and owners understand the potential benefits of low-impact logging.

This type of effort leveraged by some traditional banking institutions' recognition of forest certification can provide much needed bridge financing. This innovative financing was front-page news in the *Wall Street Journal* in 1997 (Ulman 1997). A logger received bank financing to purchase new low-impact equipment needed to secure logging contracts on an FSC-certified million-acre forest owned and managed by Seven Islands Corporation in Maine. The bank's decision to float the loan was due in part to its determination that a certified forest afforded the logging operation much more stabilized job security, therefore ensuring timely payments (Ulman 1997).

Technology and new markets leveraging–A fundamental tenet for success is the identification and implementation of new techniques, technologies, and markets to match buyers with resources generated from sustainable ecosystem management. Some that deserve immediate attention: log sort and sales yards and small-scale, small-diameter processing. Maximizing the value of every log sustainably harvested from public and private forestland is the fundamental benefit of well-designed log sort and sales yards. Log sales traditionally are accomplished through larger-scale contracts: standard log sorts based on standard log grades, with the majority of the sorting done by the logger. This approach often causes more damage to the forest floor due to the size and weight of the harvesting equipment, results in more waste left in the forest as unprofitable product, and, as a result, often achieves lower value per unit sold. A log sort and sales yard necessitates minimal sorting in the forest, uses more of the logs in the yard, and increases the opportunity to sell traditionally lower grade logs to "characterwood" or specialty markets.

For example, the Lumby Log Sort and Sales Yard project in Vernon, British Columbia, was initiated by the Canadian government in 1996 as an experiment to determine if "alternative" logging practices (such as single-tree selection, group selection, small-scale clearcuts with reserves) could be more profitable than traditional large-scale clearcuts. A log sort and sales yard was also established to determine whether logs from the project sites could demand a higher price. Within the first two years the operation accrued between 18 percent to 30 percent higher costs due to alternative logging techniques, but offset that expense with 20 percent to 40 percent higher sales value of logs more efficiently sorted and marketed from the yard. Because of the ecological and aesthetic benefits of the alternative harvesting methods, the project was also allowed to harvest on up to 20 percent more land than was originally anticipated–with the blessing of surrounding communities. The end result was a million-dollar higher profit generated to the government over what would have been obtained through traditional logging and timber sales techniques (Mater and Mater 1998).

The secret to the project's financial success was in the log sort and sales yard, where logs were sorted to serve the region's multiple value-added product manufacturer needs. Allowing for both large volume sales and single log sales, the sort yard adapted its quality and quantity parameters to meet buyer needs in the region. The project began with seven grade sorts, and increased to more than 40 sorts after only one month of operation. Large-scale log buyers continued to be interested in the standard log sorts, and smaller value-added product manufacturers were pleased to have a sort yard where they could hand-pick the logs they needed for their furniture, flooring, cabinetry, fencing, musical instruments, bird and bat house manufacturing, etc. Even logs with significant twist and turn (known as "pistol-grip" logs and usually considered waste) were in demand from log home manufacturers in the area–turning a 25 cents per cubic meter stumpage cost into an $80 per cubic meter log sale (Mater and Mater 1998).

Forest restoration projects working with small-diameter (2 to 10 inches) "thinnings" have little chance of sustained financial success without having clear options for adding value to the resources they harvest. Forest products and logging operations experiencing a significantly reduced supply of raw materials face certain closure without adequate consideration of value-added options, turning wood "trash" (e.g., defect, heavy taper, twist, cutoffs, and lumber jackets) into "cash." And all these specialty market options must be predicated on new product marketing and distribution systems. Progressive forest products companies like Vaagen Brothers Lumber in Colville, Washington, are employing some of the innovative investment tools presented here in an attempt to do just that.

An effort is currently underway in California and Arizona to integrate federal forestland restoration contracts and new small-diameter processing technology into log sort and sales yards there (Mater 1998b). New small-scale, small-diameter processing technologies such as The Economizer (which converts small-diameter limbs and logs into marketable cants instead of chips), coupled with newly-employed federal restoration contracts allowing for zero value on stumpage, can produce solid financial options for forestry and forest products operations committed to restoring ecosystems (Mater 1997b, 1998b).

Innovations in Jurisdictional Boundaries

The second strategy is a broad package of innovative jurisdictional alignments that create mechanisms for beneficiaries to pay for the ecosystem services they want and need. Political and social jurisdictional boundaries often hinder ecosystem health and restoration efforts. These boundaries, created as we settled this land, were crucial to supporting social needs at the time. With the expansion in population and improvements in technology and science, it is possible to adjust or even ignore these lines for ecosystem management. New cross-jurisdictional arrangements can overcome the separateness between ecosystem services, parts, and products, the people who care for them, and the people who benefit from them (refer to Figure 1). New cross-jurisdictional arrangements incorporate the managers and the beneficiaries of ecosystems in exchanges that increase ecosystem investment to the advantage of both parties.

Jurisdictional lines may block a financial exchange between those who benefit from ecosystem services and those who might otherwise invest in the capacities to provide them. For example, people who work in watersheds have no authority to share the benefits of actions that improve water storage, flow, and quality, thus there is no financial motive to take these actions, while those who use the water have no means of sharing their benefits to provide incentives for such investments. Conversely, there is no incentive *not* to harm ecosystems when effects are not realized by those who cause damage or degradation.

Jurisdictional lines may restrain access to some areas and services in ways that would create investable financial values. For example, protected areas and streams increase property values along their boundaries, but there are few means to capture the increased value in order to sustain the ecosystem qualities that create it. Similar patterns appear adjacent to sustainably managed forests and tree-lined urban areas, as compared to the degraded forests or asphalt urban areas. (See The Trust for Public Land's 1994 *Healing America's cities: How urban parks can make cities safe and healthy.*)

Jurisdictional lines may promote competition among jurisdictions in ways that deplete investment in the ecosystem capacities they all require. For example, adjoining federal and private forest owners, although affecting one another's prices and costs in a common timber market, lack a collaborative way to overcome the seesaw impacts one stock-harvest decision has on the other. The result is detrimental to the sustainable qualities of the forest region-wide. Counties and developers who compete for new residents lack the means to cooperate in conservation investments that would be attractive to potential residents. Metropolitan regions create greenbelts without being able to manage either disinvestment and degradation within the belt or the ecological disruptions sprawl beyond it causes.

The very problems these boundaries create suggest opportunities for changing them to generate investments people would willingly make to sustain ecosystem services. Governments are recognizing the fiscal gains from redrawing lines to facilitate exchanges that promote ecological investments and maintain property values derived from ecological quality. Public-private collaborations are developing with the same intent. This section describes some of these innovations, including:

1. Fiscal/ecological zoning
2. Watershed organizations

3. Forest economic regions
4. Cooperation among counties
5. Greenfield-greenbelt-brownfield exhanges

Fiscal/ecological zoning–Local governments zone land with their fiscal needs in mind to encourage, discourage, and distribute uses to achieve a desired tax base. In recent years, rising public environmental expectations have enriched zoning rules, with differential effects on property values. In some instances, explicitly setting aside natural areas through zoning has been used to attract commercial activity that generates revenues for the locality. The amount of revenue generated from creating desirable work settings near natural areas can be greater than the loss of the set-aside area for commercial use itself. This revenue, stemming from the value of that natural area, can then be used to fund other public needs.

In California, for example, urbanized Santa Cruz and San Mateo counties expansively zoned commercial timberland, committing to the lower tax potential as a means to secure the economic viability of forest areas and higher property values outside the zone. The tax gains more than compensated for the losses from precluding development in the protected areas. In southern California, gains in residential property values more than compensated for the cost of wildlife set-asides within specially zoned regions.

The general idea has great potential. Where subdivisions develop to the edge of farmland protection zones, their values may well reflect the pastoral virtues that agriculture protects even while their effects may undermine agricultural sustainability. There are opportunities to capture the financial values of open-space for investment in the activities that sustain it. An analogous example is the development of subdivisions along the boundaries of national parks and forests and their private or local equivalents, where the attraction's property value could be captured to rectify the under-investment causing its deterioration.

Watershed organizations–Most water-use regimes provide no direct incentive for upstream watershed management that benefits downstream water quality, timing of water flows, and riparian health. The owners and/or managers of the watershed, public or private, rarely charge for improvements created for downstream interests, since there is little precedent for this and few mechanisms to enable such charges. In the absence of taxation support, there is no financial mechanism for downstream water users to improve water-source areas.

In response to this problem, watershed organizations are emerging throughout the U.S. to facilitate negotiations that lead to benefits for both upstream and downstream interests. While many of these originated from urban criticisms of rural land management practices, growing appreciation for the financial realities is producing exchanges that raise the level of watershed investment.

Perhaps the best-known example of this development is New York City's agreement to pay for modified practices by farmers upstream from its water supply (Nickens 1998). Upstream-downstream exchanges are developing throughout the West, quietly or through increasingly prevalent watershed councils. Regional institutions, such as the Chesapeake Bay Commission, have developed a similar facility to overcome weak upstream incentives to manage resources in ways more favorable to downstream interests.

Forest economic regions–When public and private forestland owners participate in common forest products markets without cooperation, the independent decisions each makes cuts through prices and politics to have a direct effect on the others. In the West, for example, earlier overcutting on private land resulted in pressure to overcut public forests; logging restrictions on public forests for endangered species habitat protection subsequently initiated a new cycle of overcutting on private lands. This pattern is not beneficial from a regional economic, social, or ecological perspective, nor is it advantageous for the public and private owners involved. Its effect is disinvestment in the forest and the social and economic capacity that sustains forest viability. The Cooperative Sustained Yield Units and the Grazing Districts are examples of early attempts to stabilize public-private relations and strengthen reinvestment. Current innovations contain some of their elements but incorporate environmental and social interests these earlier approaches did not.

Of particular note is the Quincy Library Group which began as private discussions between a county supervisor, an environmental attorney and a timber industry forester frustrated by the damage being done by the "timber wars." Their exchanges expanded to include others interested in finding common ground on the interdependent goals of forest health and community stability. The group's work included making and promoting proposals for management of the three local National Forests to conserve wilderness and enhance environmental protection, stabilize forest commodity production and jobs, and improve forest health and public safety, by choosing actions on the basis of ecological and community processes rather than administrative traditions (www.qlg.org August 1998). The Northern Forest Lands Study, the Greater Yellowstone Ecosystem, and the Adaptive Management Units of the Pacific Northwest are other examples that draw different institutional innovations from their own contextual possibilities.

Cooperation among counties–The real estate market drives public policy in many parts of the U.S. Local governments compete in this market and have little market power when the attributes of their areas are relatively similar. This competition eliminates financial surpluses that might otherwise be used for investment, including ecological conservation, and forces development patterns that are disadvantageous for all in the long run. Confronting this situation, counties have sought cooperation, often with state and federal guidance and support, to rationalize development on a regional basis and protect collective environmental assets. What is emerging is a fiscal capacity to use conservation to generate the revenues that pay for it.

The Pinelands National Preserve may be the classic expression of this mechanism. Faced with the probability of population sprawl throughout New Jersey's unique and relatively pristine pinelands, federal, state, and local governments developed a system of financial exchange that used transferred development values to concentrate settlement adjacent to the preserve rather than within it. In exchange for lost development options, landowners inside the core preserve were given tradable rights for development in adjacent areas. An initial government subsidy catalyzed the Pinelands project, and land use planning and regulation were used to balance the purposes of the preserve and the financial requirements for its integrity. Seven counties participated in governance, planning, and financial decisions that secured gains for all of them. In effect, they used cooperation and planning to achieve a desired distribution of demographic, ecological, and financial conditions.

Groups of counties throughout the United States are pursuing similar ventures, most often with less visibility than the Pinelands project. While government subsidies launched the Pinelands effort, in other situations private banks and land trusts have played an analogous role. Public development corporations could do so as well.

Greenfield-greenbelt-brownfield exchanges–Metropolitan growth creates competing policy interests among city centers, and the surrounding suburbs and rural areas. Cities contend with deterioration of infrastructure and services, as well as a growing number of brownfields–unused sites contaminated by previous uses that complicates their reuse. Suburban residents establish greenbelts or conservation zones that enhance land values, forcing less-affluent urban migrants to the cheaper land–the greenfields–of the rural counties. The cities become segregated between those who can afford to isolate themselves and those who cannot. Regional opportunities arise to use greenbelt benefits and open space development charges and credits as incentives for urban reinvestment.

Although they may exist, we are not aware of metropolitan exchange mechanisms that use growing suburban values to prevent or reduce rural and urban disinvestment. Federal and state fiscal redistributions tend to fund activities such as agricultural conservation and urban forestry, stream recovery, and brownfield restoration. Nevertheless, the indirectness of regulatory and tax or subsidy approaches suggests opportunities for enterprising brokerage among rural, suburban, and urban interests. Swaps at the rural/suburban interface have become common, brokered by non-governmental and business enterprises rather than public entities. However, the connection with the inner city has yet to be made.

Greenfield-greenbelt-brownfield exchanges have the potential to help alleviate environmental injustice. Groups that lack mobility, whether for racial, financial or educational reasons, sometimes find their fate controlled by decisions about the location, environmental health impacts, and employment level of the factories, mills, and natural resources that affect them on a daily basis. Disinvestment is a consequence of powerlessness and isolation that can be overcome by empowerment and networking around financial and ecological goals.

Non-governmental organizations have emerged as significant intermediaries for cooperation that improves community conditions and opportunities while increasing industrial appreciation for the benefits provided by supportive and effective communities. They have made the point that investing in people improves the environment for business. Throughout the United States, thousands of groups are working to bring people together around environmental issues such as toxic wastes and hazardous pollution to empower people and provoke increased urban investment. Empowering devalued groups is a crucial part of ecosystem investment, and the results are one of the most significant sources of innovation occurring in the growth of this field.

Innovations in Government Funding of Ecosystem Management

Innovation in federal, state, and local governmental approaches to financing ecosystem services includes reforms that overcome the gross inefficiencies that developed from treating ecosystem finance as an administrative/budgetary problem rather than an investment issue.

Federal funding of ecosystem management–Funds for land and resource management activities on federally managed public lands are appropriated annually by Congress to the four land management agencies. Authorization for the expenditure of funds is either discretionary or mandatory. Discretionary appropriations are funded from the Treasury, mandatory appropriations from trust funds and special accounts into which receipts from various activities (some on public lands, some not) have been deposited. Virtually all the funds directed to ecosystem protection and restoration on federally managed public lands are discretionary appropriations. As such, the amount of money available is subject to congressional decisions and can vary from year to year, and often is influenced more by political factors than agency needs. For example, direct program expenditures in 1998 dollars for National Forest System land and resource management varied from 0.1 percent to 40 percent annually between fiscal years 1980 and 1997, fluctuating 12 percent annually on average (OMB 1982-1998). By contrast, funds used to oversee the sale or use of resources (e.g., timber or recreation) are both discretionary and mandatory appropriations. Some mandatory appropriations are used for restoration needed as a result of the sale of forest resources. Timber sale receipts deposited to the Forest Service Knutson-Vandenberg Fund, for example, pay for reforestation of the sale area.

Consistent annual funding levels would give public land managers more confidence that resources will be available for annual expenditures, as well as for multi-year ecosystem restoration projects. In addition, incentives for diminishing natural assets (quantity or quality) should be eliminated to ensure that restoration decisions are based on ecological rather than financial considerations. To achieve these goals, Congress could annually appropriate a constant level of funding specifically for ecosystem management. These funds could represent an increase over total agency funding and/or a reallocation between agency programs that better reflects legislatively delegated agency stewardship responsibilities. Money would be appropriated from the Treasury, with the amount (indexed to inflation) legislatively directed. One example is the Forest Service's Northern Spotted Owl Guarantee Fund, which provides payments to Pacific Northwest counties affected by decisions related to conservation of the northern spotted owl (USDA FS 1998). Appropriations are mandatory, with a guaranteed annual minimum. Direct funding of ecosystem management and restoration activities would then allow the elimination of special accounts funded by receipts from resource sales. Future receipts would be deposited to the Treasury. Existing fund balances could be consolidated and transferred to an interest-bearing account, with the proceeds used to further investments in ecosystem restoration.

Land stewardship contracting–Stewardship contracts can help achieve management goals on public lands while promoting the long-term involvement of community members. This innovative contracting mechanism, which evolved from the service contract, has been tested on national forests since the early 1980s. Under these contracts, stewardship objectives such as reforestation, wildlife habitat improvement, and soil and water conservation are performed by a contractor paid from appropriated funds. Recently included as key components are local small business participation, alternative commodity sales strategies, and site-specific planning by local stewardship councils (see Ringgold 1998 for case examples describing existing land stewardship projects). Although materials removed in the process of achieving stewardship objectives are considered by-products, the opportunity may exist in these contracts to capture their commercial value. These materials could help offset the cost of ecosystem management services while helping to supply new and growing markets for specialty materials.

International trade reform–An increasing percentage of timber sales in the United States now include requirements that the timber operator achieve specific ecosystem goals such as leaving wildlife trees, removing small trees to change the stand structure, and reducing historic sources of sedimentation. The bid price for the right to harvest timber and the final production cost now reflect both the cost of the product and an investment made in the ecosystem itself. This timber must compete in the open market against that produced in countries without mandatory forest ecosystem investments. This additional financial pressure may promote unsustainable practices that ultimately affect ecosystem health and resiliency (Nyland 1992; Wagner, Luzadis and Floyd 1999). One approach to ensuring that the desired ecosystem investments are made in all forests producing wood for the U.S. market would be to assign a fee on wood not produced by sustainable operations. The collected fees could then be distributed to forest product producers for ecosystem investment.

Steady harvesting levels on public lands–An important aspect of investing in ecosystems is the harvest and sale of wood products. Steadying harvesting levels would allow for better planning and decision-making for both natural resource managers and forest products businesses. In addition, rural community development and ecological goals could be supported through a program that required a minimum percentage of the sustainable harvest be made accessible to small businesses through a substantial (50-75 percent) Small Business Administration component. Forestry restoration work could be part of such a program and could create jobs at manufacturing facilities, improve forest health and watershed condition, and increase opportunity for recreational activities by reducing the chance of catastrophic fires and excessive sedimentation.

State leverage financing–Much of the emphasis on gaining community-based ecosystem management has focused on federal intervention, overlooking state financing mechanisms available through economic development assistance. All states have departments of economic development which leverage surprising amounts of state and federal dollars for economic development projects. Leverage partners at the federal level include the Department of Housing and Urban Development, through Community Development Block Grants, and the Environmental Protection Agency. These funding entities rank projects according to ability to provide matching grants, initiation of public-private partnerships, and the increasing of community values and benefits. For example, using a set-aside amount allowed under Community Development Block Grants, funding could be granted for projects that promote local employment and economic restructuring consistent with ecosystem conservation and restoration.

Leveraging these finance options using the state as both funding coordinator and one of the investors can entice private investment. An example might be combining investment in private or public land restoration and certification using an investment pool created by watershed beneficiaries including downstream water users. Manufacturer bridge-financing could be leveraged to create a community-based resource clearinghouse and log sort and value-added sales yard.

Innovative Community-Based Ecosystem Management

Communities of place are fundamental to solving the ecosystem investment problem because they are in a position to capture and reinvest the resulting benefits to sustain ecological improvements. Daly (1996) suggests maintaining each component of total capital, natural and human-made, as a tenet of strong sustainability and one viable approach to sustainable development. Communities of place have the persistence, knowledge, and skills needed to meet the long-term challenges of managing ecosystems and maintaining and improving natural and human-made capital. They are the deal makers, doers, and educators in a world of infinite possibilities.

Communities of Performance

Strengthening investment in ecosystems requires communities of place to set goals, plan, and convene the appropriate people. We refer to the relationships needed to achieve and sustain a desired outcome as a community of performance. Essentially, it is an ad hoc task group initiated by a community of place to unite people and groups with complementary knowledge, talents, and resources at all levels of influence toward a particular goal. Communities of performance also forge relationships with external groups when needed to help meet local goals. We use the term "communities of performance" to distinguish from communities of interest and to focus on performance. Communities of performance are a tool to be used by communities of place and as such are fundamentally linked to place and outcome, whereas communities of interest tend to focus on an issue. While effective process is essential to performance, the concept of communities of performance focuses on outcome (see Moote et al., this volume, for further treatment of process).

Ideally, healthy communities of place would strive for the capacity to build communities of performance that address local goals as needed. Every place faces different possibilities and pressures. Communities of performance integrate opportunities and resources in ways that suit the place and its people. They coordinate a unique mix of market, political, and governmental opportunities to fit their ecological and social realities by organizing people, knowledge, and effort to make these opportunities work. The complex and dynamic nature of investing in ecosystems suggests that communities of performance grow, change, and shift focus throughout their existence.

Communities of place are the primary architects of communities of performance for ecosystem investment. Well-developed connections and networks provide the paths that link end users to investment opportunities. Developing a plan, setting goals, and establishing the relationships necessary to perform the desired ecosystem activities are the steps that communities of performance must take to attract investment in ecosystems. They demonstrate community capacity and as such are the partners and innovators needed to leverage funding and investment needed by communities of place. Initiating and growing communities of performance will help to stabilize the long-term financing of ecosystem management.

Key features of communities of performance that make them attractive for improving ecosystem investment include:

1. Communities of performance *assume* partnerships such as public-private and urban-rural. Rather than differentiate by these features, they draw together in common goal the necessary parts to achieve the desired outcome.
2. Communities of performance focused on attracting ecosystem investments may include as essential partners small and large businesses, ecosystem workers, communities of interest, local, state and/or federal governments and non-governmental organizations. Working together through communities of performance allows for planning and innovation that draw on a wider breadth of skills than is typical if decisions are made solely by a single business, community or bank.
3. Communities of performance leverage efficiency by addressing the solution at appropriate economic, social, and ecosystem scales. Because they are created by communities of place, local and regional scales are integrated into the solution instead of the locality reacting or adjusting to decisions that are beneficial for larger entities.
4. The established relationships within a community of performance will help balance equitable, respectful processes and achievement, mediating conflict over the short- and long-term.
5. The relationships in communities of performance help build political power through networks at the appropriate spatial and temporal scales for ecosystem investment. This, in turn, flows back to maintain and improve the community of place's capacity to sustain itself.
6. With communities of performance for ecosystem investment in place throughout the United States, the potential exists to enhance the health and resiliency of ecosystems while establishing a clear role for community-based ecosystem management.

Forging effective relationships within communities that mobilize the skill, effort, capital, technology, markets, governmental facilities, and political support will advance local efforts and break through existing barriers. Local strategies, for example, may focus on opportunities to improve the economy, the environment, and the social community over time but they must also focus on relationships within larger circles of support and opportunity. Communities of performance identify these larger circles, and collect and organize what is needed to achieve tangible improvements. They also develop processes and potential strategies that fit their own place needs and possibly those of a broader range of diverse communities of place and potential investors. In other words, to achieve their ecosystem goals, communities of place and performance may find themselves working with larger, non-local groups to lower barriers to their goals. Communities of place through development of communities of performance are crucial in the process of stabilizing and growing investment in ecosystem services.

WEAVING SOLUTIONS FROM STRATEGIES

Understanding the Investment Tools

Not all financing mechanisms to stabilize, cultivate and grow ecosystem investment are the same. Communities must make the most of their potential. In dealing with dynamic strategies chosen for what they might trigger in addition to what they might do directly, planning criteria are best focused on rates of change rather than static standards. For example, considerations might include the rate of change in participation, or the rate of diversification of institutional resources, or the rate of change in productivity. Several of the investment tools presented above are characterized in Table 1 using the following categories to enhance understanding of the tools from a variety of perspectives.

Financial–Does the investment tool provide an opportunity for long-term financial stability? Is it accessible for a range of community-based ecosystem management efforts? Can the tool be used to leverage both funding and investment opportunities? Can it be used to leverage public dollars with private dollars?

Ecological–Is the tool based on scientifically defensible environmental performance criteria? Are those criteria uniformly recognized as solid environmental standards? Are those standards recognized and approved by the community? Are those environmental performance standards recognized and valued in the marketplace? What is the rate of change in productivity?

Political/Institutional–Would implementation be feasible in the short-term given existing political and institutional structures? Does the tool have an established, positive track record financially and environmentally with political and institutional organizations? Is it likely to achieve that track record within a short period of time? What is the rate of change of diversification of institutional resources? Can we respond more quickly to changing ecosystem understanding and community needs using this tool?

TABLE 1. Description of Investment Tools

Tool	Financial	Ecological	Institutional Feasibility	Equity
Innovations in Private Investment: Third-Party Certification Forest Stewardship Council Principles and Criteria	• Economic considerations are included in Principles and Criteria for Forest Management. • Long-term certification contract (five years, renewable) provides stability for forest owners and buyers • Market access: Certified Wood Product Buyers Councils connect resources with buyers. • Retailer and consumer demand is growing; financial premium until all forests are certified	• Economic considerations are included in Principles and Cirteria for Forest Managment. • Certification contract renewal provides opportunity for adaptive management • Annual certification audits provide compliance information • Price premium provides incentive for ecosystem management; forest owner has flexibility in use of funds	• Independent certifying organization is already established (FSC) • Currently limited to private lands and two state's public forests (MN and PA) • Support of industry and major environmental organizations; expansion of support needed • Regional principles and criteria are being developed	• Has been applied primarily to rural settings, but could be used in urban areas • Initial captial for certification costs may be a barrier to small forestland owners • Information about certification not yet available to all potentially qualified forestland owners • Profits locally retained if forestland owner and resource buyers are local
Carbon Sequestration Leveraged with Certification	• Cost-sharing: Investor provides initial capital for carbon sequestration application; forestland owner pays annual costs • Contract term matches certification (five years)	• Incentive for conversion to forestlands	• Applicable to private and public lands • Supported by some governments and private institutions; pilot studies needed	• Most appropriate for large, rural lands where major landuse conversions is feasible; less so for smaller areas where good ecosystem management practices occur • Profits locally retained if forestland owners and investor are local
Forest Banks	• Forestland owner receives dividends on principal deposit • Bank's return depends on management of forest	• Forestlands managed in accordance with certification principles and criteria • Incentives for conserving forestlands when owners need access to capital	• Currently implementable; private sector initiative does not require legislation • Pilot projects would expand visibility and acceptance	• Targeted to private non-industrial forest landowners • Forestlands with high commercial value likely to be favored by bank • Not suitable for forestlands to actively managed or clear-cut
Partnership Financing (Optional leverage with certification.)	• Bridge financing provides initial capital for investment and assistance in certification; later returns depend on land management • Bridge financing can be used as matching funds to leverage bank loans or government assistance • Lending institutions may perceive certification contract as indication of stability and timely repayment	• Forest management likely to cause less ecological harm with incentive for purchase and use of low-impact equipment	• Currently implementable; as private sector initiative does not require legislation • Acquisition of public funds requires working with government agencies • Pilot projects would expand visibility and acceptability	• Most beneficial to small forestland owners lacking initial capital • Profits likely to stay local if forestland owner local

TABLE 1 (continued)

Tool	Financial	Ecological	Institutional Feasibility	Equity
New Technology/New Markets Leveraging	Provides incentive for maximizing financial return from management	• Depends on initiative	• Projects may require cooperation with federal and/or state forestland managers • Requires new distribution systems and access to markets	• Applicable to rural and urban areas; large- and small-scale, public private • Success depends on access to information
Innovations in Jursidictional Boundaries: Fiscal/Ecological Zoning Watershed Organizations Forest Economic Regions Cooperation Among Counties Greenbelt/Brownfield Exchange	• Value of ecosystem services captured by investors with capacity to improve/maintain services, and beneficiaries • Discourage depletion of natural capital not in jurisdiction; stabilize markets and supply (and information) • Remove barriers to equalize property values and tax rates within ecosystems	• Facilitates the management at the broader, more relevant scale (e.g., watershed) • Provides beneficiaries of ecosystem improvements with incentive for better stewardship	• Requires cooperation between jusidiction officials: federal, states, counties, communities • Links managers and beneficiaries • Facilitates negotiation and exchange between investors and beneficiaries	• Opportunities for open space development charges/credits as sources for urban reinvestment • Applicable to public or private; small- and large-scale (but favors larger)
Innovations in Government Funding: Federal Funding	• Guaranteed stable level of funding annually	• Eliminate incentive to liquidate resources to pay for restoration and management	Requires federal/state legislation	Applies only to publicly-managed lands; large- or small-scale
International Trade (Import Fee on Non-Certified Imports)	• Financial return to domestic producers only if import process exceeds non-certified US • Magnitude of funds from import fee depends on market conditions; vary annually	• Improved forestland conditions, if funds directed to restoration	• Would require federal legislation • complies with NAFTA	• Does not provide a disincentive to purchase non-certified US lumber
Stabilization of Public Timber Supply Thinning, Small Business, etc.)	• Consistent volume for local businesses • Receipts would vary with market conditions (stumpage prices)	• Potential for improved stand structure and watershed conditions; reduced fuel loads • Predictability may encourage community investment in better stewardship	• Would require federal or state-level authorization (e.g., Congress and FS)	• Favors communities near public forestlands with local processing capacity
State Leverage Financing (Matching funds and partnerships)	Source of initial capital for investment; long-term funds from partnership	• Ecological improvements if funds effectively invested	• Cooperation with other public and require private funding sources	• Stability of funds in long-term depends on successtul negotiation

Equity–Is the investment tool affordable and available to diverse groups (e.g., rural and urban projects, large and small projects)? Can it be combined with other investment tools to reduce inequities (e.g., a short-term loan to cover initial capital costs)? Would benefits accrue to the local community? What is its potential for affecting the rate of change in concentration and dispersion of wealth?

Although far more difficult to predict, the unintended results of utilizing new, innovative investment tools should also be considered when determining "best bets" for community-based ecosystem management.

Community Enterprise: Putting the Deal Together

Throughout the United States there has been a surge of innovative community of place approaches for investing effectively in ecosystem management. These innovations are diverse. They include the development of ecological enterprises; providing the means to support people who sustain ecosystems, such as small-scale forest products harvesting, processing and sale; riparian restoration and tree-planting; fire-safing and road stabilization; and eco-tourism. These innovations include the development of stewardship institutions, i.e., the means to motivate, inform and support local efforts to protect and restore ecosystem qualities. Sometimes non-governmental organizations take on the role of catalyst or leader to help communities of place develop the capacity to create communities of performance. The Institute for Sustainable Forestry (ISF), the Applegate Partnership, and the Hayfork Watershed Research and Training Center (WRTC) are characteristic of how communities of place put together communities of performance to achieve ecosystem management and investment goals. They include a wide variety of planning, educational, conflict resolution, certification, and financial initiatives that help to develop the context for effective improvements in a specific place.

The Institute for Sustainable Forestry in California began as a highly localized effort to "certify" sustainable forestry among non-industrial forest owners, then move restoration forestry materials into a small hardwood mill and kiln so that sustainable forestry practices would pay from value-added processing and marketing. As they worked, the members began to realize the immense problems they faced in marketing, in finding investment financing for the mill, and in inventory development. ISF also realized that certification could not work on a small scale, because specialty or niche products must be marketed regionally or globally to cover costs and be profitable. The group took the lead in moving certification to the regional, national, and international scale, then began to link with large-scale industrial and large-scale environmental interests. ISF is now a member of the Forest Stewardship Council Board of Directors, operates the SmartWood certification organization, and has certified a local mill and professional forester with whom it had once been at odds. Local certified wood sources have expanded to create the basis for obtaining mill and inventory financing, and the larger dynamics of their effort have made a tremendous impact worldwide as indicated by a recent EPA grants program for sustainable development pilot projects.

The Applegate Partnership began as an effort to coordinate public and private owners in the Applegate watershed in southern Oregon. This group developed a consultative planning approach that was adopted as the model for landscape assessment under the Federal Ecosystem Management Assessment Team (FEMAT) framework. This model expanded Applegate's influence nationally; locally, it began adjusting the model to more dynamic forms that can work at a smaller community scale. Among community-based groups interested in the long-term health and resiliency of surrounding natural communities, Applegate leads regional educational, planning and research efforts by participating in the Lead Partnership Group and the Northwest Sustainability Working Group. Applegate has also developed a newsletter and reaches out to national environmental organizations for dialogue on projects and policy issues.

The Hayfork Watershed Research and Training Center in Trinity County, California began with a very loose organizational structure built solely around common interests and needs of differing groups. It quickly moved into creating diversified economic and job opportunities, developing increasing capacity to influence public agencies and foundations on a local, then national, scale. WRTC gained national influence through its leadership in the Seventh American Forest Congress in 1996 and now is central to a strategy that brings groups together around common interests at a regional scale. It participates in regional and national networking efforts. Locally, economic endeavors are beginning to take off as national attention has helped Hayfork improve return on investment in the restoration and maintenance of local ecosystems through value-added enterprises for small-diameter softwoods.

These examples illustrate communities of performance–groups of local people working together and networking with investors and beneficiaries to creatively find the tools they need to attract investment in ecosystems in their areas. Imaginative actions like these linked throughout the world through peer networking and innovative government policy will secure continued investment in ecosystems and cultivate healthy, resilient human communities and ecosystems.

CONCLUSIONS

The problems of inadequate ecosystem investment and the resulting strain on community well-being can be addressed by communities of place. As communities become more involved in ecosystem management they focus on improving investment in ecosystems, thereby strengthening the weak links of positive feedback from our economic system into ecosystems and local communities and local, value-added processing (see Figure 1). Communities of performance, ad hoc task groups, can become the focal point in this effort to meet ecosystem investment goals that are locally and regionally appropriate socially, ecologically, and economically. Using a variety of investment tools and strategies, several of which are outlined in this paper, communities can become the catalysts for establishing and strengthening the all-important positive feedback loop from economic systems to ecological systems while improving their own weak link to local value-added markets.

The financial viability of community action increases when we consider ecosystems as assets that must increase in total value over time (characterized by their health and resiliency) and be valued as producers of a wide range of goods and services. This also ensures that current production is sustainable. Using communities of performance to stimulate markets provides a revenue stream that becomes an important feedback loop by being the basis for new investment. When ecosystems produce a sustainable stream of revenue and increase the underlying asset value, they can attract more of the necessary financial capital that has not historically accompanied our greater social concern for ecosystems. Uncertainties in federal ecosystem management appropriations, as well as the market conditions for single commodities have deterred significant new investments and the development of ecosystem management skills.

As noted in the introduction and evidenced by the selected examples, flexibility is key to success. Communities must be able to establish a goal or set of goals appropriate to their ecosystems and social structures. Adjustments to broader social and economic institutions will likely be addressed along the way, as they have been with the Institute of Sustainable Forestry, the Hayfork Watershed Research and Training Center, and the Applegate Partnership. Establishing clear goals within a strong social and ecological context allows for these institutional adjustments to occur and raises questions about necessary changes in our social, political, and economic institutions to improve our overall capacity to manage ecosystems for the long-term.

The complex relationship between people and nature cannot be characterized as simply investment in ecosystems and communities. Incorporating environmental values into decision-making must be coupled with a commitment by each generation to provide to the next necessary and sufficient natural resources (Howarth and Norgaard 1992). Communities can respond innovatively to these challenges, the combined problems of under-valued ecosystem services and the need for local employment and economic restructuring, helping to put into action this commitment. It is through this type of community-level action that dynamic, creative, and integrated solutions to ecological and economic challenges will be discovered. Through commitment to future generations and innovative investment strategies, we can blaze the trail into the new territory of long-term economic and ecological well-being.

REFERENCES

Arrow, K. 1965. Criteria for social investment. *Water Resources Journal*. 1(1):1-8.

Case, K.E. and R.C. Fair. 1996. *Principles of microeconomics*. Fourth Edition. Prentice Hall, Inc: Upper Saddle River.

Daly, H.E. 1996. *Beyond Growth: the economics of sustainable development*. Beacon Press: Boston.

Daly, H.E. 1977. *Steady State Economics*. W. H. Freeman: San Francisco.

Environmental Business News. 1997. Wood products certification: a progress report. Vol. 6, No. 10. Brattleboro, VT.

Hall, C.A.S., C.J. Cleveland, and R. Kaufmann. 1986. *Energy and resource quality: the ecology of the economic process*. John Wiley & Sons: New York.

Howarth, R.B. and R.B. Norgaard. 1992. *Environmental valuation under sustainable development*. American Economic Association Papers and Proceedings: *Economics of the Environment*. 82(2):473-477.

Mater, C. and S. Mater. 1998. The Business of Sustainable Forestry. MacArthur Foundation Case Studies: Lumby Log Sort and Sales Yard. There are several.

Mater, C. 1998a. Project report for public lands certification pilot projects in the United States. Mater Engineering, Ltd. Corvallis, OR.

Mater, C. 1998b. Grand Canyon Trust business plan. Mater Engineering, Ltd. Corvallis, OR.

Mater, C. 1997a. The Nature Conservancy Clinch Valley Forest Bank business plan. Mater Engineering, Ltd. Corvallis, OR.

Mater, C. 1997b. Hayfork Watershed Center business plan. Mater Engineering, Ltd. Corvallis, OR.

Office of Management and Budget (OMB). 1982-1998. Budget of the United States Government: Appendix, Fiscal Years 1983-1999 (annual). Washington, DC: Executive Office of the President. [note: this source is actually 17 separate volumes, one for each year, all with the same title.]

Nickens, E. 1998. A watershed paradox. *American Forests*. 103(4):21-24.

Nyland, R. 1992. Exploitation and greed in Eastern hardwood forests. *Journal of Forestry* 90(1):33-37.

Ringgold, P.C. 1998. Land stewardship contracting in the National Forests: a community guide to existing authorities. Washington, DC: Pinchot Institute for Conservation.

Sampson, N. 1998. Personal communication with C. Mater.

Ulman, N. 1997. Going green: a Maine forest products firm prospers by earning eco-friendly label. *The Wall Street Journal*. November 26, 1997.

United States Department of Agriculture Forest Service (USDA FS). 1998. FY 1999 budget explanatory notes for the committee on appropriations. Washington, DC.

Wagner, J.E., V.A. Luzadis, and D.W. Floyd. 1998. A role for economic analysis in the ecosystem management debate. *Landscape and Urban Planning* 40:151-157.

PART THREE: INDIVIDUAL PAPERS

Inclusive Community Forest Management: Lessons from Mali, West Africa

Rebecca J. McLain

SUMMARY. Proponents of community-based natural resource management often use definitions of community that implicitly, if not explicitly, favor resident forest users over migrant forest users. This paper explores the shortcomings of the "fixed-in-place" model of community, using examples from ongoing community-based management projects in Mali. The author then summarizes strategies that these projects are using to expand migrant forest user participation in decisonmaking. The paper ends with a brief discussion of how these experiences in West Africa can enrich community-based natural resource management efforts in the United States. *[Article copies available for a fee from The Haworth Document Delivery Service: 1-800-342-9678. E-mail address: <getinfo@haworthpressinc.com> Website: <http://www.HaworthPress.com>]*

Rebecca J. McLain is affiliated with the Institute for Culture and Ecology, 7337 20th Avenue NE, Seattle, WA 98115.

The author thanks Beverly Brown and Jonathan Kusel for their encouragement and review comments that made this article possible.

[Haworth co-indexing entry note]: "Inclusive Community Forest Management: Lessons from Mali, West Africa." McLain, Rebecca J. Co-published simultaneously in *Journal of Sustainable Forestry* (Food Products Press, an imprint of The Haworth Press, Inc.) Vol. 13, No. 1/2, 2001, pp. 195-203; and: *Understanding Community-Based Forest Ecosystem Management* (ed: Gerald J. Gray, Maia J. Enzer, and Jonathan Kusel) Food Products Press, an imprint of The Haworth Press, Inc., 2001, pp. 195-203. Single or multiple copies of this article are available for a fee from The Haworth Document Delivery Service [1-800-342-9678, 9:00 a.m. - 5:00 p.m. (EST). E-mail address: getinfo@haworthpressinc.com].

KEYWORDS. Community-based forestry, natural resource management, international forestry (Mali), migrant forest users

The rhetoric of community-based natural resource management gained a strong foothold in much of French West Africa during the late 1980s as it became increasingly apparent that there were weaknesses in the strong centralized state management of natural and plantation forests. By the early 1990s, a variety of government and nongovernmental organizations (NGOs), both international and local, were engaged in translating that rhetoric into practice. This movement was part of a larger political trend in West Africa toward democratization, including the strengthening and creation of local-level governance structures (Thompson and Coulibaly 1995). Community-based management proponents argued that the prevailing state management regimes were major contributors toward the ecological degradation and poverty that exists throughout rural West Africa. They advocated the transfer of management authorities to local communities as a way to increase stewardship incentives for local forest users and encourage management responsive to local conditions and needs.

The arguments used in favor of community-based management in West Africa a decade ago bear a striking resemblance to those put forth in the United States today. Although there are important ecological, social, and political differences between the two areas, community-based management proponents can benefit from exploring some of the lessons learned during West Africa's decade of experience with such efforts. In the following paragraphs, I draw upon community-based natural resource management project documents and my own involvement with several innovative experiments in community-based forest management in central Mali to illustrate one key lesson: the critical importance of utilizing a broad definition of "community" that encompasses both migrant and resident forest users. I also provide some examples of community-based management projects in Mali that have incorporated migrant forest users in forest management.

CATEGORIES OF COMMUNITY-BASED RESOURCE MANAGEMENT INSTITUTIONS

Efforts to develop community-based resource management institutions in central Mali fall into three broad, often overlapping categories. In some areas, such as the western fringe of the Seno Plain adjacent to the Dogon Plateau, traditional forest police institutions continue to function even after decades of state efforts to strip them of their powers. However, most of these traditional policing systems, many of which have a religious basis, have lost important decisionmaking and enforcement powers. One strategy has been to revive these institutions by restoring some of their powers while broadening membership to account for new concerns about democratic participation in rulemaking and administrative entities (Touré et al. 1993). In other areas, such as the Samory Forest that borders the frontier with Burkina Faso, either there were no traditional systems or such systems disappeared almost entirely after Independence (Konaté and Tessougué 1996a). In these areas, community-based management efforts have supported the development of entirely new institutions, weaving new alliances across political, social, and economic boundaries. As with the revitalized traditional systems, principles of broad-based and democratic participation often underlie the new community-based institutions, at least in rhetoric if not always in practice. In still other areas, such as the forest of Boré situated at the base of the Dogon Plateau near the Niger River, community-based management has focused on developing hybrids that encompass both traditional and new institutions (Winter 1994). For all three categories, considerable attention has been paid to how these institutions function internally and how they intersect and interact with each other, and with state management systems, local NGO and private sector institutions, and the international aid and environmental communities (Winter 1995; Near East Foundation 1996; McLain et al. 1996; Diallo 1993; Konaté and Tessougué 1996a and 1996b; Touré et al. 1993).

DEFINING COMMUNITY: THE SEDENTARY VILLAGE MODEL

Community-based management in Mali accelerated rapidly after the overthrow of the Traoré dictatorship in 1991 and the installation of a multi-party democracy in 1992. The fall of the Traoré regime left a temporary vacuum in forest management as rural residents unleashed their anger on any national forest service agents who dared to enforce the country's forest laws. The forest service had a widespread reputation as a harsh and arbitrary enforcer of laws few rural residents understood and many considered inappropriate or unnecessary (McLain 1992). As efforts to fill this vacuum with community-based management systems or state-community hybrid systems got underway in the early 1990s, the topic of debate quickly became what constitutes a "local community," and therefore to whom management authorities would be shifted (Painter 1993).

Central Mali is located in the ecological zone known as the Sahel, a region where sedentary agriculturalists and migratory livestock herders have long cohabited and shared natural resources such as land, water, trees, and forage. Many early community forest proponents, often Westerners working for international NGOs or bilateral and multilateral aid organizations, assumed that villages were the logical management units for natural resources. Indeed, the bias toward sedentary populations of these early efforts at community based forestry is reflected in the term used to describe them: *gestion de terroir villageoise,* or village land use management. The tendency to identify communities with "fixed-in-place" villages was exacerbated by logistical pragmatism: people who live year-round in an area were much easier for the forest service and international NGOs to identify and contact on a regular basis. From an implementation standpoint, the fixed-in-place village model had the advantage of relying on management structures composed of people present in the area throughout the year, thus increasing the chances that critical functions, such as monitoring and enforcement, would take place on a regular basis.

Flaws in the Sedentary Village Model: The Importance of Migrant Forest Users

The community-as-sedentary-village approach, however, fits poorly with the ecological and social realities of the Sahel. In this region of highly variable rainfall and plant productivity, rural inhabitants have survived for centuries by putting together "resource portfolios" including patches of arable land, watering points, pasture lands, and small stands of wild fruit and nut trees. The spatial extent of these resource portfolios varies from a distance of a few miles for some sedentary villagers to thousands of miles for some livestock herders. Over the course of the year many types of migrants pass through central Malian forests, harvesting a variety of resources and occupying space in diverse ways and for varying lengths of time.

Transhumant cattle herders, such as the Fulani and the Fulankyrabé, who follow more or less regular routes from wet season to dry season pastures on a semi-annual basis, are perhaps the best known examples of migrant users of forest resources in central Mali. In many areas their claims to forests as browsing sites for livestock and as camping and watering points for themselves and their animals predate by centuries claims by many sedentary agricultural villages to those same resources. But the Fulani and Fulankyrabé are not the only migrant forest users. Women from villages near the Dogon Plateau travel to the Samory Forest to gather nuts and fruits on a seasonal basis, often staying with relatives in the villages near the forests. Woodcutters from distant towns and villages travel to the region's forests to cut firewood and lumber during the dry season. Young men from the villages of the Dogon Plateau have recently begun making a seasonal trek to the Samory Forest near the Burkina Faso border, using it as a feedlot to fatten goats for sale in Ivory Coast markets to the south. The word "migrant forest user" thus encompasses a broad variety of mobility strategies associated with forest resource use. These range from the daily five-mile treks a village woman makes to draw water in the neighboring village's forest to the three-month sojourn of the Dogon goat herder and woodcutter in a forest 50 miles from his home village to the 3 to 6 month-long encampments of Fulani cattle herders as they move along their annual route. In short, a diverse mix of resident and migrant users constitutes the norm for these forests.

Many of these migrant users return again and again to harvest fruits, cut wood, or pasture their animals. Some Fulani transhumance routes are centuries old, seasonal woodcutters have been operating in some areas for at least 30 years, and seasonal fruit and nut gatherers have been going to some groves on a regular basis since the early 1970s. In some cases, migrant claims to forest resources are formally recognized, such as the rights of Fulani herders to certain camping and watering sites and pasture lands. Even where migrant users' rights to resources are not formally recognized, they may still have relatively secure informal usufruct claims to the resources they utilize. Often these claims are solidified by kinship ties to residents or by the reluctance of residents to shut migrant users out since they may at some point also require access to resources located elsewhere. Reciprocity in resource access is crucial to Malian forest tenure and use systems.

Aside from the political, economic and social arguments for incorporating migrant users into community-based, there is an ecological rationale as well. Studies by range ecologists and social scientists working in dryland environments suggest that resource use strategies based on mobility and flexibility may allow people to utilize the highly variable Sahelian ecosystems in ways that are less likely to lead to long-term system collapse (Behnke and Scoones 1992; Turner n.d.).

Incorporating Migrant Forest Users into Community-Based Management Efforts

In a region where Fulani herders have long-standing claims to key resources and Fulani merchants and political leaders constitute a powerful economic and political force, organizations supporting community-based management were forced early on to consider how to incorporate migrant herders as well as sedentary agriculturalists into new management structures and processes. A brief overview follows of strategies that have brought migrant herders and, to a much lesser extent, other migrant forest users into community-based management discussions and decision-making venues.

Participatory On-Site Research to Identify Migrant Users and Their Concerns–Migrant forest users often fall through the cracks in forest management simply because their presence is ephemeral and often they camp or work in remote locations. To better understand the range of forest users in an area at various times of the year, most community-based management projects in central Mali have developed and refined intensive applied participatory research programs that are directly incorporated into ongoing management and implementation efforts. Rather than limiting the research to village visits, as was often the case in the past, project researchers now incorporate visits to seasonal encampments, watering sites, and harvesting areas for various resources and at various times throughout the year.

Migrant User Group Presence in Management Councils–Community-based management efforts in central Mali encourage herder representation within decision-making entities, regardless of whether those entities are traditional organizations that have been revitalized, such as the Ogo Kana in the Seno Plain or entirely new organizations, such as the Waldé Kelka in the forest of Boré. Finding herders to take part is easiest in areas where herding groups have established some year-round presence, even if limited, in the forest. This phenomenon has become more widespread during the past thirty years as herders seek to strengthen claims to pasture lands that are sedentary agriculturalists are appropriating for farm land. In areas where herders are not present year-round, community management councils use the Malian host family system, in which herder groups (indeed all migrant users) generally establish close links to a resident family. In such cases, the leader of the host family is considered to "speak for" his "guests."

Meeting Migrant Users on Their Turf–To increase opportunities for migrant forest users to participate in critical discussions, community-based management efforts in central Mali emphasize holding key meetings when migrant herders or woodcutters are in the area, or holding meetings in different locations to ensure nonresident users have a voice in such decisions. For example, an SOS/Sahel-sponsored community-based management project in the Samory Forest held stakeholder meetings in a town adjacent to the forest and in a town some distance away where most of the woodcutters and wood merchants lived.

Making Information Available in Appropriate Formats and Languages–In some cases migrant forest users in central Mali come from different ethnic and linguistic backgrounds than their host villagers. Community-based forestry project staff have paid considerable attention to ensuring the presence of translators at forest management meetings, and to producing informational radio broadcasts and audio cassettes in a variety of languages. At least one community-based management project evaluation (McLain et al. 1996) has cited a need for better understanding of how information flows among migrant users and between migrants and residents.

Importance of Key Players Who Understand and Respect Migrant Users–Village leaders, acutely aware of their interdependence with herders, sometimes have insisted that herders be represented in meetings to discuss forest management issues. In areas with large numbers of migrant woodcutters and fishermen they have noted the importance of including these groups as well.

The personal links of several key NGO administrators and forestry service officials to herding groups and wood merchants in the region has also heightened awareness of herder and woodcutter concerns in local and regional policy debates on community-based management.

Most efforts to incorporate migrant users into community-based management efforts have focused on Fulani herders who have a strong economic and political presence in central Mali. Woodcutters, with their connections to wealthy and powerful wood merchants in the region's towns, have also played a prominent role in some areas in forest management discussions. The incorporation of less powerful migrant users, such as Bella sheep and goat herders, young Dogon goat herders, and women fruit and nut gatherers has been much more difficult, and community forest project staff are still exploring strategies to expand their participation in forest management.

LESSONS FOR THE UNITED STATES

Central Mali's experience with community-based management provides a potentially useful alternative to the fixed-in-place model that underlies much of the community-based management literature in the United States. As in Mali, a variety of migrant forest users, ranging from hikers to loggers to mushroom pickers, inhabit and use our nation's forests. Some follow regular circuits, returning year after year to harvest products like mushrooms, beargrass, and salal. Often they develop extensive knowledge of a variety of ecosystems stretching across hundreds of miles. Others travel to the same places for a few weeks or months during the year to plant trees, conduct vegetation surveys, hike, or fish. Still others commute on weekends from distant cities to camps and cabins in the forest to ski, hunt, and pick mushrooms. All have an interest in how the forests they inhabit and use are managed; they should not be dismissed as "outsiders" with less say in forest management decisions than people who live in or near the forests year-round (McLain and Jones 1997). The difficulty in the United States, as in Mali, is how to ensure that those who do not remain conveniently "fixed-in-place" still have a seat at the table.

Many of the strategies utilized in Mali are applicable to U.S. contexts. Indeed, some are already being used here; such as providing information in multiple languages and incorporating nonresident environmental and industry groups in community-based management discussions. What the Malian experience has to teach us, however, is the importance of taking a harder look at the range of migrant users in our forests, gaining a better appreciation of their knowledge and concerns about forest management, identifying ways to meet less powerful migrant groups on their own turf, and figuring out how to develop and solidify reciprocal relationships among migrant and resident users.

REFERENCES

Behnke, R. H. and I. Scoones. 1992. Rethinking range ecology: Implications for rangeland management in Africa. Overseas Development Institute Dryland Networks Programme. Issues Paper 33. London: Overseas Development Institute.

Bocoum, Aly. 1993. Etude de cas de Boré: Décentralisation de la gestion forestière. Paper presented at the Regional Forest Code Workshop, Bobo-Dioulasso, Burkina Faso.

Diallo, Yamadou. 1993. Décentralisation foncière à Boré: Etude des instances locales et des options juridiques. Near East Foundation, Douentza, Mali.

Konaté, Aly Bacha and Martin M. Tessougué. Janvier 1996a. La gestion des ressources renouvelables dans la forêt du Samori. Volume 1: Elements de reconnaissance de la forêt du Samori. SOS Sahel/Great Britain, Bankass, Mali.

Konaté, Aly Bacha and Martin M. Tessougué. Janvier 1996b. La gestion des ressources renouvelables dans la forêt du Samori. Volume 2: Analyse Institutionelle. SOS Sahel/Great Britain, Bankass, Mali.

Konaté, Aly Bacha and Martin M. Tessougué. Janvier 1996c. La gestion des ressources renouvelables dans la forêt du Samori. Volume 3: Quelques réflexions: Que faire? SOS Sahel/Great Britain, Bankass.

McLain, Rebecca J. 1992. Recommendations for a new Malian forest code: Observations from the Land Tenure Center's study of land and tree tenure in Mali's Fifth Region. LTC Research Paper 109. Land Tenure Center, University of Wisconsin-Madison.

McLain, Rebecca, Ousmane Maiga, and Mamdou Konaté. July 1996. Environmental Protection Project (Bankass) Final Evaluation. SOS Sahel/Great Britain, Bankass, Mali.

McLain, Rebecca and Eric Jones. 1997. Challenging 'community' definitions in sustainable natural resource management: The case of wild mushroom harvesting in the USA. Sustainable Agriculture Programme Gate Keeper Series no. 68. London: International Institute for Environment and Development.

Near East Foundation. July 1996. Etude de cas: Waldé Kelka. Paper presented at the Reflecting on the Future of Natural Resources in the Mopti Region Workshop. Bankass, Mali.

Painter, T. M. 1993. Getting it right: Linking concepts and action for improving the use of natural resources in Sahelian West Africa. Dryland Networks Programme Paper 40. Drylands Programme, London: International Institute for Environment and Development.

Thompson, J. T. and C. Coulibaly. 1995. Common property forest management systems in Mali: Resistance and vitality under pressure. *Unasylva* 46:16-22.

Touré, Mahamané Dédeou, Alassané Kanouté and Ibrahim Sangaré. November 1993. Analyses institutionelles: Prise de responsabilité des institutions paysannes dans la gestion des ressources naturelles renouvelables. Etudes de cas des: Ogokana de Koro, Alamodiou de Bankass, Waldé Kelka de Douentza. Care International–Mali, SOS Sahel/Great Britain and Near East Foundation. Bamako, Mali.

Turner, M. n.d. The role of social networks, indefinite boundaries, and political bargaining in the ecological and economic resiliency of the transhumance systems of Sudano-Sahelian West Africa. Draft manuscript for Niamir (ed). The Role of Transient Institutions in CPR Management.

Winter, Mike. August 1995. Institutions locales et la gestion des ressources naturelles renouvelables: Notes méthodologiques. SOS Sahel/Great Britain, Bankass.

Winter, Mike 1994. La gestion decentralisée des ressources naturelles dans l'Arrondissement de Boré: Quelques observations. Near East Foundation, Douentza.

The Evolution of Watershed Councils and the Oregon Plan: Tempering the Populist Tradition

Lance R. Clark

SUMMARY. Revisiting the history of Oregon politics, especially environmental policy, shows that populist tradition and innovative government are tied together. Political innovation through massive reform led to the Oregon System, a governmental approach that allows a good deal of direct, or participatory, democracy. The degree of direct democracy in Oregon, or at least the threat of voters bypassing the legislature, led to a series of environmental laws passed in the 1970s. Along with similar federal laws, these set a regulatory baseline from which to build a watershed protection strategy. Key to this Watershed Health Program is the role of community-based watershed councils, which provide a good example of local, inclusive, participatory democracy. Utilizing watershed councils as the key component of a state watershed restoration strategy within the context of a regulatory baseline is an example of a hybrid centralized/decentralized approach to environmental policy. This hybrid approach is a political innovation that serves as an incremental response to competing environmental and anti-government populism. This approach will be severely tested by the 1997 Oregon Plan for Salmon and Healthy Watersheds, a state conservation plan designed to head off a listing of salmonid species under the Endangered Species Act. If this approach is successful, community-based ecosystem management groups and states that wish to copy Oregon's example will have new ideas to foster maintenance and restoration of ecosystems

---

Lance R. Clark is Resource Planning Analyst, Oregon Department of Forestry, 2600 State Street, Salem, OR 97310.

[Haworth co-indexing entry note]: "The Evolution of Watershed Councils and the Oregon Plan: Tempering the Populist Tradition." Clark, Lance R. Co-published simultaneously in *Journal of Sustainable Forestry* (Food Products Press, an imprint of The Haworth Press, Inc.) Vol. 13, No. 1/2, 2001, pp. 205-221; and: *Understanding Community-Based Forest Ecosystem Management* (ed: Gerald J. Gray, Maia J. Enzer, and Jonathan Kusel) Food Products Press, an imprint of The Haworth Press, Inc., 2001, pp. 205-221. Single or multiple copies of this article are available for a fee from The Haworth Document Delivery Service [1-800-342-9678, 9:00 a.m. - 5:00 p.m. (EST). E-mail address: getinfo@haworthpressinc.com].

through state-administered conservation plans. *[Article copies available for a fee from The Haworth Document Delivery Service: 1-800-342-9678. E-mail address: <getinfo@haworthpressinc.com> Website: <http://www.HaworthPress.com>]*

KEYWORDS. Community-based forestry, watershed councils, participatory democracy, state conservation plans

Oregon is well known for its history of political innovation, including its populist tradition. In its newest political innovation, the state in 1993 passed legislation encouraging establishment of community-based watershed councils that would implement the restoration work of Oregon's Watershed Health Program. In 1997 that idea was put to the test as watershed councils formed the basis of a formal state conservation plan to restore salmonid species.

Oregon's history of populism and innovation provided the political conditions needed to implement these new policies. The management of watersheds by councils is made possible by tempering the episodic populist fervor that historically accompanies major sociopolitical/environmental problems in the state with innovative policy mechanisms. In the case of watershed councils and the Oregon Plan for Salmon and Watersheds, comparison to environmental policy theory is useful.

Oregon's idea of state-sponsored watershed councils may be copied by other states, as so many other of its political innovations have been. Conversely, by learning from the examples of other community-based ecosystem management groups, Oregon can refine its use of these councils as the primary vehicle for watershed management policy and restoration.

THEORETICAL CONSTRUCTS FOR STATE-SPONSORED WATERSHED RESTORATION

Cortner et al. (1994) states that, "To fully understand the implications of alternative resource policies, there must be an examination of the theoretical principles which undergird resource policies. Too often the link between theory and the development of policy is overlooked" (p. 28). This paper discusses two recent political theories and compares them with policy development for the Oregon Plan for Salmon and Watersheds.

Environmental Theory

Political scientists often discuss competing theories of centralization versus decentralization to solve environmental problems (Press 1994, John

1994). These theories apply to all facets of government in the US, but the environmental arguments are only a few decades old. Early environmental theorists favored the centralized government approach. They claimed that if democracies could not marshal the foresight and consensus needed to solve problems, then perhaps centralized power would cut through delays and small-minded policymaking (Press 1994, p. 12). From a similar perspective, "the increasing scarcity will therefore create overwhelming pressures toward political systems that are frankly authoritarian (Ophuls 1976, p. 163, cited in Cortner et al. 1994, p. 29). This view seems to have been widely held in the late 1960s and 1970s, as evidenced by federal laws (and ensuing regulations) passed during that time that were centralized and prescriptive in nature.

Theorists arguing for decentralization believe local control and flexible management are key to solving environmental problems. Rodman (1980, cited in Cortner et al. 1994) advocates a decentralized and participatory society organized around local ecological conditions. Press (1994) further states that "resistance to the inertia of the status quo and the logic of existing economic and political paradigms is the key to the decentralist position" (p. 17). In contrast to centralization, decentralization clearly calls for more direct democracy.

Press also discusses hybrid regulatory/nonregulatory approaches that mix the centralist and decentralist positions. He believes the difficulties presented by the hybrid approach tend to achieve the most common denominator–the status quo. Further, these policy experiments fail when they intersect with partisan and regional legislative politics. Press cites as an example hazardous waste management authorities in California (pp. 104-105). State agency regulators repeatedly vetoed management plans that reflected regional, decentralized decisionmaking because they could not get past their own regulatory agency culture.

Press does not believe all hybrid approaches are doomed to failure and lays out three conditions for success. He claims that innovative policy and institutional designs must consider (1) how specific issues constrain policy choices, (2) how participants and nonparticipants react to technical information, and (3) how jurisdictions and rules of inclusion and exclusion can be agreed upon (p. 135).

Related to the idea of a hybrid approach is the concept of civic environmentalism, which is a bottom-up approach to environmental protection (John 1994). Civic environmentalism as described by John allows more citizen participation and state and local governmental control in environmental policymaking as an alternative or complement to federal regulation. The five distinctive features of civic environmentalism are: (1) a focus on the unfinished business of resolving nonpoint problems, pollution prevention, and protecting endangered ecosystems; (2) the extensive use of nonregulatory tools; (3) interagency and intergovernmental cooperation; (4) a search for alternatives to political confrontations; and (5) a new role for the federal government as a participant in decisions made at the state level (p. 260).

HISTORY AND POLITICAL CONTEXT

The historical and political context illustrates how the populist tradition set the stage for later innovation in Oregon government. Some argue that this tradition of political innovation has become so prevalent it is considered part of the Oregon psyche (Mason 1994).

The Rise of Populism in Oregon

In the late 1880s, politics in Oregon, as well as in most other states, was characterized by widespread political corruption. Most citizens tolerated these abuses as a by-product of the industrialization of Oregon's natural resources, which fueled growth. Nevertheless, the populist movement was about power and economic self-determination for those who did not share equally in this accumulation of wealth. Nationally, the movement was heavily invested in and suffered with the 1896 defeat of Populist Party candidate William Jennings Bryan (Thompson 1929), but the movement remained alive and well in Oregon.

Within one decade, Oregon passed Initiative, Referendum and Recall (called the populist holy trinity, also known as the Oregon System), and established a presidential primary, voter registration, and secret ballot laws (Thompson 1929, O'Donnell 1988, Burnett 1915). The initiative process, which allows citizens to introduce ballot measures for a statewide vote, is popular among Oregon citizens. When the initiative to the Oregon Constitution came up for vote in 1902, the people approved it by a ratio of 11 to 1 (Thompson 1929).

Other states saw the popular appeal of the Oregon System, and many in the West soon copied its reforms. For most of this century, Oregonians have channeled their populist sentiments through the initiative and referendum of the Oregon System.

Current use of the initiative has been problematic for the state. Political scientists are beginning to argue that the Oregon System is backfiring on a populace that does not understand the aggregate effects initiatives have on their lives. That's particularly true, they say, when tax cuts or spending programs initiated by single-interest groups severely restrict the legislature's ability to make sound decisions that balance social programs with available budgets (Mason 1994). Some believe this tool, once used to force the legislature to consider the ideas of the people, is now a tool used by affluent special interests to get their way. It seems that current use of the initiative is not really populism, but special-interest use of a populist tool. In a way, this is "fringe" populism, where interest groups use the initiative process to set a populist agenda of their own making as opposed to an agenda based on shared sentiment of the majority.

Early Conservation in Oregon

In one of Oregon's first conservation triumphs still celebrated today, Governor Oswald West's 1913 bill declared the state's beaches public highways in order to protect them from development (Walth 1994, Mason 1994). Conservation in the early part of this century was far different than it is today. Early conservationists such as President Theodore Roosevelt (1901-1909) and Gifford Pinchot (first Chief of the US Forest Service, 1905-1910) believed that scientific management could marry industrialization of forests, farms, and streams with protection of natural resources (Botkin 1990, Robbins 1997). According to Robbins (1997), "conservation ideology preached virtues that were consistent with the modernizing world of industrial capitalism: efficiency, the elimination of waste, and the development and scientific management of resources" (p. 239).

From the WWI era through President Franklin Delano Roosevelt's New Deal, populist sentiment in Oregon waned. Building the Oregon economy through industrialization and "engineering" nature took center stage. The centerpiece of this effort was the taming of the Northwest's great rivers for hydropower, flood control, and agricultural development (Robbins 1997). Ironically, the need to reverse the effects of this aggressive taming of natural systems probably led to the state's environmental ethic and is a primary reason why restoration-oriented watershed councils exist today.

Water pollution was one of the first environmental problems addressed by Oregon's citizens. As to raise awareness, newspapers early as 1910, and for decades after lamented the deplorable condition of the Willamette River. Reports described lumber and paper mills dumping untreated waste directly into the river a few miles upstream from Portland (Walth 1994). Through most of the first half of this century, many state and citizen-based attempts to control water pollution in Oregon were defeated or diluted by powerful development interests.

Adler (1995) has tracked the history of federal attempts to manage the complexities of water planning and watershed protection. Echoing the taming nature theme, Adler says, "In part, this history is explained by the prevailing Progressive era view that we could build our way out of almost any problem, with engineers revered in American society as only rock stars and sports heroes are today." Understanding Progressive era ideas–particularly the unintended consequences of those ideas–helps to explain why watershed restoration is such an important issue today. A critique of federal flood control programs states, "The massive number of dams, reservoirs, and impoundments, and thousands of miles of levees and channelization projects have severely impaired the hydrological and biological integrity of the nation's rivers, streams, lakes, and estuaries" (Reuss 1993, cited in Adler 1994). The Pacific Northwest is a textbook example of the effects of such policies.

Still, Adler (1994) argues that the U.S. turned the corner on watershed protection with passage of the Clean Water Act and related environmental protection statutes, regional watershed protection programs, and federal land management statutes. A 1973 report of the federal Water Resource Council (established in the Water Resources Planning Act of 1965) made recommendations now being implemented in Oregon (Adler 1994). Some of those recommendations were: priorities should shift from water resources development to restoration; water resources planning should be more closely tied to land-use planning; and the development, management, and protection of water resources should be controlled by the level of government closest to the problem and most capable of representing the interests involved.

Environmental Populism

Throughout the country in the 1960s, race, war, women's rights, and the environment spawned a new populism. Post-WWII optimism was slowly replaced by a society that began to question both its social values and the sustainability of engineering nature.

As the unintended effects of building the Pacific Northwest economy began to emerge in the late 1960s, an environmental populism emerged in Oregon. As the timber industry responded to the post-WWII-housing boom, rapid timber harvesting ensued along with the associated impact on the landscape. The Willamette River became increasingly polluted as industrial growth exploded new residents swarmed into the Willamette Valley after the war.

Predictably, Oregon always had a land ethic, but secondary to commerce as natural resources dominated the regional economy from the time of white settlement until the 1980s. Although early conservation efforts are noteworthy, Oregon's environmental ethic really took root during the late 1960s and early 1970s.

In 1966 Oregon elected a maverick Republican governor, Tom McCall, who made his name as a Portland television newsman. In the, his environmental leanings were apparent from a series of early 1960s "60 Minutes"-style documentaries on pollution in the Willamette River. Dealing with a less than reform-minded legislature, McCall took environmental issues directly to the people (much the same as original populists acted in spite of the legislature), sometimes coalescing populist fervor, sometimes creating it. Interactions between public opinion and legislative compromise were complex. With others' help, McCall led the charge as Oregon passed a wave of innovative environmental legislation. A 1967 beach bill increased protection of Oregon beaches. The relatively weak and ineffective Oregon Sanitary Authority, which dealt mostly with water pollution, was upgraded to a Department of Environmental Quality and given strong enforcement powers to address a wide range of environmental issues. The 1973 bottle bill initiated a national wave of recycling. Finally, a controversial land-use planning law passed in 1973 that is still is considered a national precedent in land-use planning (Mason 1994, Walth 1994).

Typical of that era, Oregon's environmental laws represent official regulatory, centralized approaches to environmental protection. During this same period the federal government passed such regulatory environmental laws as the Clean Water Act, the Clean Air Act, the National Environmental Policy Act, and the Endangered Species Act. One Oregon environmental law moved in another direction, however, and it paved the way for future watershed councils.

Oregon's land-use planning law, enacted primarily to preclude unrestricted development of forest and farmland, represented a whole different way of thinking. The law did not focus on centralized regulation. Though the law set clear statewide goals and gave a state commission and appeals board veto power, it handed off authority for site-specific planning to local governments (Oregon Department of Land Conservation and Development 1995). Watershed councils in Oregon operate much like that land-use planning law. Some decision making is passed on to community groups, while the state determines statewide goals and restoration priorities.

MODERN OREGON ENVIRONMENTAL POLICY AND THE RISE OF WATERSHED COUNCILS

In the 1990s, political thinking in Oregon further explored the concept of the state and local governments sharing power and responsibility as expressed in the 1973 land-use planning law. That power sharing represented an early attempt to address environmental issues while bridging the urban/rural split that Mason (1994) claims is a fundamental characteristic of Oregon politics. Urban residents tend to favor government spending for education, welfare, and mass transit, while rural citizens tend to favor it for agriculture and highways. This comparison is important in natural resource issues because state government is usually identified as having an urban perspective and local governments a rural one.

The Establishment of Watershed Councils

In 1987, the Governor's Watershed Enhancement Board (GWEB) was created to provide lottery-funded state grants for fish habitat-improvement projects (Oregon Watershed Health Program 1995). GWEB would undergo many changes over a few years.

In 1993, the legislature passed a law creating state-sponsored watershed councils, an expansion of GWEB, and enhancing a state interagency management focus into the comprehensive Watershed Health Program. In 1995 the role of watershed councils expanded and program management was transferred to GWEB, building on its successes of GWEB and those of earlier state agency programs that used volunteers for fish habitat-improvement projects. It has been suggested that the rationale of the Governor and the legislature for creating the new program was partly due to evolving public policy, and, more importantly, a fear of looming endangered fish listings in Oregon's south coast streams and in the northeast corner of the state.

By legislative definition, a watershed council is a locally organized, voluntary, nonregulatory group established to assess the condition a watershed and create a plan for enhancement and protection activities. The council becomes official when a local government such as a county Board of Commissioners recognizes it (Oregon Watershed Health Program 1995, app. G). Official recognition makes the councils better candidates for federal and private grants (GNRO 1997). Joint recognition by the state and counties is another indicator of this power-sharing "partnership" between state and local governments.

Local watershed councils tackle a number of issues such as fish and wildlife habitat; fish passage; water quantity and quality; land use planning, including rural and urban development; recreation; and resource extraction. There are currently 81 watershed councils in Oregon (GNRO 1997). Watershed councils are intended as a balance of perspectives, so their membership could include private citizens and landowners as well as representatives of local government, state and federal agencies, nongovernmental organizations, commissions, tribes, public interest groups, industry, and the academic, scientific, and professional communities. The councils represent the community's interest and help build trust, both within the local communities and between the communities and governmental entities (Oregon Watershed Health Program 1995).

Policy Adjustments in the Kitzhaber Administration–Governor John Kitzhaber, elected in 1994, brought a new political perspective to his office. As president of the state senate for much of the 1980s, Kitzhaber was well known for seeking collaborative, bipartisan solutions. Noted statewide for his environmentalism, he also had specific ideas about using innovative collaborative solutions to solve environmental problems.

Kitzhaber catalyzes environmental populism's energy and interest of to facilitate state approaches to resource management. With an approach hinting at direct (or participatory) democracy, community-based watershed councils are highly compatible with his philosophy of a government that utilizes numerous citizens' advisory panels and task forces. His reliance on citizens supplements the expertise of state agency staff, builds credibility for his policies, and gives Oregonians a sense of empowerment.

Kitzhaber's collaborative style is important in 1998. It can be argued that due to the complexity of federal laws and modern politics, former Governor Tom McCall's mastery of Oregon populist sentiment could not work today. Government and political organizations in Oregon are larger, more complex, and more contentious. In addition to the difficulty of mastering the political media machine, leaders today must also wade through a maze of federal regulations to make significant policy changes that are broadly supported by Oregon citizens.

Watershed Councils as a Political Force–The Oregon Plan for Salmon and Watersheds (formerly known as the Oregon Coastal Salmon Restoration Initiative, hereafter referred to as the Oregon Plan) was written primarily in an attempt to head off the listing of coho salmon as a threatened species along Oregon coastal streams. Under the Endangered Species Act (ESA), federal agencies issue regulations to prohibit the "taking" of a species but cannot mandate habitat restoration. Landowners and local governments that support the Oregon Plan generally want to enhance fish habitat but fear additional regulations or fines if they unintentionally violate the ESA. Additionally, listing regulations could affect the land's production of commodities or the flow of water that supports agriculture, thereby endangering the livelihoods of private landowners.

After the plan was introduced, the National Marine Fisheries Service (NMFS) decided not to list coho under the ESA for the upper two-thirds of the Oregon Coast–the Oregon Coast Ecologically Significant Unit (ESU–a unit used in ESA regulations to identify discrete populations of threatened or endangered species). NMFS cited as its rationale the belief that coho would not become endangered in the interval between its decision and the adoption of improved habitat measures under the Oregon Plan (Oregon Natural Resources Council v. Daley 1998).

Coho was listed in southern third of the state, primarily because two-thirds of that ESU was in California, which has no state coho conservation plan. In southern Oregon, NMFS will use the Oregon Plan as the required species recovery plan.

The state committed $15 million to the Plan in the first two years, matched by $15 million from the timber industry from an industry-endorsed timber harvest tax. The industry also pledged in-kind support in the form of voluntary conservation measures on private lands. The tax money was approved unusually early in the legislative year, due to hints from NMFS that it would list the coho without some guarantee of conservation funding.

In March 1998, the Oregon Plan was expanded to include steelhead, and other fish are expected to be added in the future, especially east of the Cascade Mountains.

The Role of Watershed Councils–Voluntary watershed councils can be found in the northeast and the Lake States (Doppelt et al. 1993). Literally hundreds of cooperative watershed-based ventures have been springing up throughout the West over the past few years (Getches 1996). What is unique in Oregon is the state legislature encouraged sponsorship of the coordination of voluntary efforts to restore threatened species and avoid or mitigate listings under the ESA (GNRO 1997).

The Oregon Plan's comprehensive management and funding directives establish restoration priorities within and above a regulatory baseline. That baseline encompasses laws such as the Oregon Forest Practices Act (the 1994 revision of the Act included tougher streamside rules to protect riparian and aquatic habitat) (Oregon Department of Forestry 1994, GNRO 1997). The Oregon Plan consists of four essential elements: coordinated agency programs; community-based action; monitoring; and appropriate corrective measures. This is clearly spelled out in a summary known as the "Oregon Approach":

> In contrast to many endangered species recovery plans that rely primarily on regulatory approaches, this plan represents a new way of restoring natural systems . . . the Oregon Approach. This approach meshes scientifically sound actions with local watershed-based public support. It relies on teamwork among the various levels of government and is dependent on monitoring and accountability for results. Strong enforcement of existing laws and regulations are a foundation upon which voluntary and cooperative actions can be built. We believe that this is the only approach–one that will generate the support and commitment across all sectors, from landowners and industry to government agencies–to restore salmon and their natural systems. This plan will require an unprecedented level of cooperation and coordination among local, state, and federal agencies. It represents the commitment of all Oregonians to the fish, the watersheds, and our children. (GNRO 1997)

The Oregon Plan also relies heavily on federal conservation plans. The aquatic conservation strategy and the commitment to monitoring in the federal Northwest Forest Plan provide a cornerstone to the Oregon Plan (GNRO 1997). Similarly, as the Oregon Plan expands to address fish issues in the eastern part of the state, the multi-state Interior Columbia Basin Ecosystem

Management Project is expected to serve as a cornerstone to conservation efforts east of the Cascade Mountains.

Conceptually, watershed councils drive the Oregon Plan. The Plan recognizes that the conservation and restoration of fish and other aquatic and riparian species must be accomplished by local communities and landowners, who have local knowledge of problems and ownership in solutions. The plan further acknowledges that state and federal agencies will provide the regulatory foundation and technical support for the watershed councils, but the bulk of the work needed to conserve and restore watersheds will be accomplished by local people (GNRO 1997).

Judging Progress of the Oregon Plan–It is far too early to determine whether the Plan is a success. While some councils have been operating for years, have completed watershed assessments, and are aggressively implementing restoration projects on the ground, others are barely past the initial organizational stages. Some councils are a model of citizen cooperation, while others have failed to make notable progress due to distrust and disagreement between environmental groups, commodity group interests, and landowners over traditionally contentious natural resource issues. Some councils lack broad-based representation from stakeholder groups (GNRO 1998). In others, some natural resource professionals are hesitant to accept the decisions of ordinary citizens, concerned that councils dominated by laymen will become "organized stupidity" (Barsotti 1998).

The plan will require considerable patience. The first year's work was mostly procedural, so few treatments on the ground can be monitored. Current habitat conditions indicate that it will take many decades for riparian conditions to develop at desirable levels for salmon freshwater habitat (GNRO 1998).

Political problems on the horizon may unravel the Oregon Plan. Although NMFS has endorsed the Oregon Plan, it has given the state only three years to show results and has reserved the right to negotiate with the state over changes in the Oregon Forest Practices Act.

A recent federal court decision (Oregon Natural Resources Council v. Daley 1998), currently under appeal, barred NMFS from not listing coho salmon as threatened solely on the promise of an unproven state conservation plan. The magistrate's decision interpreted the ESA to read that NMFS, when making a listing decision, could not consider voluntary restoration measures and/or the prospect of a future conservation plan (the Oregon Plan, which was not yet fully implemented).

The 9th Circuit Court of Appeals rejected a motion to postpone the requirement for NMFS to reconsider its listing decision by August 3, 1998 until after the appeal was heard. As a result of the two decisions, on August 3

NMFS listed the coho salmon as threatened under the ESA. NMFS's "conditional" decision not to list is moot since the court decisions forced a listing.

The Oregon Plan funding package approved by the state legislature in 1997 included a private timber tax (estimated at $15 million/biennium) would be rescinded if coho were listed. The industry wanted leverage to persuade the state, and indirectly, NMFS, to stick to a voluntary restoration plan. NMFS is expected to use the Oregon Plan as the basis of a recovery plan as mandated by the ESA. However, the lack of funding may seriously jeopardize the Oregon Plan's viability as the cornerstone of a recovery plan. On a positive note, soon after the listing occurred the Oregon Forest Industries Council announced it would try and convince the legislature to reinstate the timber tax in spite of the listing. This suggests the industry is heavily invested in the state's plan, and less concerned about the potential economic pitfalls of an ESA listing, which prompted the "rescission" language in the 1997 legislative session.

THEORY AND THE OREGON PLAN

The Oregon Plan, as implemented through community-based watershed councils, has a chance for success. The plan meets most of Press and John's criteria for innovative policy. In short, the authors of the Oregon Plan have done their homework in assembling a detailed but flexible plan that considers a broad spectrum of issues related to fish recovery and overall management of watersheds.

Whether deliberate or not, the Oregon Plan addresses many of the theoretical issues or dilemmas associated with modern resource management. In theory and practice, the Plan makes a strong case for utilizing a combination of regulatory and nonregulatory approaches to resource management. It also demonstrates how state leadership can facilitate flexibility between centralized regulation and local control.

The Oregon Plan seems to meet Press's criteria for a hybrid centralized/decentralized resource policy. The Plan for the most part discusses how specific issues constrain policy choices by the Plan's comprehensive, strategic nature. Through its use of an independent scientific advisory committee, the full weight of scientific capabilities of state agencies, plus a strong education component, the plan anticipates how participants and non-participants will react to technical information. Finally, the Plan considers how jurisdictions and rules of inclusion and exclusion can be agreed upon using examples from its flexible collaboration between citizens and state and local governments. In sum, most of Press's criteria for a hybrid centralized/decentralized resource policy are met because the Plan's authors spent a great deal of time on strategic thinking. In other words, the state did its homework in formulating this policy.

The Plan also seems to match up well with John's five criteria of civic environmentalism due to its focus on the "unfinished business of resolving non-point problems, pollution prevention, and protecting endangered ecosystems" through various mechanisms. Nonpoint source problems are being addressed by the state through a parallel, complementary effort to improve "water quality-limited" stream segments as mandated under the Clean Water Act. Pollution prevention is addressed by including the participation of all river users descending from the upland federal forests (usually on Forest Service land), to upland industrial forests, to valley farmers and ranchers, to urban dwellers and homeowners, to sport and commercial fishermen in the bays and ocean. Endangered ecosystems are protected through holistic ecosystem planning. In this case, the Oregon Plan complements the federal Northwest Forest Plan implemented to conserve the habitat of spotted owls and other late-successional forest-dependent species. The extensive use of nonregulatory tools is highlighted in the Plan, as is interagency and intergovernmental cooperation. The Oregon Plan, an alternative to traditional natural resource political confrontations, proactively involves citizens in restoration work rather than merely allowing them limited opportunities to comment on agency plans. In addition, the plan offers citizens the opportunity to influence the actions of private landowners who traditionally do not seek the community's advice. The plan involves the federal government in a new role as a participant in decisions made at the state level. Instead of reacting through formal channels, under the Oregon Plan NMFS and other federal agencies are encouraged to work informally and collaboratively with state representatives. This is intended to be synergistic, as relations between federal regulatory agencies and state agencies are traditionally strained, cumbersome, and time-consuming.

The state of Oregon has found a middle road between voluntary conservation approaches and command and control regulation, which the state believes will lead to better natural resource management, more efficient government, and local empowerment. The ideas that relate to the protection and restoration of aquatic habitat and large watersheds have shown progressed constantly over the past 15 years, but the ideas have taken off and found more celebrity and political endorsement in the current state administration. Some of this change is due to leadership, some to incremental differences in the management philosophy of state and federal agencies, and some to a conscious strategy to avoid the regulatory burden of a species listing under the ESA.

The substance of the most recent Oregon environmental doctrine can be found in a recent policy position of the Western Governors Association (WGA 1998), which states:

> The old model of command and control, enforcement-based programs is reaching the point of diminishing returns. Successful environmental policy implementation is best accomplished through balanced, open, and inclusive approaches at the ground level, where interested public and private stakeholders work together to formulate critical issue statements and develop locally-based solutions to those issues. Collaborative approaches often result in greater satisfaction with outcomes, broader public support, and lasting productive working relationships among parties. Additionally, collaborative mechanisms may save costs when compared with traditional means of policy development, and can lessen the chance that an involved party will dispute a final result.

Many environmentalists, among others, believe the Oregon Plan inappropriately hands power off at the local level to citizens who may be more interested in economics than environmental stewardship. However, the WGA doctrine suggests that sharing power while maintaining existing environmental regulations will yield greater environmental stewardship than by regulation alone.

Through its salmon plan, Oregon is acting on the words of the WGA. Since this resolution has been adopted, it seems likely other western states will follow Oregon's example. Conversely, Oregon will likely follow specific policy examples from other western states that contributed to the creation of the resolution.

CONCLUSION

This paper has tried to demonstrate how states are best positioned to manage broad-scale conservation plans. The Oregon Plan is designed to be the most efficient watershed restoration strategy ever attempted. But to do that the state must manage some of the populist traditions that have fostered an atmosphere of political innovation in Oregon.

It can be argued that the Oregon Plan takes an incremental approach in responding to populist demands for reform. This response is probably tied to both the environmental populism of many of Oregon's citizens (Mason 1994) and conservative "anti-establishment" or anti-government populism evidenced by middle-class voters embracing of property tax limitations that severely curtail funding for state and local governments (Schneider 1987, Mason 1994). Environmental polls routinely show Oregonians are opposed to clearcutting even though their consumption of wood products continues to increase. In spite of this apparent conflict in values, citizen initiatives to modify forest practices focus on the visceral reaction to viewing forest harvesting. Similarly, citizens almost always vote for tax cuts, while expecting increased social services. An elected legislature balances the conflicting demands of society; initiatives sometimes fail to see the big picture.

Through the Oregon Plan, the state encourages citizen involvement through local participation in resource planning. This is done in the context of well thought-out state government policies instead of through indiscriminate ballot measures. Through experiments in hybrid approaches to governing (combining centralization and decentralization) in both the Oregon Plan and the legislation that helped to create watershed councils, Oregon is tempering "fringe" populism through innovative government. Although the current approach is different than the result of populist reforms that led to regulatory laws of the 1960s and 1970s, it is yet another example of the political innovation for which Oregon is famous.

The development of watershed councils through state leadership and funding is a novel approach to environmental policy. Other states are observing Oregon's progress. Some political conditions for the emergence of watershed councils are unique to Oregon and may not apply to other states, but imitating a program that works may not require similar political conditions–it may simply require enough political will to take a chance.

Conversely, Oregon is not the only state to experiment with watershed councils or other types of community-based ecosystem management. The state has made it clear that its resource policy is intended to be flexible and adaptable. Oregon clearly should look for examples from other states or countries to develop better policy models for community-based groups.

Watershed councils, by their very nature, tend to focus on riparian and aquatic habitat and fish survival at the expense of upland ecosystems. Other community-based projects that focus on upland forest and range ecosystems will allow Oregon to improve and expand the watershed council model without sacrificing the inclusive local nature of the councils. However, that idea is a topic for future study as the Oregon Plan evolves.

REFERENCES

Adler, Robert W. 1995. Addressing barriers to watershed protection. *Environmental Law* 25(4): 973-1106.

Barsotti, Mike. 1998. Watershed councils: Oregon's solution for today's resource concerns. *Northwest Woodlands* 14(1): 16-17. Portland, OR: World Forestry Center.

Botkin, Dan. 1990. *Discordant harmonies: a new ecology for the 21st century.* New York, NY: Oxford University Press.

Burnett, James D. 1915. *The operation of the initiative, referendum, and recall in Oregon.* New York: McMillan Co.

Burton, Robert E. 1970. *Democrats of Oregon: the pattern of minority politics 1900-1956*. Eugene, OR: University of Oregon Books.

Cortner et al. 1994. *Institutional barriers and incentives for ecosystem management: A problem analysis. Issue Paper #16*. Tucson, AZ: Watershed Research Resource Center.

John, DeWitt. 1994. *Civic environmentalism: alternatives to regulation for states and communities*. Washington, DC: Congressional Quarterly, Inc.

Doppelt, Bob et al. 1993. *Entering the watershed: a new approach to save America's river ecosystems*. Washington, DC: Island Press.

Getches, David H. 1996. Changing the river's course: Western water policy reform. *Environmental Law* 26(1): 157-174.

GNRO (Governor's Natural Resource Office). 1998. *The Oregon Plan for Salmon and Healthy Watersheds: 1998 annual report summary*. Salem, OR: GNRO.

GNRO. 1997. *Oregon Coastal Salmon Restoration Initiative (OCSRI) Conservation Plan*. Salem, OR: GNRO.

Mason, Thomas L. 1994. *Governing Oregon: an inside look at politics in one American State*. Dubuque, IA: Kendall/Hunt Publishing Co.

O'Donnell, Terrence. 1988. *That balance so rare: the story of Oregon*. Portland, OR: Oregon Historical Society Press.

Ophuls, William. 1976. Reversal is the law of the Tao: the imminent resurrection of political philosophy. In Stuart Nagel (ed.), *Environmental Politics*. pp. 34-48. New York: Praeger Publishers.

Oregon Department of Forestry. 1994. The Oregon Forest Practices Act Water Protection Rules: scientific and policy considerations. Forest Practices Policy Unit. Salem, OR: Oregon Department of Forestry.

Oregon Department of Land Conservation and Development. 1995. Oregon's statewide planning goals and guidelines. Salem, OR: Oregon Department of Land Conservation and Development.

Oregon Natural Resources Council v Daley, No CV-97-1155-ST (D. Or. June 1, 1998) (opinion on cross motions for summary judgment).

Oregon Watershed Health Program. 1995. *Oregon's Watershed Health Program: Volume 1*. Salem, OR: Governor's Watershed Enhancement Board.

Press, Daniel. 1994. *Democratic dilemmas in the age of ecology: trees and toxins in the American West*. Durham, NC: Duke University Press.

Reuss, Martin. 1993. Introduction. In Reuss, Martin (ed.), *Water resources administration in the United States: policy, practice and emerging issues*. Bethesda, MD: American Water Resources Association

Robbins, William G. 1997. Landscapes of promise: the Oregon story 1800-1940. Seattle, WA: University of Washington Press.

Rodman, John. 1980. Paradigm change in social science. *American Behavioral Scientist*. 24(1):49-78.

Schneider, William. 1987. The new shape of American politics. *The Atlantic*, 259:39.

Thompson, Cecil T. 1929. The origin of direct legislation in Oregon: how Oregon

secured the initiative and referendum. Master's thesis. University of Oregon, Eugene, Oregon.
Walth, Brent. 1994. Fire at Eden's gate: Tom McCall and the Oregon story. Portland, OR: Oregon Historical Society Press.
Western Governors Association. 1998. Enlibra: a new shared doctrine for environmental management. Policy Resolution 98-001. Denver, CO: Western Governors Association.

The Great Lakes Forest Alliance: Building Bridges to Raise Collective Wisdom Across Governments and Institutions

Wendy Hinrichs Sanders

SUMMARY. The Great Lakes Forest Alliance, created by charter in 1987 at the direction of the governors of Michigan, Minnesota, and Wisconsin, is a mutual aid, public/private partnership that integrates global, national and local interests by bridging the gap at a regional level. It expanded in 1997 to include Ontario. Trustees include key leaders of government and industry and citizens from a broad range of forest interests. It was designed to be as learning environment to address the resurgence of forest growth and the increasing demand for conservation, wood products and recreation. The need for the Alliance resulted in part from a perceived underrepresentation of regional forest-related issues in the national arena. The Alliance attempts to consider leading-edge strategies over the long-term in a pro-active manner, and trustees recognize the need to build respect, trust, information exchange, cooperation, coordination and collaboration among diverse interests. Among the projects that demonstrate the bridge role played by the Alliance: a regional forest resources assessment, public and private funding that supports research toward a more frequent forest inventory process, training for communities to use the collaborative learning process to address economic prosperity and environmental protection strategies and the development of sustainable forest management criteria

---

Wendy Hinrichs Sanders is Executive Director, Great Lakes Forest Alliance, P.O. Box 722, Hayward, WI 54843.

[Haworth co-indexing entry note]: "The Great Lakes Forest Alliance: Building Bridges to Raise Collective Wisdom Across Governments and Institutions." Sanders, Wendy Hinrichs. Co-published simultaneously in *Journal of Sustainable Forestry* (Food Products Press, an imprint of The Haworth Press, Inc.) Vol. 13, No. 1/2, 2001, pp. 223-236; and: *Understanding Community-Based Forest Ecosystem Management* (ed: Gerald J. Gray, Maia J. Enzer, and Jonathan Kusel) Food Products Press, an imprint of The Haworth Press, Inc., 2001, pp. 223-236. Single or multiple copies of this article are available for a fee from The Haworth Document Delivery Service [1-800-342-9678, 9:00 a.m. - 5:00 p.m. (EST). E-mail address: getinfo@haworthpressinc.com].

and indicators for the region. A continual challenge is relationships among diverse forest interests across jurisdictional and institutional boundaries in a manner that promotes exchanges that build collective wisdom. *[Article copies available for a fee from The Haworth Document Delivery Service: 1-800-342-9678. E-mail address: <getinfo@haworthpressinc.com> Website: <http://www.HaworthPress.com>]*

KEYWORDS. Community-based forestry, regional forest alliance, public-private partnership, forest conservation

Wanted: Professionals and citizens with vast, regional forest resource knowledge, experience, vision, wisdom, discretion, and the capacity for global and systems thinking with the capability to act locally. Must have credibility to make decisions on behalf of their peers in their absence and with their ensuing support. Must have integrity and be willing to be open, honest, and fair. Must be poised to take great risks to raise collective wisdom in the region. Thick skin preferred. Only those with countless time available when crises occur need apply. No pay or remuneration other than that which is intrinsic, but in-kind contributions will be greatly appreciated. Must be able to fit this demand on time and wisdom into an already overloaded work or volunteer schedule.

Some may argue such people do not exist. However, it is exactly such people with all but the "countless time available" who serve as trustees for the Great Lakes Forest Alliance (GLFA), a multinational, collaborative partnership spanning multiple levels of government and interest. It has the flexibility to evolve as needs and issues change. It was created by charter under the direction of the governors of Michigan, Minnesota and Wisconsin in 1987 and expanded in 1997 to include Ontario. The Alliance is a forum to foster and facilitate cooperative efforts that enhance management and sustainable use of the region's public and private forest lands. It develops the tools necessary to build a framework of technical expertise that supports community-based ecosystem stewardship across jurisdictional boundaries and institutional sectors. It helps to bridge the communication gap between agencies and citizens (Cortner et al. this publication). It supports sound forest management in the region, including both use and conservation.

The Alliance serves as a springboard from which communities of place and interest can thrive. To understand this framework, an understanding of the cultural context of the region is helpful. For the scope of this document, the description focuses on the United States partners, Michigan, Minnesota, and Wisconsin, unless otherwise indicated.

UNDERSTANDING THE REGION

The Forests

More than 60 percent of the forests in the Michigan, Minnesota and Wisconsin region are privately owned (GLFA 1996). Forests are in relatively good shape: Following extensive harvest in the latter half of the 1800s, major public and private regeneration efforts have resulted in a resurgence of forest growth. Climatic change and fragmentation due to the development of second homes for recreation or retirement in remote forested areas are potential threats to forest health (Stearns 1997).

The People

With a distaste for public, contentious debate and impasse, people in the region are inclined toward consensus and pragmatism. An Oneida Nation tribal leader who served as president of the Wisconsin Rural Partners vividly describes how elders traditionally sat and *listened* until consensus was reached among participants. In fact, there may be a reluctance to institutionalize local community-based ecosystem management efforts in some areas of these states since many rural leaders already follow the informal structure when participating in county and private forest land management.

Schools in the region are of high quality and serve as the backbone for a preference for education, which is viewed as a catalyst for change. There is a commitment to research, but residents want research that supports the needs of the people and their work. Outreach from the three land grant university Extension services has provided a variety of private landowner education efforts in an array of media formats. With increased recognition of the acreage held by private forest landowners, demand for landowner education has increased significantly in the region.

Timber and tourism have, for the most part, emerged as complementary rather than competitive industries, as they offer better employment opportunities than communities that have either one or the other or neither (Webster, Chappelle, and Andrews 1997). The timber industry has been rebuilt based upon forest regeneration efforts, technology and retooling that has kept pace with the changing diameter and species of harvestable trees. The tourism sector was created as logging efforts in the last century brought news of the beautiful lake and wooded areas. It has equal tenure with the timber industry and current demands for travel related to education, cultural, historic and recreation have strengthened this sector (Chappelle 1997).

What Is the Alliance?

The Alliance was envisioned in the mid-1980s by forestry professionals (namely research, academic, and government staff) who saw long-term benefits to institutionalizing cooperation and coordination across governance levels before a crisis over forest uses prevented a cooperative attitude in the Lake States. These individuals recognized that the region's forests were in a major phase of regeneration, a renaissance some say, based upon more than 100 years of restoration after extensive harvest and intense fires. They understood the increase in regional population, the increase in forest-based recreational travel, and the uncertainty of timber supply in other parts of the country. They sensed this would place increasing demand on the Lake States' forest resource.

At the same time, they felt the region to be underrepresented, lacking political visibility at the national level. The three states together have a smaller congressional delegation than California. Perceiving the potential for a crisis, they created a vehicle they thought would at least mitigate the crisis and at best yield constructive alternatives to improve the forest resource condition.

A three-state governors' conference on forest resources was hosted and the Alliance emerged. The Board of Trustees believed it might best meet the needs of the forest resource over the long-term in a proactive, nonreactive manner by focusing on teaching professionals and the public to make informed and reasoned policy choices. It recognized the need to build respect, trust, information exchange, cooperation, coordination, and collaboration among diverse forest interests (Shands 1988).

Trustees represent the following groups: USDA Forest Service; the Canadian Forest Service; each state or province's chief forester; and gubernatorial/ministerial appointees as well as citizen representatives of tourism and timber industries; the universities; research, community, and economic development organizations; private woodland owners; conservation groups; and local government (GLFA 1997). The Alliance Board of Trustees meets three times each year in a forum based on an emerging issue featuring as speakers regional experts who reflect diverse viewpoints. Several hours of discussion ensues. A Resource Issue Paper, a thought piece, is usually written afterward to put forth the collective wisdom of the trustees and provide a factual and theoretical framework for potential policy and practice. The Alliance disseminates a quarterly newsletter on partnerships, collaboration, and situation improvement to more than 3,500 citizens and business and government representatives. Conferences are held as issues of regional significance emerge and conference proceedings are usually published.

The Alliance is funded through several sources, mostly Northeastern Area State and Private Forestry (USDA), each state/province's natural resources agency, small project grants, and a number of private contributions.

When the Alliance first formed, this type of collaborative partnership was a relatively new concept. Much has been learned about building trust and exchanging of resources by facing opportunities and challenges while addressing emerging issues. Some have brought allegiance because sharing strategies and building a base of knowledge helped individual groups and institutions meet their own constituent needs. Some activities have brought disenfranchisement because of growing pains associated with new collaborative processes. Ultimately, the effort has brought a greater understanding of the region's forest resource, an understanding that has been made accessible to all government, business, interest groups, and citizens. This has led to a greater commitment by the trustees and the groups they represent to enhance forest management to meet a variety of forest values. This is most apparent in the commitment to work together on sustainable forest management criteria and indicators, an effort described later in this paper. Examples of cooperation across institutions and interests have helped create a positive climate for community-based ecosystem management. The following section describes some of the initiatives in which the Alliance has played a major role.

The Work of the Alliance

From its inception, the Alliance was respected for its leading-edge concept of bringing diverse interests to the table. The work of the Alliance, an ebb and flow similar to that experienced by many organizations breaking new ground, continually forces rethinking and course correction to sustain that leading-edge capacity. Process, timing, civility, and trust-building are key factors as evidenced by the Alliance's work.

Lake States Regional Forest Resource Assessment–To build upon a tradition of deliberate and careful decisionmaking, trustees felt it imperative to study and analyze broad forest resource data ranging from the ecological forest history, to the role of private landowners, industrial technology, the interface of timber and tourism industries, and the condition of the forest. Each state provided significant funds, and universities, state agencies, and the region's Forest Experiment Station contributed much of the research. A technical review team and an overall study team guided the project of twenty-one technical papers and an analytical report. These papers represent a body of shared knowledge that public and private forest land managers use in management and planning. The key finding identified the opportunity for a forest resource that could provide areas for environmental protection, recreation and timber harvest, all emphasizing the need for cooperative sustainable management (LSFA 1995).

Some on the project advisory committee challenged the content of the assessment analysis and suggested gaps in the environmental research. At a Board of Trustees meeting, concerns were raised by representatives of that viewpoint. The trustees listened and recommended changes in semantics, research citations and analysis. The document was published but some still were unhappy and distributed contradictory press releases. The three state governors cohosted a conference to release the report, which was disseminated extensively throughout the region to build an understanding of the forest resource conditions.

The process forced the Alliance to reexamine its purposes and action steps. It also heightened awareness of the Alliance's need to demonstrate its commitment to openness and transparency of process. The results emphasized the importance not only of convening advisory groups that represent diverse forest interests but connecting them on a regular and meaningful basis. It also raised the challenge of timing: When dissension exists within a diverse group, at what point can the group say minority issues have been accommodated as much as is feasible without overly delaying the product.

Incidental Activities–Highlighting a couple informal activities of the Alliance demonstrates its capacity to influence regional action. The Alliance's trustees are of a caliber that allows them to challenge current systems through the influence of partnerships and collaboration. Forestry education surfaced as a need at the Governors' Conference on Forest Resources at the end of 1994. The Alliance hosted a regional forum on private land ownership and its board made a commitment for each state delegation to meet with the respective Extension Services deans at the land grant universities. One state Extension service later convened a meeting of stakeholders related to private lands and sought input on how to better serve landowners. The meeting offered a good exchange of existing programs and prompted new ideas as well. In another state, additional Extension Foresters were hired to help with landowner forest management plans.

The Alliance staff serves on the Steering Committee of the Communities Committee of the Seventh American Forest Congress. That group position statement has been shared across the region and the Michigan Sustainable Forest Development Roundtable resolved to take it to Michigan state government, asking its agencies to include communities of place as standard practice in decision-making processes.

Another project of the Alliance demonstrates a different collaboration of resources to meet the information needs of diverse communities of interest with relevance to communities of place.

Continuous Forest Inventory Project–Forest inventories are valuable tools to provide citizens, as well as forest managers, with accurate and current information on forest conditions. Forest inventories in the Lake States had stretched to a frequency of fourteen to sixteen years and Alliance trustees perceived a need to increase their frequency to check potential shifts in harvesting, regeneration, or forest health which might otherwise go unobserved.

The North Central Research Station which conducts the inventory and the staff of two state natural resource agencies suggested that enhanced technology could allow a continual inventory process and approached the Alliance trustees to convene a planning meeting. Industrial forest managers, universities and state forest management agencies formed a task force to raise more than $240,000 to support a research effort to test these new concepts. The Alliance facilitated the process, holds the money in an interest-bearing account, and is instrumental in evaluating progress.

There was some initial difficulty as the Research Station struggled to meet the expectations the private sector held for its proffered funds. The Alliance stepped up the communication process with an advisory group, co-chaired by an Alliance trustee from the academic community and an industry staff person, which helped clarify project expectations and bring about a Cooperative Agreement. The project exemplifies the way government and industry can share resources to achieve mutual goals that benefit individual companies, communities, and citizens with environmental concerns.

It is the Alliance's goal to promote this kind of cooperation and collaboration. The Alliance has also facilitated dialogue between the forest resource-based economic sectors of timber and tourism.

Tourism and Timber Opportunities for Communities–The Resources Assessment emphasized the benefits of strong and diverse forest-based economic sectors in a community. On a simplistic basis, a diverse economy may offer a community more opportunities than an economy based solely upon one industry. On another level, there is opportunity for complementarity between one industry, timber which grows and harvests some trees and pays good salaries and another, tourism which thrives when the forest is diverse and in a healthy and aesthetically pleasing condition with vast opportunities for recreation. Residents in such communities learn from each other as they regularly discuss the level and location of timber harvesting and recreation trail paths in informal conversations on the Post Office steps, before school board meetings or along cross country ski trails through the forest. Many people in such communities admit that they value the forest for wood products as well as for recreation, spiritual endeavors and ecological integrity. Their conversation keeps all values in the forefront, and the level of trust and respect enables creative alternatives able to surface and potentially satisfy more values.

The Alliance committed itself to extending these informal conversations across the region. A task force guided work between the tourism and timber sectors. A recent conference sponsored by the Alliance and a variety of partners from the university Extension services to the Towns Association (local government) provided an opportunity. Speakers exchanged their perceptions of the need and value of cooperation in communities, and participants discussed strategies to enhance such cooperation across the region. Ideas focused on the need for dialogue. The strategies suggested included forestry and tourism field tours, focused discussion among tourism and timber operators, and travel promotion films to include reference to sound forest management practices. Participants also suggested ways to diffuse conflict such as holding autumn meetings for forest industries and snowmobile groups to discuss potential logging sites, alternative trail routes, and opportunities for public education about forest harvesting. The proceedings from the conference will further this exchange of ideas (GLFA 1998).

The most current Alliance project demonstrates an evolving organizational culture. As the Alliance develops sustainable forest management criteria and indicators, the process has been open and well-defined. Citizen participation has been invited early on and throughout the process at interactive meetings, through media presentations, and by presenting the draft work on the Alliance's web site and inviting feedback.

Sustainable Forest Management Criteria and Indicators–Responding to a request from the Michigan Department of Natural Resources, the Alliance sought federal and state funds to develop sustainable forest management criteria and indicators relevant to the region. In autumn 1996 the Alliance hosted a forum on environmental issues of the forest. To broaden accessibility, a teleconference on sustainable forest management was conducted in 1997 with speakers on emerging demographic and economic insights, sustainability work at the global level, and social concerns relative to forest management. Two public input sessions were held for professionals, and citizens in the region and individual interviews were conducted with representatives from several environmental organizations. Draft criteria and indicators were presented for public comment and sent to more than 400 leaders in the region in June 1998. The document was also made available on the Alliance's web site and through media promotion. A unique feature was the identification of indicators on three scales: statewide, countywide and by private woodlot.

The Alliance intends to share the criteria and indicators with professionals and citizens in Michigan, Minnesota, Ontario and Wisconsin to help them understand the value of sustainable forest management. A regional conference of leaders later will be convened to address data collection, monitoring processes, research necessary to help forest managers implement sustainable forest management, opportunities to integrate this regional work with other public and private efforts, issues of actual implementation, and potential public education efforts to convey the importance of sustainable forest management. This project will take the global and national perspectives of sustainable forest management criteria down to a region-specific set of indicators useful to professionals and local communities, their landowners, and citizens.

Each state/provincial agency has been at a different level in the process of developing criteria and indicators. The Province of Ontario has the Canadian Forest Management criteria and a set of 25 indicators applicable to forests on Crown lands. It is beginning to calibrate the measurement of these indicators. In Minnesota an interagency group, the Environmental Indicators Initiative, has set broad criteria and management responses across all natural resource areas. Wisconsin has just begun the process to identify criteria and indicators for state forest lands. Michigan is poised to work with citizens to prioritize applicable indicators from the menu developed by the Alliance. In all states, the private woodlot scale is unique and private woodland owners have been eager to be included in this process.

While these support mechanisms build a knowledge base of sound forest stewardship accessible at the community level, the Alliance has launched a project to directly initiate community-based ecosystem management groups.

Collaborative Learning and Community Ecosystem Management–The Alliance has helped residents across the states see the region as a community of place by drawing attention to the similarity of the forest across the region. This has resulted in more coalitions and exchanging of information. The Alliance became interested in promoting community-based forest efforts because of the National Rural Development Partnership's Systemic Change Initiative on Forests and Land Use Planning and its participation in the Seventh American Forest Congress. Both released reports noting the importance of ensuring local citizens are not disenfranchised and have a place at the table in forest planning processes (Bouchard, Archey and Danison 1997; Giebler and Sanders, 1996). Lake States participants in these processes perceived a new voice in forest resource use and protection coming from the communities themselves.

The Alliance invited Drs. Gregg Walker and Steven Daniels of Oregon State University to present their collaborative learning concept to Lake States regional leaders from diverse forest interests (Walker and Daniels, this publication). The collaborative learning concept suggests that if invited to participate, local citizens with diverse perspectives can learn from each other and create meaningful ways to improve the forest resources.

The Johnson Foundation hosted the session for 40 regional leaders in 1997 at its Wingspread Conference Center in Racine. Leaders nominated by state agencies, Alliance trustees and representative groups were considered people of influence who were committed to cooperation and learning new ideas. Following the two intense days of study and deliberation, the leaders of each state group left the meeting with a plan to bring the concept to their states and province.

The Alliance received a small grant in 1998 from the Wisconsin Environmental Education Board and co-hosted a one-day orientation session on collaborative learning with the Wisconsin Rural Partners (WRP, the State

Rural Development Council facilitated by USDA Rural Development, National Partnership Office). Community representatives who attended were eligible to later seek a volunteer facilitator to help them experiment with the collaborative learning process in their area and one proposal was chosen. The Alliance and WRP will staff collaborative learning sessions among citizens and local government officials in a northeastern Wisconsin township to develop community strategies for economic prosperity (including forest-based tourism and/or timber industries) and environmental protection related to land use planning. Up to four community meetings will be advertised and open to the public. The focus will be on learning from each other as critical issues are addressed and strategies are suggested.

Drs. Walker and Daniels also conducted three days of facilitator training for individuals from the state Department of Natural Resources, county Extension services, the Forest Service, Resource Conservation and Development groups (RC&D's) and non-profit organizations. The planning staff of the Chequamegon/Nicolet National Forest participated and may use the process to engage communities of place and of interest to address one forest planning issue as a trial.

The Alliance will seek additional funding to make this a pilot project to entice community groups in the other states to coalesce around sustainable forest management goals. Ontario has been a model for this learning process as it has utilized local, community-based decision groups to plan how best to assign Crown forest lands to tourism, harvest, and protection (Wianecki 1998). Canada has also used a community-based ecosystem management process to manage special forest areas, called the Model Forest Programme (International Model Forest Network, 1996).

CHALLENGES FACING THE ALLIANCE: IMPLICATIONS FOR FUTURE WORK

The Alliance's culture and structure have evolved to meet new challenges the organization itself has helped produce. These challenges, in turn, generate new opportunities to enhance its relationships with government, the academic community and the private sector. The proof will ultimately lie in the level of commitment to sustainable forest management across the region and the resultant level of public trust. Perhaps the greatest opportunity rests in the potential to build stronger relationships with other forest resource interest groups and with local communities. To build stronger relationships, the Alliance must take advantage of opportunities and continue to address certain issues. Several emerging challenges are presented here.

Leaders of the Alliance with diverse backgrounds will help generate better suggestions for enhancement of the forest resource. Although that is articulated in a white paper on the selection of Alliance trustees (GLFA 1996), the Alliance must convey that conviction consistently and meaningfully.

Environmental organizations were a part of the Alliance at its inception, have participated as trustees, and have been presenters at conferences and forums. Many were concerned about the implications in the analysis of the Regional Forest Resources Assessment alleging their views had not been adequately represented. Since that time some have been reluctant to participate in Alliance activities, particularly in serving as trustees. With limited time available, they may perceive their interests to be better served in other forums.

Tribal members and representatives have participated in Alliance activities, are willing to comment and provide feedback, speak at forums, and share their work with the Alliance. However, not one has yet committed to serve as a trustee. The Alliance must continue dialogue with tribal leaders to ascertain what changes in format, structure or issues are necessary for their greater participation.

Certainly, there are many other groups that might enhance the work of the Alliance from environmental education teachers to county government officials and other tourism, private woodland owner, and secondary wood industry representatives. It does matter who is involved and just as importantly, it matters how well how respectfully they perceive themselves as being treated.

> If they (citizens of communities near the forest) were allowed to solve their problems and manage their resources, they would soon discover that no one wants the local sawmills closed, and no one wants wildlife habitat annihilated. They would learn to inhabit the place on the place's own terms better than any regulatory bureaucracy will ever accomplish. (Dan Kemmis, *Community and the Politics of Place*)

Trust is earned by an accumulation of positive work together. It can be lost in one careless or surprising move.

Decision makers, wise and prudent, knowledgeable and experienced, must be at the table. When a proxy without decisionmaking authority is present, little is said, no action can be taken and a sense of unimportance is cast on the group and its activities. In order for leaders to participate consistently, positive working relationships must be actively and meaningfully cultivated.

The Alliance must conduct its business grounded in a learning climate, emphasizing civility. Trustees place the highest value on the Alliance as a forum to exchange strategies, lessons and emerging issues. Its work must remain open, honest, and of value, as well as sensitive to cultural differences.

The Alliance must continue to educate while allowing individual jurisdictions and institutions to customize to meet the values of their constituents as well as the specific needs of the forest resource they manage. With its regional scope and the capacity of its trustees, the Alliance has been able to attract experts on emerging forest interests and create an environment for cross-fertilization to bring the broad world view to the local level.

The Alliance will continue to reach out to communities to build capacity and bridge the communication gap that exists between government, agencies, interests, and citizens. A variety of projects must be implemented to build community capacity for ecosystem management in the Lake States region. A greater inclusion of local citizens on its board and better tools to exchange information between the regional and local levels may prove valuable. The Alliance may spur pilots in individual states, a systemized linking of community-based forest groups or the provision of support and technical expertise to help CBEM groups. Cooperation with other groups, from the State Rural Development Councils to Cooperative Extension and the Forest Service, will ensure greater impact across the region. The Alliance can use its arsenal of promotional efforts to encourage community-based forest groups. Further efforts must be made to instill support for community-based ecosystem management within institutional cultures. State and federal agencies, nonprofit groups, economic development associations, trade associations, and the academic and research communities are all directly involved in the Alliance and trustees from those organizations are in a position to diffuse these ideas.

Funding an interstitial forum such as the Alliance is a continual challenge when the outcomes are indirect: it is not the Alliance itself, but its trustee groups that implement actual change. Such groups are enticed by grantors, and the lure of future funding, to find ways to document specific outcomes yet the real measure is the often intangible building of learning networks and relationships.

The process of problem solving, issue resolution, and situation improvement still does not seem to receive the same social and political attention as efforts that generate conflict. Finding private foundations that support work in intermediary foundation building is one option, but it is an incomplete one. Other mechanisms to generate a solid base of operational funds must be identified. In either case, a way is needed to track these indirect outcomes that result from exchanges of information to verify the value of the work to funding sources.

CONCLUSION

There are wise people of integrity who are committed to building innovative bridges across governments and institutions and through collaboration to raise collective wisdom within the Great Lakes region. They have woven a rich tapestry of sound forest management upon which individuals, agencies, groups, and communities can work together and thrive while the forest resource is sustained. The citizens and professional leaders who head up the Alliance are more than capable of addressing the challenges it faces: building diversity, creating a learning climate with global implications, instilling an organizational culture that facilitates collaboration, drawing decisionmakers to the table, respecting the manner in which various groups implement the work of the Alliance, infusing a commitment to community-based ecosystem management that pervades through all efforts, and securing appropriate funding. After 10 years of groundbreaking efforts, adjustment, and evolution, the work may just be beginning but the mounting successes are leading the commitment to proceed.

REFERENCES

Bouchard, Carl E., Bob Archey and Kurt Danison. 1997. Sustainable rural development and natural resource land use: A model for effective change. A paper of the National Rural Development Partnership, Systemic Change Initiative. Washington, DC.

Chappelle, D. 1997. Interaction of tourism and forest products sectors on community employment/unemployment in the lake state region. In M. Vasievich and H. Webster (eds.), *Lake States Regional Forest Resources Assessment: Technical Papers*, pp. 206-227. St. Paul: USDA Forest Service North Central Forest Experiment Station.

Chappelle, D. 1997. A detailed examination of economic effects of growth in tourism and forest products sectors in the lake states. In M. Vasievich and H. Webster (eds.), *Lake States Regional Forest Resources Assessment: Technical Papers*, pp. 228-248. St. Paul: USDA Forest Service North Central Forest Experiment Station.

Giebler, M., and W. Hinrichs Sanders. 1996. Sustainable rural development and land use planning: Is the American land use process broken? A paper of the National Rural Development Partnership, Systemic Change Initiative. Washington, DC.

Giebler, M., and W. Hinrichs Sanders. 1996. The impact of FACA on shared decision-making. A paper of the National Rural Development Partnership, Systemic Change Initiative. Washington, DC.

Great Lakes Forest Alliance. 1998. *Proceedings: Timber and Tourism Opportunities for Communities, A Conference for Leaders*, Rhinelander, WI.

Great Lakes Forest Alliance. 1997. Charter. Hayward, WI: Great Lakes Forest Alliance.

Great Lakes Forest Alliance. 1996. A picture of the lake states forests. *Lake States Forest Facts*. Hayward, WI: Great Lakes Forest Alliance.

Great Lakes Forest Alliance. 1996. Guidance for selection of trustees. A white paper. Hayward, WI: Great Lakes Forest Alliance.

International Model Forest Network. 1996. Model Forest Programme Information packet. International Model Forest Network Secretariat, International Development Research Centre, Ottawa, CA.

Kingsley, Neal P. 1997. Resource availability from nonindustrial private forest land in the lake states. In M. Vasievich and H. Webster (eds.), *Lake States Regional Forest Resources Assessment: Technical Papers*, pp. 39-71. St. Paul: USDA Forest Service North Central Forest Experiment Station.

Lake States Forestry Alliance, 1995. *Forest Resource Trends and Opportunities in the Lake States, A Continuing Resources Renaissance*. Hayward, WI: Lake States Forestry Alliance. (Soft cover, offset printed guidebook, not published in a book with hard or soft cover by a book publisher.)

Mielke, Manfred E. 1997. Health of the lake states forests–1992. In M. Vasievich and H. Webster (eds.), *Lake States Regional Forest Resources Assessment: Technical Papers*, pp. 30-38. St. Paul: USDA Forest Service North Central Forest Experiment Station.

Shands, William E. 1988. *The Lake States Forests: A Resources Renaissance*. Washington, DC: The Conservation Foundation. (Paper back, published book.)

Stearns, Forest W. 1997. History of the Lake States forests: Natural and human impacts. In M. Vasievich and H. Webster (eds.), *Lake States Regional Forest Resources Assessment: Technical Papers*, pp. 8-29. St. Paul: USDA Forest Service North Central Forest Experiment Station.

Stone, Robert. 1997. Great lakes states forest trends, 1952-1992. In M. Vasievich and H. Webster (eds.), *Lake States Regional Forest Resources Assessment: Technical Papers*, pp. 39-71. St. Paul: USDA Forest Service North Central Forest Experiment Station.

Vasievich, J. Michael, K. Potter Witter and L. Leefers. 1997. Timber supply and market trends in the lake states. In M. Vasievich and H. Webster (eds.), *Lake States Regional Forest Resources Assessment: Technical Papers*, pp. 290-310. St. Paul: USDA Forest Service North Central Forest Experiment Station.

Webster, Henry H., Daniel E. Chappelle and Stephen C. Andrews. 1997. *Tourism and forest products: twin resource sectors for effective community development in the lake states*. College of Natural Resources and the Agricultural Experiment Station, University of Minnesota.

Wianecki, Karen. 1998. *Proceedings: Timber and Tourism Opportunities for Communities, A Conference for Leaders*. Ontario Ministry of Natural Resources. Rhinelander, WI. pp. 23-30.

Public Involvement in Forest Management Planning: A View from the Northeast

Martha Twarkins
Larry Fisher
Tahnee Robertson

SUMMARY. In 1997, the US Forest Service initiated public involvement processes in three national forests in the northeastern United States–the Finger Lakes National Forest in central New York, the Green Mountain National Forest in Vermont, and the White Mountain National Forest in New Hampshire and Maine. Citizens' perspectives were sought on forest management prior to determining the changes needed for revising the Land and Resource Management Plans (Forest Plans) as well as to exchange information on management of these three forests. These processes represent a pro-active effort by the US Forest Service to engage communities of interest in dialogue about the management of these national forests. The Forest Service has emphasized the principles of ecosystem management and community partnership in developing these plans. The paper describes the policy framework for forest planning and an outline of the Plan Revision process as developed in the Northeast. It further offers a preliminary analysis of this approach, including lessons to date, their implication for subsequent phases of plan revision processes in the Northeast, and possible rele-

Martha Twarkins is District Ranger, Finger Lakes National Forest, 5218 State Route 414, Hector, NY 14841.

Larry Fisher and Tahnee Robertson are Director and Associate Director, respectively, Cornell Program on Environment and Community, 112 Rice Hall, Ithaca, NY 14853.

[Haworth co-indexing entry note]: "Public Involvement in Forest Management Planning: A View from the Northeast." Twarkins, Martha, Larry Fisher, and Tahnee Robertson. Co-published simultaneously in *Journal of Sustainable Forestry* (Food Products Press, an imprint of The Haworth Press, Inc.) Vol. 13, No. 1/2, 2001, pp. 237-251; and: *Understanding Community-Based Forest Ecosystem Management* (ed: Gerald J. Gray, Maia J. Enzer, and Jonathan Kusel) Food Products Press, an imprint of The Haworth Press, Inc., 2001, pp. 237-251. Single or multiple copies of this article are available for a fee from The Haworth Document Delivery Service [1-800-342-9678, 9:00 a.m. - 5:00 p.m. (EST). E-mail address: getinfo@haworthpressinc.com].

vance for agencies embarking on similar public planning initiatives. *[Article copies available for a fee from The Haworth Document Delivery Service: 1-800-342-9678. E-mail address: <getinfo@haworthpressinc.com> Website: <http://www.HaworthPress.com>]*

KEYWORDS. Community-based forestry, national forest planning, public involvement, community partnership

There are three national forests in the northeastern United States: the Finger Lakes National Forest in central New York, the Green Mountain National Forest in Vermont, and the White Mountain National Forest in New Hampshire and Maine (see Figure 1). In 1997, the US Forest Service initiated a process designed to gain citizens' perspectives on forest management in these three forests prior to determining changes needed in revising the Land and Resource Management Plans (Forest Plans) for each national forest. The planning process involved a proactive effort to engage communities of interest in dialogue about the management of these national forests, and the Forest Service has emphasized the principles of ecosystem management and partnerships in plan development.

While these efforts are in various stages (ranging from on-hold to in-prog-

FIGURE 1. Map of National Forests in the Northeastern United States

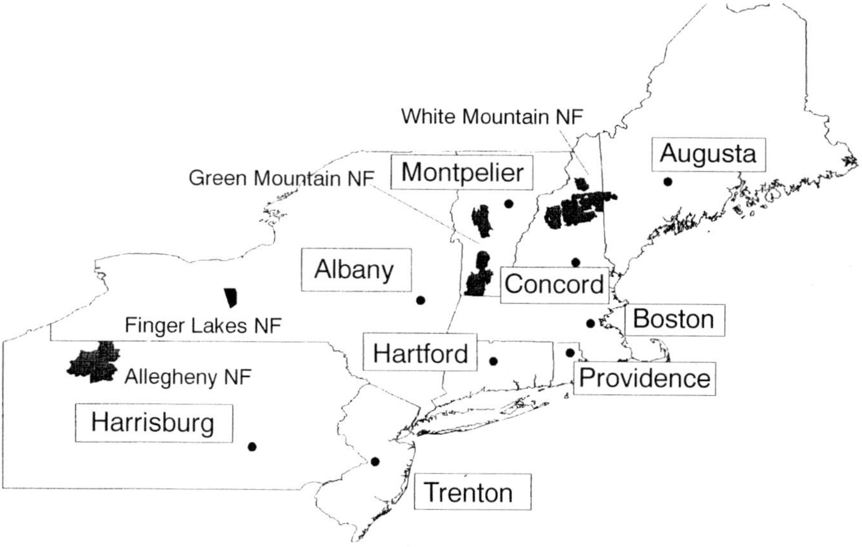

ress), a preliminary analysis of the planning process offers some important insights into opening dialogue among stakeholders about land and resource management of national forest lands, as well as engaging the public as partners in forest management. The discussion below highlights the lessons to date, their implications for subsequent stages of the Plan Revision process, and possible relevance for agencies embarking on similar planning initiatives. While the information presented is from all three forests, we rely heavily on our direct experience in the Finger Lakes, as well as on numerous discussions with the Forest Service Joint Core Planning Team and an evaluation workshop comparing experiences across the three forests.

BRIEF DESCRIPTIONS OF THE NATIONAL FORESTS IN THE NORTHEAST

The Finger Lakes National Forest (FLNF) is the only national forest in the state of New York. At 16,000 acres, it is also the smallest forest in the National Forest system. Despite its limited size, there are more than 2 million people living within a 100-mile radius. Its primary use is recreation: principally hiking, camping, cross-country skiing, horseback riding, driving for pleasure, fishing, and birding. The North Country Trail, a national scenic trail, traverses the Forest as does a portion of the Finger Lakes Trail system. While recreation is important, the FLNF is also a working forest that employs multiple-use principles to create and/or maintain essential wildlife habitat, graze cattle, provide clean water, and supply trees for wood products such as furniture, paper, and baseball bats. Key issues surrounding management planning include concerns over the proper mix of uses for the trail system, the management of recently acquired lakeshore property, conservation and configuration of potential old-growth areas, and the desirability of further land acquisition.

The Green Mountain National Forest (GMNF), located in central and southern Vermont, consists of more than 367,000 acres of lush woodlands and streams. This forest constitutes approximately 5 percent of the state's land area and hosts more than 2.5 million visitors each year. Recreation plays an important role on the GMNF, with 550 miles of hiking trails, which include the Appalachian/Long Trail, and six Wilderness areas. The forest also provides habitat for wildlife, clean water, thriving fisheries, and wood products. Fifty-seven communities are located within the proclamation boundary of the Green Mountain National Forest, and the economic viability of many of these communities is tied to its management. Key issues include forest health and biodiversity conservation, achieving a desirable mix of recreational uses, addressing the role of timber harvest, and assessing management's impacts on the economic viability of the surrounding communities. Public planning group participants also suggested more systematic monitoring and evaluation than is currently practiced.

The White Mountain National Forest (WMNF) covers almost 780,000 acres in New Hampshire and western Maine and contains some of the highest mountains in the Northeast, including the 6,288 foot Mount Washington. Within easy driving distance of Boston and other East Coast cities, the White Mountain National Forest receives more than 7 million visitors a year. Twenty-three campgrounds, numerous trails and streams, the Appalachian Trail, and extensive Wilderness areas all lie within the forest's boundaries. Forest products are also offered with an average annual sale program of 20 million board-feet (MMBF). Currently, the major issues on the White Mountain National Forest involve questions about recreation (mix of uses and the quality of the recreational experience), biodiversity and wildlife habitat conservation, and federal policies on clearcutting.

Although the types of issues being addressed in the Plan Revision process are similar for these three national forests, they are articulated somewhat differently for each site. All three have acquired additional land in recent years, drawing a mix of responses from local residents and elected officials. In the Finger Lakes, for example, some individuals and local governments have expressed concern over the loss of tax revenues to local communities, while others would like to see the forest acquire substantially more land. In Vermont, land acquisition has caused a variety of reactions, even among local governments. On all the three forests, varied concerns have been raised about the management of trails, use of motorized vehicles, and general public access. While timber harvesting did not prompt much discussion on the Finger Lakes National Forest, it remains a volatile issue on both the Green and White Mountain National Forests. Biodiversity and wildlife conservation are also viewed somewhat differently at the three sites.

THE FOREST PLAN

Systematic planning on the national forests has been proscribed by federal statute for more than 20 years. In the mid-1970s, two pieces of legislation, the Forest and Rangeland Renewable Resources Planning Act (RPA 1974) and the National Forest Management Act (NFMA 1976), addressed concerns over resource management. The RPA requires land and resource management plans for each national forest, while the NFMA describes a process to guide the Forest Service in multiple-use management. Major features include integrated resource management plans for each national forest, compliance with the National Environmental Policy Act (NEPA), identification of suitable timber land along with criteria for harvesting, consistency with other plans and permits, and a requirement that the public be allowed to participate in the planning process.

Several laws have impacted how forest planning is executed. One such law with implications for public involvement in land management is the Federal Advisory Committee Act (FACA) passed in 1972. FACA sets requirements for advisory committees and defines a set of open government procedures. While its intent was to promote more balanced participation and transparency in decision making, the net effect has been to narrow the range of options for (non-designated) advisory committees that assist federal agencies (Rieke 1998).

Forest Plans are ordinarily revised on a 10 to 15 year cycle according to USFS NFMA regulations. Revisions are based on changes in understanding about forest management, new trends and opportunities, and the social and economic changes that take place in communities (USFS 1997).

Forest Plan Revision Processes at the Three Sites

Forest Plan Revision processes in the Northeast were initiated in 1996, with the USFS Joint Core Planning Team (USFS staff representing the three northeastern Forests) outlining basic principles and procedures for completing the process within the allotted time frame. The planning process pivoted on two primary tenets: partnerships and ecosystem management. The public would be involved from the very beginning, all information would be widely shared, dialogue and learning would be essential, and joint problem-solving would be emphasized.

NFMA requires public participation. In the previous round of Forest Planning, participation consisted primarily of soliciting the public's concerns on proposed management and incorporating the Forest Service's solutions into the final Plans. As a result, some people felt alienated from the process and unhappy with the decisions made by the Forest Service. Many communities remained detached from the process and "took what they got" in goods and services from the forests. During this plan revision effort, the Forest Service strove to engage the public as partners in management from the beginning of the process rather than asking for reactions to proposals. To engage those interested in forest planning, the Forest Service has created a variety of participation formats including one-on-one interactions with interested people and local governments, public meetings to work on issue development, working groups to answer technical and/or management questions, and contracts for surveys and interviews to ascertain public perceptions and desires. As the process evolves in the future, the Forest Service will monitor which techniques work best and modify their approach. The techniques used during the planning effort and their effectiveness are discussed in greater detail later in this paper.

This premise of partnership versus traditional public involvement is new for forest planning in the Northeast. The desired outcome is for the Forest Service to better meet the needs of the "community of interest" (Sirmon, Shands and Liggett 1993) and for the community to better understand the Forest Service mission and how the USFS community partners can work together on mutual goals. Another desired outcome is for the partnerships to continue not only through the revision process, but into all aspects of forest management. The Planning Team outlined a five phase process for revising the Forest Plans:

> *Public outreach*–The Forest Service develops information about current plans and issues through discussions and public meetings with Forest Service employees, the public and groups currently engaged in forest management; develop a list of issues to consider during Plan Revision.
>
> *Public Planning Group (PPG)*–The Forest Service hosts Public Planning Group meetings; disseminates packets of Forest Planning regulations, past management plans, and other pertinent information; the PPG then reviews performance of current plans and raises further issues for the revision.
>
> *Collect information to evaluate revision needs*–The Forest Service and PPG form technical working groups (TWGs) to collect and analyze information on specific issues raised by the PPGs.
>
> *Need for change*–TWG members work with the PPG to document areas of possible change to the existing Forest Plan.
>
> *Formal NEPA process to revise Forest Plan*–The Forest Service submits a Notice of Intent to prepare an Environmental Impact Statement (EIS) for Forest Plan Revision. The Notice of Intent is followed by the scoping and preparation of the EIS, a Record of Decision (ROD), and publication of the Revised Forest Plan. The public will be involved throughout the entire NEPA process, providing comment to proposals, ideas for management, solutions to problems, and concerns to be addressed.

Comparisons Between the Three Forests

While the general Revision framework was adhered to for all sites, each followed somewhat different approaches (and schedules) in facilitating public involvement. For example, the Finger Lakes National Forest held two public outreach sessions, which generated more than 600 comments. The public planning group (PPG), created after these outreach sessions, worked to sort and further define these issues in a series of meetings. A field study tour was also held on the Finger Lakes National Forest, which helped clarify specific issues.

The Finger Lakes PPG clustered issues into the following action categories: (a) issues to be addressed via the Revision process, (b) issues best addressed by amending the current Forest Plan, (c) issues outside the jurisdiction of the Forest Service, and (d) issues already covered by the Forest Plan which could be addressed immediately by changing Forest priorities. The PPGs prioritized their interest in these to help the Forest Service decide which projects to consider for next year's funding. Finally, the PPGs developed questions to be addressed by the technical working groups (TWGs).

The Green Mountain National Forest held four public outreach meetings, followed by five PPG meetings which led to the identification and discussion of 22 management issues. The White Mountain National Forest held three outreach sessions for stakeholders reaching out to urban centers as distant as Boston. The three PPG meetings held on the WMNF were structured to cover more issues in a shorter time frame than either the Finger Lakes or Green Mountain sessions. Currently, the revision efforts of the Green Mountain and Finger Lakes National Forests are under a congressionally mandated hold. The next step for the White Mountain National Forest will be to form technical working groups.

RESULTS AND OBSERVATIONS

While premature and mixed, the results from these initial phases of the planning process have been encouraging. Active participation by stakeholders–local and state government and a variety of concerned organizations, agencies, and individuals–in all three sites has helped clarify the range of forest management issues and identify key questions for further analysis. Facilitation by neutral third parties (the Cornell Program on Environmental Conflict Management for the FLNF and independent contractors for the GMNF and WMNF) encouraged participants to view the meetings as balanced and open with no predetermined results. External facilitation also allowed Forest Service staff to interact informally with the participants, allowing better communication and exchange among all parties. The participants were able to see Forest Service employees more as people than bureaucrats or experts, so communication was increasingly open and honest. On the WMNF, however, some PPG members declined to participate, claiming the independent contractor who acted as facilitator was not objective. This points to the need for a clearly defined role for the facilitator and for diligence by the

Forest Service to ensure that the facilitator encourages an atmosphere of openness, fairness, and joint learning, free of predetermined results.

The outreach sessions, PPG meetings, and other public forums have led to a more engaged and aware public, increased volunteer support, and a better understanding of issues on all sides. This is evidenced, for example, by the increased number of people who have signed up to volunteer for trail maintenance, bird surveys, and other forest activities on the FLNF, several of which are a direct result of the PPG meetings. Also on the FLNF, citizens are becoming more involved in environmental issues pertaining to both the national forest and the community. Local government is becoming more engaged in efforts Forest staff have undertaken outside the planning process, including using the national forest as an anchor to expand ecotourism in the area. The US Forest Service has benefitted too, gaining new insights into its constituency and their diverse views. Now the Forest Service is hearing the reasons behind citizens stated concerns rather than citizens just stating demands and complaints. This has opened the door for increased understanding among the various parties, and has established a more creative and collaborative approach to problem solving.

As community participants have become more educated in aspects of forest management, they have become more involved in the planning process. They are meeting with the Forest Service to help plan better trails and wildlife areas, as well as assisting with historic restoration. They are helping to explore future management options for potential old-growth forests and the best use of recently acquired lands. These meetings are outside the Forest Planning process but were spawned by the enthusiasm generated through those meetings. A collaborative research effort is also underway, as a result of the FLNF PPG process, to help clarify the range and intensity of views on land acquisition, and further engage the USFS and local communities in constructive dialogue (Jirka 1998).

The participants have also gained an understanding of each other's views on forest management options, which has led to greater cooperation and joint problem solving among participating groups. For example, some members of the hiking and equestrian trail user groups in the FLNF had been feuding over which trails to use for which activities. They are now working together and with other user groups on a management scenario to best meet their varied interests and concerns. They're even planning a joint trail cleanup day to get to know each other better.

These results are based on observations and have not been statistically measured. Future planning efforts perhaps should include a research component that specifically measures these outcomes.

While the experience in the Northeast has revealed the strength and potential of collaboration, it also exposed weaknesses that must be addressed in subsequent phases of the planning process. The type and structure of meetings was ineffective at engaging all stakeholders in the entire process, and the more structured periodic public meeting did not meet everyone's needs or sense of opportunity for active involvement. Some participants wanted to be able to react to a specific proposal and were frustrated with the open format, while others appreciated the opportunity to form and clarify issues. Some participants were also frustrated by the lack of a concrete product at the end of each session.

One of the objectives of the PPG format was to provide an opportunity for people to see the interaction of different management activities, such as recreational use and timber management. Nevertheless, several participants chose to engage in only one topic area. This single-issue orientation limited the kind of exchange that has proven so beneficial in other dialogues. Time was another factor that prevented many people from becoming involved in multiple issues. The Forest Service must develop a system that will allow for multi-issue discussion without an overwhelming time commitment by participants. Organizers must also determine the best setting, time, day of the week, and location for meetings.

The facilitators and Forest Service strove to achieve participation by all primary stakeholder groups. One of the activities in the first meeting of both the FLNF and GMNF PPGs was to have participants create a stakeholder map to identify all stakeholders and determine which were present and which were missing. Then the participants actively recruited some of those absent stakeholders. This met with varied success. Discussions with Forest Service staff and PPG members indicated that the reasons for lack of full stakeholder participation could include: lack of perceived benefits, time commitment, undesirable venue, tension with the Forest Service over certain issues, and lack of interest in the collaborative approach. The lack of consistent attendance and the absence of certain stakeholders should be seriously scrutinized to determine its impact on the goals of the process and to identify effective ways to involve hard-to-reach stakeholders.

Few participants had a broad understanding of forest management issues and the institutional constraints of the Forest Service, and thus were limited in their ability to integrate planning within a multiple-use, ecosystem-based framework. In addressing these educational challenges, the Forest Service frequently erred in extremes–providing too much or too little background material. In many cases, the material provided was too technical and detailed for quick comprehension by PPG participants.

Many challenges remain to improve collaboration in forest planning: (a) identifying stakeholder groups and gaining their consistent representation throughout the process, even in the face of limited concrete achievement and difficult time commitments; (b) finding ways to involve more distant stakeholders; (c) integrating forest planning with other local and regional planning efforts.

Lessons Learned

In April 1998, Cornell University hosted a collective evaluation of the Plan Revision process in the three National Forests of the Northeast. Participants included US Forest Service staff, citizen members of the PPGs, researchers and extensionist educators from Cornell and Syracuse Universities, as well as other observers. Their analysis highlighted the following key lessons learned from the planning process thus far.

The importance of transparency–An open and sincere interest in collaboration on the part of the Forest Service helped create an atmosphere of frankness that fostered greater public involvement in the planning process. From the beginning the Forest Service view that "all comments are valid" helped encourage a stronger sense of trust, which improved dialogue and enhanced opportunities for participants' contributions. In the Finger Lakes, tangible results can already be seen in the increased role of volunteers and continued engagement and follow-up on key issues.

The value of partnership–Throughout the process, the Forest Service relied on considerable public and private support. External facilitators helped design the process and organized and led most of the meetings. Community volunteers helped prepare for the meetings by contacting participants and providing food, logistical support, and valuable insight. This involvement has also extended beyond the plan revision process, as citizen members have continued to meet to address issues important to the community as well as the Forest Service. This emerging volunteerism will become increasingly important during the subsequent "need for change" phase of planning, as plans are translated into actions, which will require the strong support of community-based groups. The concept of partnership has included cooperation with local universities, such as Cornell, which has offered assistance in process design, facilitation, cross-site analysis, and research support, and Syracuse, which conducted research on the local economic impacts of the Finger Lakes National Forest. Linking the three northeastern forests in the planning process resulted in the sharing and comparison of important information.

Stakeholder analysis and involvement–Identifying and including of key stakeholders is a fundamental part of the planning process. Stakeholder maps served as reference points to identify and seek out various perspectives on forest management. While it was clear that not all stakeholders were represented, or represented adequately, the principle of inclusiveness was an important element in the collaborative process. Thoughtful, participatory stakeholder analysis is important to identify and involve key groups, gain varied insight and expertise, and ultimately reach some form of consensus for planning, implementing, and monitoring the Forest Plan. As the PPG processes progressed, stakeholder "involvement" became "engagement," and now is moving towards an active, ongoing "partnership" on key management issues.

Multiple methods–It has become increasingly obvious that no single time, place, or approach works for all constituencies. In early surveys, potential Finger Lakes PPG participants were split almost equally between preferring weekday, weekend, or evening meetings, and all options were tried during the PPG phase. Since the national forests are used by people of all classes and vocations, often traveling great distances, multiple venues were used for the meetings. Nevertheless, participation was generally uneven, and there remains a sense that not all stakeholders were able (or willing) to attend all meetings and that some key stakeholders were still unrepresented. National interest groups were represented to a small degree, sometimes by local constituents attendance, however more effort will be made in the future to encourage their participation. As this experience shows, varied approaches are needed to successfully involve the public, meetings at different times and in different places, plus other activities and approaches that will gain input from the varied constituencies with a stake in forest management. Additional methods discussed for future planning include: public surveys (individual and group interviews, questionnaires), more directed bio-physical or socio-economic research, internet-based approaches such as bulletin boards and on-line discussions, field days and other forest events, public opinion polls, and write-in opportunities.

Dialogue and education–Education and joint learning are critical to the planning process; this education is reciprocal between the Forest Service and the public. Community representatives learned a great deal about the technical and administrative aspects of forest management, while the Forest Service gained new insights into local community dynamics and the experience, concerns, and ideas of many national forest users. This kind of joint learning requires both understandable, accurate information and an atmosphere of trust in which to convey and listen to concerns and ideas. It can be achieved only through open dialogue and exchange. New, practical, and innovative ways of sharing information, integrating concise reports, brochures, and presentations into the structure of the meetings (including distributing materials in advance), and reducing technical terminology, jargon, and acronyms will make forest management information more accessible to a wider community.

Managing the planning process–Both structure and flexibility proved important. The general framework for planning provided an overall sense of the process; nevertheless, frequent adjustments were necessary based on input from the meetings or to accommodate new elements (e.g., field visits). Independent facilitation was useful in reducing the perception that the Forest

Service was directing the outcome, and it reduced the responsibility (and visibility) of Forest Service staff. Informal settings reinforced opportunities for fun and relationship-building and affirmed that quality communication often occurs outside the formal meeting setting. This can be contrasted with the more formal public hearing format in which participants take turns stating their positions and demands without the opportunity for dialogue and constructive joint problem solving.

CHALLENGES AHEAD

The Public Planning Groups in these sites completed their tasks of sorting through and clarifying public comment, reviewing current Forest Plans, and developing Plan Revision issues. They also identified key questions for the technical working groups to study. On the White Mountain National Forest, technical work with the public will begin soon to address issues raised in aquatic and terrestrial biological diversity, recreation, social and economic conditions, and education. On the Finger Lakes, small working groups continue to meet on issues of trail use and maintenance, old growth forests, and potential uses of a former Boy Scout camp property. During the remaining phases of the planning process, the primary tasks for community participants and Forest Service staff will include: (a) the collection and analysis of relevant information to address questions raised by the PPGs, (b) documentation of the need for changes to the current Forest Plans, and (c) the analysis of proposals and alternatives to meet needed changes through the NEPA process. The Forest Service will continue to face challenges through the next phases of the planning process. Those challenges are:

Balanced stakeholder involvement–Finding the proper balance and means for stakeholder representation is important, as is maintaining the momentum of participation over time. There is a need for creative new ways to expand and deepen participation: (a) by finding innovative methods and venues for reaching all stakeholder groups as well as for incorporating new people who enter part way through the process, and (b) by broadening the range of issues discussed in these forums. The planning process must seek effective ways to engage people critical of Forest Service policies (e.g., opponents of land acquisition or timber harvest), along with urban and national constituents who represent a growing interest in National Forest management practices. Furthermore, building a sense of stakeholder ownership and commitment in forest planning and management will continue to be challenging, since both the process and decision making still rests with the USFS.

Integration with local and regional planning procedures–Local, ecosystem, and larger regional/national scales should be linked to position forest planning in the context of these broader efforts. Expanding partnerships with local communities requires closer institutional links with local and state government. As the national forests are increasingly seen as an integral part of local communities and ecosystems, there is an obvious need to incorporate forest planning within these local, ecosystem, and larger regional/national scales. This task will be challenging as governmental bodies reduce staff and in many instances have yet to see the connection. There are exceptions to this as evidenced by greater state employee participation in Vermont; examining the reasons for this greater integration into state-level planning would be helpful for other forest planning processes.

Managing and sharing information–An informed public can better contribute to the planning process. Written materials should be brief and accessible to all interested participants, be provided prior to meetings to give participants time for study, and be targeted for specific issues and constituencies. Presentations should reinforce key points and provide the opportunity for questions. Field visits and other information-sharing venues should be encouraged throughout the process.

National politics–Congress and the administration have increasingly become involved in national forest management decisions and issues have become polarized in the general public and in Congress. This is evident by the increasing number of decisions being questioned by both the Congress and the courts. This polarization will make plan revisions more challenging.

In several instances Congress passed bills that directly affect Forest Management. One of the latest was the decision that until the Forest Service revised and adopted new regulations to address the National Forest Management Act (NFMA), Congress would not to authorize the Forest Service to proceed with Forest Plan Revision on forests that had not yet issued a Notice of Intent. None of the northeastern forests has yet issued a Notice of Intent (although the White Mountain National Forest has congressional permission to proceed with revision). Because of the inability to move forward with revision, many participants have lost enthusiasm for the process and faith in the Forest Service to carry out its mission.

Attitudes, values, and culture–There is a need to change institutional culture from within in order to allow the Forest Service to move towards partnership and joint management approaches. These changes are essential for changing public perceptions which will in turn encourage greater public involvement in forest planning and management. During formation of the current Forest Plans, the Forest Service engaged public review but for reaction to a proposal rather than formation of that proposal. The people both within and outside of the Forest Service that participated in that process need to understand the new process for changing the Forest Plans. The current and former Chiefs of the Forest Service have stated and reiterated the importance of public involvement and engagement in national forest management. This means understanding the diversity of values and preferences, and seeking ways to resolve and/or accommodate different perspectives, including the varied class and ethnic backgrounds of stakeholders. The Forest Service and the citizens participating in the process need to listen to each other and try to understand all points of view.

Integrating existing laws–The Federal Advisory Committee Act (FACA) controls federal agency use of advisory committees and sets direction for public participation and agency decision making. This law directs how the public can interact with the federal government while allowing for open and active participation by all interested parties. Yet it also limits the creation of advisory committees and constrains opportunities for developing partnerships with various communities.

CONCLUSIONS

The US Forest Service's efforts in these northeastern forest units are still initial and imperfect. While the Forest Service has moved beyond conventional public hearings and agency-centered technical forest management plans, they are not yet fully rooted in community-based institutions nor well integrated with local and regional planning processes. These initiatives in the Northeast are a sincere effort by the Forest Service to engage the public early on in an extended planning process, before positions and plans become too fixed to be negotiable. At each of the sites, the Forest Service has begun to identify and convene a community of interests focused on multiple-use management, creating a working environment for more open, engaged, and better informed decision making (Sirmon, Shands and Liggett 1993). The next phase of planning–identifying critical gaps in technical information–will enable more in-depth analysis and discussion of key issues and involve a widening circle of participants in the dialogue over forest management. This process, while occasionally contentious, is a critical investment in sustainable stewardship of forest and community resources (Richards and Burns, 1998).

With Chief Mike Dombeck's pledge "to achieve quality land management under the sustainable multiple-use management concept to meet the diverse needs of the people," the mission of the Forest Service is clear although the means to achieve it are not. The challenge facing the Forest Service is how to effectively respond to this diversity of interests and needs. This means finding an approach to constructively engage the public, one that emphasizes joint learning and problem solving, that forges the best way to achieve sustainable multiple-use management, and that can adapt over time. The process outlined in this paper illustrates a set of sincere efforts to involve citizens in forest management planning. Much can and should be learned from the strengths and weaknesses of these, and many other, innovative public involvement efforts.

REFERENCES

Jirka, A. 1998. *Finger Lakes National Forest Community Impact Study: Community Perceptions Regarding Land Acquisition and Compensation.* Masters research proposal, School for International Training, Brattleboro, Vermont.

Rieke, B. 1998. *The Federal Advisory Committee Act: Impacts on Community-Based, Collaborative Groups, Quarterly Report.* Forest Trust, Santa Fe, NM, Number 18.

Richards, T. and S. Burns. 1998. Beyond Scoping: Citizens and San Juan National Forest Managers, Learning Together. *Journal of Forestry*, 96(4):39-43.

Sirmon, J., W.E. Shands and C. Liggett. 1993. Communities of Interests and Open Decisionmaking. *Journal of Forestry*, 91(7).

U. S. Forest Service. 1997. *Public Planning Group Notebook: Finger Lakes National Forest.*

Natural Resource Policy and the Paradox of Public Involvement: Bringing Scientists and Citizens Together

Gregg B. Walker
Steven E. Daniels

SUMMARY. Immersed in natural resource policy approaches such as ecosystem management is the expectation that the best available science will be applied so that the best policy management decision will result. Citizens, like scientists and land managers, want natural resource management decisions based on good science rather than special interest group politics. Yet citizens also want to be involved in the decision process and are skeptical about the very science they claim must be the basis for policy actions.

Herein lies an apparent paradox. Citizens' want the best science to guide natural resource management decisions, but not to the exclusion of their input. Similarly, there seems to be a paradox in the sentiments expressed by natural resource management agency administrators and specialists. Agency personnel know they need meaningful citizen involvement in their management decisions, but they also want citizens to trust their scientific expertise.

This paper is about that paradox and innovative ways to work through it. We first discuss the nature of natural resource conflict, then address the paradox in some depth. A discussion of traditional public

---

Gregg B. Walker is affiliated with the Department of Speech Communication, Oregon State University, Corvallis, OR 97331-6199.

Steven E. Daniels is affiliated with the Department of Forest Resources, Oregon State University, Corvallis, OR 97331-6199.

[Haworth co-indexing entry note]: "Natural Resource Policy and the Paradox of Public Involvement: Bringing Scientists and Citizens Together." Walker, Gregg B., and Steven E. Daniels. Co-published simultaneously in *Journal of Sustainable Forestry* (Food Products Press, an imprint of The Haworth Press, Inc.) Vol. 13, No. 1/2, 2001, pp. 253-269; and: *Understanding Community-Based Forest Ecosystem Management* (ed: Gerald J. Gray, Maia J. Enzer, and Jonathan Kusel) Food Products Press, an imprint of The Haworth Press, Inc., 2001, pp. 253-269. Single or multiple copies of this article are available for a fee from The Haworth Document Delivery Service [1-800-342-9678, 9:00 a.m. - 5:00 p.m. (EST). E-mail address: getinfo@haworthpressinc.com].

participation precedes innovative methods for working through the paradox. *[Article copies available for a fee from The Haworth Document Delivery Service: 1-800-342-9678. E-mail address: <getinfo@haworthpressinc.com> Website: <http://www.HaworthPress.com>]*

KEYWORDS. Community-based forestry, public involvement, natural resource decision-making, national forest planning

Anyone who has read *High Country News* and regional newspapers in recent years is aware of the major environmental conflicts that dominate the West. Front page stories about the comprehensive nuclear waste clean-up project at Hanford in Washington, efforts to save salmon in the Pacific Northwest's Columbia River system, and the Sierra Club's proposal to remove Glen Canyon Dam in Arizona, all attest to the substantial and complex natural resource management issues confronting agencies, citizens, scientists, nongovernmental organizations, and interest groups.

Immersed in these and similar natural resource and environmental policy situations is the issue of science: the expectation that the "best available science" will be applied to the case at hand so that the best policy management decision will result. In our natural resource conflict work throughout the West, we repeatedly hear from citizens that natural resource management decisions need to be based on "good science" and not on "special interest group politics." Yet we also hear from citizens that they want to be involved in the decisionmaking process and are skeptical about the very science they claim must be the basis for policy actions.

We have noted a similar paradox in the sentiments expressed by natural resource management agency administrators and specialists. Agency personnel know they need meaningful citizen involvement in their management decisions but they also want citizens to trust their scientific expertise. Some agency managers have told us, "If the public only knew what we knew they would agree with our decision . . ." and "If we can just get the public to understand . . ."

The methods we propose to work through this paradox are grounded primarily in communities of place hopefully while being responsive to communities of interest. Natural resource management policies are enacted in many different arenas and levels, but ultimately they are implemented locally. Communities are places where science and interested citizens come together.

NATURAL RESOURCE CONFLICT

Natural resource policy conflicts are inevitable and fundamentally unresolvable (Walker and Daniels, 1997a). That inevitability stems from their complexity. A complex array of factors are typically involved, relating to people and parties, others pertaining to structure. Party-based sources of complexity include: (1) the number of parties, (2) deeply held values, (3) diverse incentives, (4) cultural differences, (5) varied forms and degrees of symbolism, and (6) perceptions of standing. Structure-based sources of complexity are (1) multiple issues, (2) power imbalances related to alternatives, (3) multiple venues, (4) scientific and technical uncertainty, (5) legal and jurisdictional constraints, and (6) an entrenched conflict industry (Walker and Daniels, 1997b). To manage natural resource conflicts and disputes well, federal and state agencies must understand their complexity. Natural resource controversies are rarely single issue or two-party; disputes typically include many interrelated matters, both tangible and symbolic. A dispute over the removal of timber salvage, for example, may involve issues related to visual impacts, fish and wildlife habitat, forest health, industry needs, occupational identity, and community dependence. Stakeholders may include local and national environmental groups, government agencies, small and large timber interests, local community officials, recreation interests, and so on.

Natural resource conflicts, although manageable, may be unresolvable. That difficulty stems from a number of factors. First, natural resource conflicts often involve diverse world views. They are frequently ongoing, reflecting different values, interests, and fundamental views of the relationship between people and the natural environment (Henning and Mangun 1989). These factors may contribute to a perception that the conflict is intractable (Kriesberg 1989). Some conflicts exist at the "ontological" level, featuring value systems diametrically opposed to the situation at hand (Hunter 1989). Parties will not negotiate agreements that conflict with their world views and may make little progress until they understand one another's values and regard them as legitimate, even though they seem incompatible. Assigning economic values to a specific western range, for example, may seem to deny aesthetic or religious values associated with that same land.

Second, natural resource conflicts are often perceived as zero-sum. Conflicts rage over natural resources that are viewed as finite; i.e., a limited amount of rangeland, old-growth forest, water, etc. Zero-sum perceptions encourage parties to adopt rigid positions and seek their goals by means other than negotiation, such as the courts or legislature. Some environmental organizations, for example, will state publicly that no policy decision situation justifies cutting any more old growth or "ancient" forest. Similarly, commodity groups will lobby to change legislation that has provided the basis for court decisions they deem harmful to their interests.

Third, traditional public participation methods have at times escalated conflict with officials acting as arbitrators as groups state extreme demands and use volatile rhetoric, hoping to sway the decision maker/arbitrator toward their preferred outcomes.

Given the complex and unresolvable nature of natural resource and environmental conflicts, methods are needed to "manage" controversial situations. By emphasizing learning, these methods can be responsive to situation complexity in general and the need for scientist-citizen dialogue specifically. Management can be achieved by focusing on of specific disputes that occur within a broader conflict context (Cormick 1982; Crowfoot and Wondolleck 1990). Regardless of the methods used, policy decisions are made both in response to and as an incentive for conflict.

THE PARADOX OF POLICY DECISION MAKING

Within the complexity of natural resource policies situations lies a paradox, a tension that might be called the "paradox of public involvement." On one hand, complex policy situations benefit from cutting-edge scientific and technical knowledge. Scientific and technical experts want this and the public at large expects it. On the other hand, citizens want a meaningful voice in policy decisions that affect the quality of their lives. Environmental policy legislation over the last three decades has institutionalized public participation but not necessarily in ways that work through the "technical expert–inclusiveness" paradox. Effective decisionmaking and conflict management for controversial natural resource management issues requires both awareness of and responsiveness to this paradox.

The Paradox of Science and Citizenry Tension

This juxtaposition between technical competence and open process is a defining characteristic of American policy formation. Citizens demand technically sound decisions, but as situations become more complex few have the technical background to either contribute or critique. These complex situations often touch people's lives in very fundamental ways and our traditions of participatory democracy imply that those people should at least be consulted, if not directly involved. This creates a compelling dynamic that cuts across public policy conflict situations such as nuclear waste disposal, health care, and land management. We desire the efficiency provided by science-based solutions while ensuring the equity associated with meaningful citizen input (Fuller 1997; Collier 1997b). Finding ways to do both is perhaps the most fundamental paradox of effective policy formation.

Several laws illustrate this paradox. The Administrative Procedures Act (APA) of 1946, the National Environmental Policy Act (NEPA) of 1969 and the Federal Land Policy and Management Act (FLPMA) and National Forest Management Act (NFMA) of 1976 all require high quality decision or planning processes, while simultaneously needing either administrative appeals, public disclosure documents, or public involvement activities.

This dilemma has captured the attention of scholars from several disciplines. Political economist Robert Reich (1985) argues that the limited success of post-World War II public administration stems from competing paradigms. Neither the paradigm of administration as technocratic analysis or the paradigm of administration as interest intermediation pay sufficient attention to social learning activities that might benefit both. Daniel Yankelovich (1991), a pollster and political analyst, sees improved political process coming from a citizenry working through the complexity of modern policy situations. According to technical writer James Collier (1997b), the public image of science as uniform and certain is central to this "tension in policy making decisions" (pp. 176-180). Kai Lee (1993), an environmental scientist, contends that sustainable development will come from the integration of adaptive management–to deal with the scientific uncertainty and complexity–and political negotiation, which meets the need for an involved citizenry. Political scientists Pierce et al. (1992) refer to this paradox as the "post-industrial quandary" and see interest groups as playing important roles as information brokers and political actors.

Perhaps it is more important to focus on these authors' diverse academic traditions than on the specific ways that they address the paradox of technical competence and inclusion. Finding an aspect of policy formation that has attracted attention from so many different fields suggests that this is not a superficial notion, but neither is it likely to be easily resolved. Progress in natural resource and environmental policy decisionmaking and conflict management, and indeed progress on many wide-ranging and complex policy questions will depend on how we resolve this paradox. We cannot view as progress wild swings from purely technocratic processes to purely inclusive ones. Our ability to develop processes that can truly identify and further the public interest hinges on our ability to enhance both.

Public Participation and the Paradox

Public participation activities have attempted to provide inclusion while maintaining scientific, technical, and administrative expertise. But they have not systematically brought scientists and citizens together in an interactive, mutual learning environment. Instead, traditional public participation methods have displayed a variety of limitations.

Citizen Input Ambiguity–One aspect of traditional NEPA-type public participation is that the participants cannot tell how their input changes the agency decision. After the comment period on a draft document ends, the agency retreats into internal deliberations, emerging weeks or months later with a final decision and supporting documentation. The public has no way of knowing how their input changed that decision, nor do they have any insight into how seriously (or perfunctorily) their comments were taken. Furthermore, citizens do not see how their ideas–their "local" knowledge–fare in comparison with the "expert" knowledge of agency scientists. And when the situation is not characterized by considerable trust and goodwill, citizens may conclude, perhaps cynically, that their comments did not make any difference.

Limited Constructive Impact–In fact, research substantiates the sense that most public involvement has few positive impacts. Surveys of participants indicate that many feel their input had little or no impact (Lyden, Twight, and Tuchmann 1990; USDA Forest Service 1990); are dissatisfied or mistrustful (Dixon 1993); and value interactive participation methods that favor two-way communication and shared decisionmaking over formal public hearings or letter writing (Force and Williams 1989). Traditional public meetings seem to do little to dispel stereotyped perceptions of disagreement with agency positions (Twight and Paterson, 1979). They may become venting sessions motivated out of generalized resentment and mistrust of public officials (Twight 1977).

The Agency as Arbitrator–Rather than fostering mutual learning among scientists and citizens, traditional public participation methods in natural resource decisionmaking have at times escalated conflict among agencies, stakeholders, and unaffiliated citizens. For example, the game theoretic incentives embedded in traditional public participation increase the likelihood of extreme behaviors. Agency decisionmaking resembles conventional arbitration with the deciding official acting as arbitrator as different public groups make their cases in an effort to affect the decision. Theoretical research shows clearly that participants state extreme demands and use volatile rhetoric because they assume the decision maker/arbitrator will somehow split the difference between the groups. To move the decision its way, each group must be more forceful and compelling than the others. The old maxim of "the squeaky wheel gets the grease" applies here. Such "squeaks" may have little to do with scientific and technical knowledge; the agency as arbitrator may respond to "pressure politics" in ways that diminish scientific as well as citizen interests. Learning-oriented scientist and citizen interaction may be critical to including meaningful knowledge–scientific, local, and indigenous–as an important part of the policy decisionmaking process.

Formal Communication Processes–The relatively formal nature of communication during public involvement processes also tends to affect the quality of the information the agency receives. To comment at a hearing, a participant must speak for the record, often by making a short speech into a microphone before a relatively large assembly. Given the number of people uncomfortable with public speaking, it is likely this formal protocol diminished the quality and quantity of the comments. Only the most motivated speak, resulting in comments that tend to be more extreme than they might be in a setting where dialogue is more natural.

There may also be substantial attitudinal barriers to participating in public involvement processes, even if the structural barriers are low. Some portions of a population may feel so alienated, frustrated, intimidated, or unempowered that they do not enter the process at all.

Highly structured, rule-governed public participation processes encourage formal communication. They foster and reward "advocacy" by offering incentives for parties to lobby and advocate their point of view, to amass evidence, to "debate" their case. Communication competence takes the form of a strategic adversarial advocate, one who can prove that his or her position is superior to others.

Need for Innovation

As natural resource management agencies embrace and implement new land management approaches such as ecosystem-based management and sustainability, they need to utilize public participation activities consistent with these approaches. Ecosystem-based management, for example, emphasizes a continual rather than static planning endeavor. Public participation processes must be adaptive and durable.

There is a trend in public participation toward practices that are more responsive to the complexity, diversity, and systemic nature of natural resource policy situations. First, one is less likely to hear that public involvement is used to educate the public so they "will understand why our proposal is the best thing to do." Second, public participation increasingly is a means to reduce the us-versus-them dynamic and create opportunities for participatory learning. An International Association of Public Participation survey of participation practitioners identified these core values of participation:

1. People should have a say in decisions that affect their lives.
2. Public participation includes the promise that the public's contribution will influence the decision.
3. The process communicates the interests and meets the needs of all participants.
4. The process seeks out and facilitates the involvement of those potentially affected.
5. Participants are involved in defining how they participate.
6. Participants are informed of how their input was, or was not, used.
7. Participants are given the information they need to participate in a meaningful way (Delli Priscolli 1997).

These core values can affect the way scientists and citizens communicate constructively with one another, as well as the design of venues and activities that foster meaningful interaction and learning.

SCIENTIST AND CITIZEN COMMUNICATION COMPETENCE

Scientists and members of the public must be willing to participate meaningfully if they are to work together to resolve natural resource management controversies. Interaction and mutual learning rely on a basic degree of communication competence. All parties must contribute constructively to the process and strive to communicate well.

Communication Competence

Competent communication is perceived as effective in achieving certain objectives while being appropriate to the context in which the activity occurs (Spitzberg 1988, p. 68). The components of communication competence include context; appropriateness and effectiveness; adaptability based on knowledge and motivation; and actions, performing communication behaviors considered appropriate and effective (Lustig and Koester 1993). In complex natural resource management situations communication competence is fostered by developing and implementing discourse and interaction guidelines (e.g., "ground rules" that value diversity), facilitating, and taking stock (ongoing process assessment). Skills associated with learning-based collaborative public participation processes include: (1) listening, (2) questioning and clarifying, (3) feedback, (4) modeling, (5) social cognition, (6) dialogue, and (7) collaborative argument techniques (Walker and Daniels 1997c).

Scientist Communication Competence

Constructive and substantive citizen-scientist dialogue begins with the scientific community. Citizens expect natural resource management decision-makers to base policy judgments on the best available knowledge, which includes the best available scientific and technical studies. Citizens also want to contribute their knowledge of the situation–their local and indigenous knowledge–and want scientists and managers to respect and be responsive to their local expertise. Scientists can foster respect and promote meaningful dialogue by communicating competently. Scientific activity is fundamentally different from many other human endeavors (Keith 1997), but knowledge must be communicated effectively to influence policy. Influence relies on communication competence. This competence encompasses, but is not limited to, the following:

1. *Assessing and adapting to the communication situation.* This includes analyzing citizens as an "audience" understanding what they know, what they want to contribute, what the scientist hopes to learn from them, and what the scientist would like them to learn (Collier 1997a).
2. *Respecting cultural preferences in learning and knowing.* Cultural communities vary in the ways they prefer to learn and in the ways of knowing that they value.
3. *Employing clear language.* Scientists should minimize jargon and acronyms. Clarity does not mean "talking down." Rather it relies on appropriate language frames, such as metaphors and illustrations.
4. *Active, open-minded listening.* Scientists need to listen openly, taking notes on the information citizens contribute. Just as citizens question scientists about what they know, scientists should question citizens about their expertise and listen actively in order to understand local and indigenous insights.
5. *Patience.* Learning about complex natural resource situations takes time. Scientists need to be patient with citizens and with themselves as a mutual learning environment emerges. Citizens and scientist often understand a situation differently. To learn, to make progress, and to discover points of agreement requires a patience and tolerance.
6. *Self-awareness and self-monitoring.* Scientists need to monitor their nonverbal and verbal behavior as they interact with citizens about natural resource management controversies. Through self-monitoring, scientists can better model communication competency behaviors that promote mutual learning and that they hope citizens will exhibit.

Citizen Communication Competence

In a meaningful scientist-citizen dialogue, citizens, also, have a responsibility to communicate competently. They need to listen actively and be open to learning. Citizens can invest in understanding the complexity and systemic nature of a situation by according scientists the appropriate space and place to share their technical knowledge. Citizens should use clear language and provide evidence to support their claims. Citizens also need to recognize science is uncertain and that there is no scientific or technical "silver bullet" where complex natural resource management issues are concerned.

As agencies embrace and implement new resource management approaches, such as ecosystem-based management and sustainable development, they need to utilize public participation activities consistent with those approaches. Ecosystem-based management, for example, emphasizes a continual rather than static planning endeavor. Public participation processes must be adaptive and durable. Given the scientific and technical uncertainty of natural resource policy situations and a motivated citizenry, venues and approaches must integrate scientific knowledge and concerns with of those citizens. Current management philosophies and democratic principles call for a new generation of public participation approaches that confront the paradox, with structures and tasks that emphasize learning and interaction between scientists and citizens.

CONFRONTING THE PARADOX: METHODS FOR CITIZEN AND SCIENTIST INTERACTION

Innovative methods of public participation provide opportunities for working through the paradox of public involvement in natural resource policy situations. These methods typically emphasize collaborative interaction: interdependent parties work together to affect the future of an issue of shared interest (Gray 1989). More specifically, Gray (1985) defines collaboration as "the pooling of appreciations and/or tangible concerns, e.g., information, money, labor, etc., by two or more stakeholders to solve a set of problems which neither can solve individually" (p. 912). Drawing upon Gray's ideas, Selin and Chavez (1995) assert that "collaboration implies a joint decisionmaking approach to problem resolution where power is shared, and stakeholders take collective responsibility for their actions and subsequent outcomes from those actions" (p. 190). This section proposes methods for bringing together science and citizenry. The list is not exhaustive; rather, it provides a starting point for creative thinking.

Study Groups

Study groups represent a recent innovation in public involvement, one that has proven conducive to the interaction of scientific and indigenous knowledge. Currently sponsored by the Community-Public Lands Partnership (a.k.a. the Ponderosa Pine Partnership) of southwestern Colorado, study groups have played a role in the San Juan National Forest's forest plan revision process (Richard and Burns, 1998). People voluntarily joined one or more study groups that correspond to the forest plan revision's major issue areas, such as fish and wildlife habitat, vegetation management, and access and travel management. Study groups include federal and state agency scientists, concerned citizens, and members of interest groups (some of whom are scientists). Study groups meet monthly (and in some cases more often) to discuss research reports, management options, and other matters pertinent to the issues at hand.

Field Trips

Natural resource management agencies use field trips as part of the planning process. Frequently, though, field trips include members of a particular interest group or compatible groups, rather than those with diverse interests who may hold very different goals and positions about the land. A field trip can provide a rich opportunity for scientists to interact with members of different interest groups in a mutual learning environment.

Many community-level collaborations use field trips to foster scientist-citizen interaction. Organizations such as the Applegate Partnership of southwestern Oregon, the Cispus Adaptive Management Area group of southwestern Washington, and the Community-Public Lands Partnership have periodically sponsored field trips to areas of their region pertinent to important management issues. Whether tied to a study group process or simply an activity related to a natural resource management planning effort, field trips are an effective way for citizens and scientists to interact "on the ground."

Science/Scientist Consensus Forums or "Science Summits"

In some complex natural resource policy areas, such as the restoration of salmon runs in the Pacific Northwest's Columbia River basin, there seems to be significant conflict over the "science of the situation." Scientists need venues in which they can interact collaboratively to work through science conflicts. In a science "consensus forum," scientists–from federal and state agencies, sovereign Indian nations, interest groups, and industry–can come together in an extended (e.g., multi-day) forum to "argue constructively" about research needs, data, and scientific interpretations. Through "collaborative argument," argument that is both inquiry and advocacy (Senge 1990), scientists can discover where consensus exists, where it is not likely, and what they can contribute toward meaningful progress.

A science consensus forum can provide the foundation for a broader policy discussion that involves citizens and policymakers. A science consensus forum could produce a report that clarifies areas on which scientists agree, the areas of disagreement, and research needs. The report could be distributed to interested parties who might then come together in a subsequent forum, such as a "science and citizenry dialogue" to discuss the findings and their implications.

Scientist-Citizen Dialogues

As a follow-up to a science consensus forum or in situations where study groups may not be feasible, organizations and agencies can sponsor dialogue venues that bring together scientific and local/indigenous knowledge. Ecosystem-based management involves a variety of systemic dimensions: physical, biological, economic, social, and cultural. Natural resource agency, interest group, university, and industry scientists generally contribute "expert" knowledge in the physical and biological areas. Within any community in most natural resource management situations, there are people with rich local or indigenous knowledge of the issues at hand. This knowledge may be key to understanding the cultural and social dimensions of ecosystem-based management; local knowledge can contribute to understanding the physical, biological, and economic aspects of the situation as well.

For scientist-citizen dialogues to be effective, both the scientist and the citizen must be respectful of one another's knowledge in an atmosphere of mutual learning. All parties must be open to sharing ideas and willing to risk that their views might change through dialogue. Given the complexity of natural resource management situations, no one discipline or knowledge source has a monopoly on understanding the situation (Daniels and Walker 1997).

Citizen-Scientist Project Collaboration

Study groups and dialogue forums emphasize social learning, learning processes in which all parties are committed to learning from each other. These venues can lead to more substantive working relationships between agency scientists, nongovernmental organization and interest group scientists, and citizens. Citizens and scientists who have participated in Collaborative Learning workshops we have conducted tell us that, while they find meetings useful, they prefer to be out, practicing stewardship rather than simply talking about it. Field projects, led by scientists and relying on citizen participation, can foster interaction that works through the paradox. Project collaboration may take a variety of forms, including research (e.g., field site development, data collection), implementation (e.g., planting seedlings in riparian area, installing fish screens in a watershed), and monitoring (e.g., fish counts, water temperature readings).

Given declining agency staff and budget resources, the need for citizen participation in science projects may be more critical than ever. On the Kenai Peninsula in Alaska, for example, the Kenai Watershed Forum, a community-based collaborative group, is leading a water quality monitoring effort that includes agency specialists training citizens in monitoring techniques and technology. A variety of state and federal agencies with Kenai River watershed management responsibilities will use the data those trained citizen volunteers gather.

Citizen Assessment and Planning

Assessment is another way to work through the paradox and foster mutual learning between scientists and citizen communities. Natural resource management agency scientists and specialists have been trained to perform analyses, but that expertise may exist outside agencies as well. For example, in Northern California, with funding from a resource conservation development council, a nonagency team of local specialists has prepared a watershed analysis that compares favorably with similar analyses developed by agencies.

Citizen Environmental Impact Statements (EIS) represent a second type of planning project nonagency experts may develop. Interest groups have frequently contributed an alternative to agency prepared environmental impact statements. Citizen EIS may address issues not featured in an agency EIS. For example, a recent feature article in *High Country News* reports that the Sierra Club has sponsored the preparation of an EIS concerning the removal of Glen Canyon Dam.

A third way for citizens to participate in an assessment is as adjunct members of planning teams. For example, a USDA Forest Service interdisciplinary team could include citizens as participant-observers in Interdisciplinary (ID) team meetings and other activities. Although care must be taken to ensure that citizen involvement is compatible with relevant statutes, such as the Federal Advisory Committee Act, having citizens on the adjunct planning team could foster trust and dialogue between scientists and members of the public.

The Chugach National Forest land management plan ID team is attempting, on a modest scale, to involve citizens as adjunct members in two ways. First, the dates and times of ID team meetings are publicized so that interested citizens can attend and observe. Although citizens cannot participate in the actual meetings; they can talk with ID team members before and after meetings. Second, as the ID team works through public comments (generated via Collaborative Learning workshops and other comment opportunities), citizens are consulted regularly for feedback on the team's ideas and interpretations (Daniels and Walker, in press).

As mentioned earlier, these activities are proposed as methods for bringing together the seemingly competitive tensions of expertise and inclusiveness. It is possible to meet the dual goal of applying the best available science to a natural resource policy situation and involve citizens meaningfully. Scientists and citizens must be brought together in venues where mutual learning can occur and both expert and local knowledge can be realized. These venues may involve the design of new, innovative methods or may draw upon existing approaches to collaboration (e.g., Burgess and Burgess 1996; Diemer and Alvarez 1995; Sirmon et al. 1993).

CONCLUSION

Given their complexity, natural resource policy decision situations are inherently controversial. The conflicts that accompany them seem manageable but not unresolvable. Ecosystem management as a land management philosophy is not immune from controversy and conflict. Its multi-systems nature, in fact, implies that conflict should be an accepted feature. This conflict is often particularly visible in communities of place, where people's concerns about work, family, the natural environment, and quality of life coalesce.

To understand the policy challenges posed by ecosystem-based management, this discussion of integrating science and citizenry should be viewed as illustrating some of the policy processes that ecosystem-based management is likely to require. Public policy processes that have a rich agency/citizenry dialogue at their core have been developed in recent years, dating back at least to Freidmann's transactive planning (1973). The social and political forces that spawn these efforts are at least as likely to increase as to abate.

Designing policy processes that can accommodate ecosystem-based management and can work through the paradox requires matching the tool to the task. We must think carefully about the fundamental attributes and challenges of ecosystem-based management, then design compatible systems that are robust in the face of challenges. Doing less is analogous to screwing in a light bulb with a hammer or painting a room with a screwdriver.

Scientist and citizen interactions seem particularly important for handling ecosystem-based management conflicts because they can be designed to specifically address the policy challenges of mixed public lands. Collaborative Learning (CL), a method we have been developing, applying, and refining over the past few years, provides a case in point (Daniels and Walker 1996; Daniels et al. 1996; Walker et al. 1996). CL has three features that make it well suited to integrating science and citizenry as a part of ecosystem-based management: (1) It explicitly adopts a systems approach to the situation and works to improve the participants' understanding; (2) It is more modest in its expectations for progress than the more frequently used rational-comprehensive models, which seek solutions; and (3) It expects and attempts to accommodate a wide range of world views about land management and the strategic behaviors that those world views are likely to generate in controversial situations.

We believe that any situation in which ecosystem-based management makes progress will be identified by at least these three characteristics: a systems approach, realistic goals, and high political acumen. Whether these attributes arise spontaneously or are carefully thought out is immaterial. If a structured method is used, it needs to be appropriate to the situation, whether it goes by Collaborative Learning, Science Consensus Forum, or any other name. Perhaps the only thing that matters, at the core, is that it make progress on the paradox of public deliberation: It must generate technically sound decisions while allowing stakeholders a meaningful voice in the process. The scientific burdens of ecosystem-based land management, combined with the range of interests in the mixed public/private lands, appear to require nothing less.

REFERENCES

Burgess, H., and G. Burgess. 1996. Constructive confrontation: A transformative approach to intractable conflicts. *Mediation Quarterly*, 13: 305-322.

Collier, J.H. 1997a. Understanding audiences. In J.H. Collier (ed.), *Scientific and Technical Communication: Theory, Practice, and Policy.* Thousand Oaks, CA: Sage.

Collier, J.H. 1997b. Participation and policy. In J.H. Collier (ed.), *Scientific and Technical Communication: Theory, Practice, and Policy.* Thousand Oaks, CA: Sage.

Cormick, G. 1982. Intervention and self-determination in environmental disputes: A mediator's perspective. *Resolve*, Winter:1-7.

Crowfoot, J. and J. Wondolleck. 1990. *Environmental Disputes: Community Involvement in Conflict Resolution.* Washington, DC: Island Press.

Daniels, S.E., and G.B. Walker. 1996. Collaborative learning: Improving public deliberation in ecosystem-based management. *Environmental Impact Assessment Review*, 16: 71-102.

Daniels, S.E. and G.B. Walker. 1997. Collaborative processes and scientific information. Oregon Chapter American Fisheries Society annual meeting, Gleneden Beach, OR.

Daniels, S.E. and G.B. Walker. In press. *Working Through Environmental Policy Conflict: The Collaborative Learning Approach.* Westport, CT: Praeger.

Daniels, S.E. et al. 1996. Using collaborative learning in fire recovery planning. *Journal of Forestry* 94(8):4-9.

Delli Priscolli, J. 1997. Participation and conflict management in natural resource decision-making. In B. Solberg and S. Miina (eds.), *Proceedings: Conflict Management and Public Participation in Land Management.* Joensuu, Finland: European Forest Institute.

Diemer, J.A. and R.C. Alvarez. 1995. Sustainable community–sustainable forestry: A participatory model. *Journal of Forestry* 93(11):10-14.

Dixon, K.M. 1993. The relationship of benefits and fairness to political confidence in the U.S. Forest Service. M.S. thesis, School of Renewable Natural Resources, University of Arizona, Tucson.

Force, J.E. and K.L. Williams. 1989. A profile of National Forest planning participants. *Journal of Forestry* 87:33-38.

Freidmann, J. 1973. *Retracking America: A Theory of Transactive Planning.* Garden City, NY: Anchor Press.

Fuller, S. 1997. Putting people back into the business of science: Constituting a national forum for setting the research agenda. In J.H. Collier (ed.), *Scientific and Technical Communication: Theory, Practice, and Policy.* Thousand Oaks, CA: Sage.

Gray, B. 1985. Conditions facilitating interorganizational collaboration. *Human Relations* 38:911-936.

Gray, B. 1989. *Collaborating*. San Francisco: Jossey-Bass.

Henning, D. and W. Mangun. 1989. *Managing the Environmental Crisis*. Durham, NC: Duke University Press.

Hunter, S. 1989. The roots of environmental conflict in the Tahoe basin. In L. Kriesberg, T. Northrup, and S. Thorson. (eds.), *Intractable Conflicts and Their Transformation*. Syracuse, NY: Syracuse University Press.

Keith, W. 1997. Science and communication: Beyond form and content. In J.H. Collier (ed.), *Scientific and Technical Communication: Theory, Practice, and Policy*. Thousand Oaks, CA: Sage.

Kriesberg, L. 1989. Conclusion: Research and policy implications. In L. Kriesberg, T. Northrup, and S. Thorson (eds.), *Intractable Conflicts and Their Transformation*. Syracuse, NY: Syracuse University Press.

Lee, K. 1993. *Compass and Gyroscope: Integrating Science and Politics for the Environment*. Washington, DC: Island Press.

Lustig, M. and J. Koester. 1993. *Intercultural Competence: Interpersonal Communication Across Cultures*. New York: Harper Collins.

Lyden, F.J., Twight, B.W., and T.E. Tuchmann. 1990. Citizen participation in long-range planning. *Natural Resources Journal* 30:123-138.

Pierce, J., Steger, M., Steel, B., and N. Lovrich. 1992. *Citizens, Political Communication, and Interest Groups: Environmental Organizations in Canada and the United States*. Westport, CT: Praeger.

Reich. R.B. 1985. Public administration and public deliberation: An interpretive essay. *Yale Law Journal*, 94: 1617-1641.

Richard, T. and S. Burns. 1998. Beyond "scoping": Citizens and San Juan Forest managers, learning together. *Journal of Forestry*, 96(4):39-43.

Selin, S. and D. Chavez. 1995. Developing a collaborative model for environmental planning and management. *Environmental Management* 19:189-195.

Senge, P. 1990. *The Fifth Discipline: The Art & Practice of the Learning Organization*. New York: Doubleday Currency.

Sirmon, J., Shands, W.E., and C. Liggett. 1993. Communities of interests and open decisionmaking. *Journal of Forestry* 91(7):17-21.

Spitzberg, B. 1988. Communication competence: Measures of perceived effectiveness. In C. Tardy (ed.) *A Handbook for the Study of Human Communication*. Norwood, NJ: Ablex.

Twight, B.W. 1977. Confidence or more controversy: wither public involvement? *Journal of Forestry* 75:93-95.

Twight, B.W. and J.J. Paterson. 1979. Conflict and public involvement: measuring consensus. *Journal of Forestry* 77:771-774.

USDA Forest Service. 1990. *Critique of Land Management Planning, Vol. 2. National Forest Planning: Searching for a Common Vision*. Washington, DC: FS-453

Walker, G.B. and S.E. Daniels. 1997a. Foundations of natural resource conflict: Conflict theory and public policy. In B. Solberg and S. Miina (Eds.), *Conflict Management and Public Participation in Land Management*, EFI Proceedings No. 14. Joensuu, Finland: European Forest Institute.

Walker, G.B. and S.E. Daniels. 1997b. The complexity of natural resource conflict: implications for conflict management. Oregon Chapter American Fisheries Society annual meeting, Gleneden Beach, OR.

Walker, G.B. and S.E. Daniels. 1997c. Communication competence and public participation in natural resource policy decisions. Paper presented at the annual meeting of National Communication Association, Chicago, Illinois.

Walker, G., Daniels, S., Blatner, K., and M. Carroll. 1996. Civic discovery and ecosystem-based management: Collaborative learning and fire recovery planning. Speech Communication Association annual meeting, San Diego, CA.

Yankelovich, D. 1991. *Coming to Public Judgment: Making Democracy Work in a Complex World*. Syracuse, NY: Syracuse University Press.

A Civic Conversation About Public Lands: Developing Community Governance

Sam Burns

SUMMARY. Community-based ecosystem management (CBEM) is being offered as an alternative to *agency*-based public land management. Its fundamental premise is connecting communities to public lands for the purpose of increasing ecosystem stewardship and community sustainability. For CBEM to appropriately serve the public interest, new social and institutional relationships will need to be formed, collaborative learning will need to occur, and capacities in community participation and ecological literacy will need to be developed. A new civic conversation about public lands is essential to these relationships, learning, and capacities. Community governance processes, which guide these collaborative and participatory activities, must correspondingly reflect the community-based nature of CBEM, while preserving the national interests in public land resources. *[Article copies available for a fee from The Haworth Document Delivery Service: 1-800-342-9678. E-mail address: <getinfo@haworthpressinc.com> Website: <http://www.HaworthPress.com>]*

KEYWORDS. Community-based forestry, public lands management, community governance, collaboration

A VOCABULARY OF SOCIAL DISUNITY

Especially in the Western United States, the debate over the future of public lands has grown ever more contentious. Its vocabulary includes the

Sam Burns is affiliated with the Fort Lewis College, Office of Community Services, 1000 Rim Drive, Durango, CO 81301.

[Haworth co-indexing entry note]: "A Civic Conversation About Public Lands: Developing Community Governance." Burns, Sam. Co-published simultaneously in *Journal of Sustainable Forestry* (Food Products Press, an imprint of The Haworth Press, Inc.) Vol. 13, No. 1/2, 2001, pp. 271-290; and: *Understanding Community-Based Forest Ecosystem Management* (ed: Gerald J. Gray, Maia J. Enzer, and Jonathan Kusel) Food Products Press, an imprint of The Haworth Press, Inc., 2001, pp. 271-290. Single or multiple copies of this article are available for a fee from The Haworth Document Delivery Service [1-800-342-9678, 9:00 a.m. - 5:00 p.m. (EST). E-mail address: getinfo@haworthpressinc.com].

phrases–range of natural variability, forest restoration, range reform, endangered species, county supremacy, communities of place, multiple use, good science, forest health, national interests, wise use, adaptive management, stewardship contracting, and ecosystem management, among numerous others. While each concept has a positive meaning intended by certain advocates, it often triggers an opposite interpretation by another interest group. These words, while well-intentioned for the most part are creating and reflecting a depth of social disunity that many citizens and policymakers believe is increasingly untenable within many regions of the United States, perhaps more so in the West, because federal land ownership is so dominant there.

For example, the use of the term "ecosystem management," considered by many as a more progressive way to care for land and natural resources in the public domain, conveys to many traditional users of public lands that they will lose opportunities for "multiple use." "Forest health," when used as a rationale by advocates of forest restoration, is viewed skeptically by environmental interests, as simply another devious method for producing timber products. Most recently, a sharp critique has evolved about resource management processes grounded in "communities of place," by proponents of "national interests," because they advocate what they believe are higher ecological standards and a reliance on federal legal mandates, and deeply mistrust "local interests."

Absent an authentic, civic conversation, courts rather than the authorized public agencies are managing public lands. Legislatively mandated or presidentially negotiated forest plans, and a moratorium on US Forest plan revisions are outcomes of a governmental and civic process to manage public lands which is fractured at key points–public awareness, participation and support. The processes of public land resource planning, management and conservation have simply not been working, whether you take an ecological perspective or that of a traditional user. In what ways is this happening?

While federal land managers must by law make the final management decisions, an absence of well-grounded consensual problem solving is harming the social and economic fabric of communities and regions. Healing public lands, desired by everyone, is clearly not occurring in any significant ways, whether measured by standards of biodiversity, a range of natural variability, vegetation restoration or retention of native species. While land resources remain in a poor state of health, the US Congress is limiting the capacity of authorized public land agencies to complete their assigned work by cutting fiscal resources needed for monitoring and restoration, by restricting planning processes, and turning resource management into a political controversy rather than a much needed formative dialogue. It is indeed the worst of all possible worlds, disempowered communities and unhealthy lands–considerable social and economic pain experienced by public land dependent communities with little if any public benefit to communities or the land as a result of proposed ecosystem related management initiatives. While there are many community-forest examples of this situation, the most notable among them is Catron County, New Mexico and the Gila National Forest.

Even though the degree of resource degradation, as well as the level of community trauma and reaction varies considerably across federal and state public land systems, the underlying problem and challenge is quite similar. Public lands and their official stewards are caught in a rhetorical crossfire, blanketed by a litany of cross cutting, polarizing words, conveying disparate social, economic and ecological values. The formal organizational structures, the legal and institutional patterns for public land resource planning and management, are not attuned to the types of collaborative discussion, learning, adaptation, and institutional change needed to move from a resource extractive economic dependency model to one that prioritizes adaptive, resource sustainability and community stewardship as the new vision for public lands.

Those persons, interests, advocates and voices that need to be talking, building relationships, conversing, and finding common ground solutions do not have an adequate chance to hold a civic conversation. They are not conversing soon enough with each other. Rather, they are forced to use means of communication of the "last resort," formalistic and legalistic governance structures of public hearings and appeals and lawsuits to polarize, delay, and ultimately derail most opportunities for seeking common ground.

This paper rests on the premise that authentically recovering the health of public lands will require a new commitment to a new civic conversation within and between both communities of interest and place. Whether one believes current patterns of *governance* and community deliberation for public land management in the United States are in deepening crisis or merely going through a much needed evolution in philosophy and direction, most agree that there are no silver bullet solutions. The basic fact that land resources are *public* sets the stage for multiple perspectives, which in turn generates a set of conflicting values and intentions that, when left unresolved, are capable of further damaging the natural well-being of forests, watersheds, rangelands, and nearby communities caught in a disunifying tidal wave.

Simultaneously, and just as destructive, are the negative impacts of conflicting claims and actions upon the social and cultural health of communities, especially those adjacent to public lands. Communities whose current and future quality of life is tied to public lands can neither plan for themselves nor provided the public land stewardship of which they are capable. Because for several decades we have not possessed socially effective, institutional means of public decision making, we have disempowered communities of place from cooperatively contributing to the healing of public lands. This contribution is increasingly a necessity with the downsizing of public land management agency resources.

Among the perspectives being expressed about the future of public lands are those that seek significant reductions in resource utilization in order to rehabilitate or restore natural ecosystems. In most regions of the country, traditional forms of commodity use, such as ranching and timbering, are being replaced by outdoor recreation activities–hiking, biking, four wheeling, fishing, mountain climbing, hunting–most of which present a new set of resource and social impacts on both the land and nearby communities. As a result, many rural communities see older and more traditional forms of resource use as increasingly more threatened and their way of life as being destroyed. Within the array of contending perspectives, some of the more vocal interests claim to speak for national interests and others for local interests, some claim sustainability, some "wise use" and others claim an ecosystem orientation. Other interests, often labeled as environmental or ecological, are characterized as claiming that local or community-based interests are economically self serving, short-term, parochial, and unconcerned with resource sustainability.

Accordingly, the public debate has become overwhelmingly defined by a polarizing, socio-political extreme vocabulary, drowning out a common ground conversation at the center. Whether these centering views were arrived at through historical tradition, formal negotiation or community debate, they have been overpowered by legal or administrative gridlock. Seeing escalating community conflict and no ecological progress, alternative voices are now arising to argue that another approach to management, planning, and stewardship is needed–one based not in ideology, but on maximizing community stewardship to further what works on the land.

CONNECTING ECOSYSTEMS AND COMMUNITIES

A new framework for public land management and decisionmaking has been emerging since the early 1990s. This alternative perspective recognizes the *connection of communities to ecosystems* in different ways than previously considered and is defined by various names–community forestry, community-based forestry, community-public land partnerships, community ecosystem stewardship, and for our purposes here will be called community-based ecosystem management (CBEM). Because CBEM is still evolving, the process has a number of sometimes indigenous and everyday, informal or vernacular meanings. Even with these variations, we can identify a number of values and conditions that set the stage for various emerging community forestry or public land stewardship strategies.

1. Many individuals, communities, and social institutions have experienced the dysfunctional, fragmented, centralized institutional patterns of managing public land resources.
2. Long-standing, so-called rational, scientific analysis does not insure that the most sustainable management solutions will be developed or selected.
3. Land management is characterized by political, cultural, and administrative gridlock.
4. Local communities believe they have not had an equitable voice in determining the future of public lands.
5. Increasing losses of economic benefits and the way of life in rural communities have occurred, particularly in those which have been most economically dependent on public lands.
6. Even with more stringent environmental protection laws, public lands still suffer from neglect due to the disempowerment of land management agency staff.
7. New interests have developed in improving reinvestment in ecological restoration, job retraining, and sustainable, cooperative, economic development processes.
8. A belief has grown that local communities need to be directly involved in land stewardship, because of their knowledge and skills, and the lack of traditional governmental resources.
9. The desire to make both communities and public lands more sustainable is increasingly more prominent, providing a new opportunity for achieving common ground alternatives.
10. Many citizens and resource practitioners see the necessity for civil balanced debate about the future of public lands, rather than a civil war of ideological extremes.

In a recent issue of *American Forests*, Gerry Gray and Jonathan Kusel outlined four essential components of the community-based ecosystem management perspective (Gray and Kusel, 1998, pp. 28-29):

Process Community-based practitioners feel–after two decades of "public involvement" in which their comments have been synthesized, coded, counted, considered too late, or taken out of context–they have had little or no impact on what happens to the forests that surround them.

What do they want? A more open process that includes local know-how–knowledge that comes from being "of" a place–and that involves a variety of perspectives, just as a community is made up of a variety of interests.

. . . What they do not want, they say, is to call the shots.

Stewardship Stewardship seeks to define an enduring, reciprocal relationship between people and the natural environment; in this case, between communities and the forests that sustain them.

Ideally, that relationship is based on respect and responsibility for forest ecosystems, allowing people to learn through both their actions and their observations. In simple terms it is "the forest sustaining the community and the community sustaining the forest," Communities Committee chair Lynn Jungwirth says.

Reinvestment Community-based activists contend that forests historically have been exploited, their resources extracted with little thought given to reforestation and maintenance. They argue that much of the economic value from these ecosystems has been transferred out of the local area. Left behind, too often, are degraded forests and watersheds that need restoration work but that have insufficient resource values remaining to pay for it.

. . . Reinvestment focuses on the need for restoration and maintenance of natural or ecological capital on public and private lands. Community-based advocates want a mechanism that allows them to use a variety of funding options for activities on public lands and incentives that entice private landowners to better maintain their land.

Monitoring Because we don't know everything about how an ecosystem works, all ecosystem management is a kind of experiment. To learn from our successes and failures requires monitoring of ecological, biological, social, and economic conditions.

. . . All party monitoring is based on the notion that various parties with different local, national, ethnic, cultural, economic, and educational orientations should share responsibility for monitoring and assessing what they deem vital. Combining their knowledge and interests, they monitor the health of an ecosystem, the health and well-being of affected workers and communities, and the quality of management.

We are able to discern in these four essential focus areas a willingness and desire on the part of many local communities to establish cooperative partnerships with public land agencies, to improve public land resources by increasing participation, stewardship, responsibility, equitable financing of reinvestment, and all-party monitoring. Considerable individual energy, community conversation, and practical experience are now being dedicated to focusing and defining this new framework for public land management. It is being offered as a better way *to care for the land and sustain communities* by many types of practitioners of community-based ecosystem management. The evolution of this work must continue in order to clarify the new stewardship enterprise itself, and to demonstrate to its critics that community-based ecosystem management has social and ecological benefits worth considering.

Among the critics, the characteristic of CBEM's communal or partnership activities that seems most problematic is the notion that public land managers would enter into a *collaborative* process with communities. Collaborative, community-based efforts are being criticized as being a co-optation of public processes by communities of place with parochial interests. Critics of the community-based approaches believe collaborative community groups will continue the same resource management practices that have previously raped and destroyed the public lands. If community-based ecosystem management is going to have an opportunity to demonstrate itself to be publicly beneficial, it's collaborative activities must be guided by governance process based in civic conversations, viewed as equitable and inclusive of all interest advocates, who begin to view themselves as community citizens in the broadest sense of this term.

COMMUNITY GOVERNANCE NEEDS

The need for a civic, accountable, sustainable, equitable structure for public land decisionmaking is revealed quite clearly in a recent article in *Chronicle of Community* (Coggins 1998), in which George Cameron Coggins addresses what he refers to as "some perils of devolved collaboration," which he observes as a central principle of community-based approaches. A series of Coggins' statements well underscore the need to reconcile concerns of equity, authority, and stability in community governance processes (Coggins, 1998 pp. 27-33):

> . . . the values and premises of participants in resource allocation controversies sometimes–perhaps most times–differ irreconcilably, and no amount of talking is going to change that.
>
> . . . allocation and conservation of federal lands and resources necessarily and properly are national, not local, questions.
>
> Whatever the faults of the BLM or the Forest Service, the appropriate remedy is neither abdication nor vigilantism, both attributes of collaborationism.

> Groups assembled to collaborate are transient entities. Without guaranteed long-term funding, they will have no professional staff, and thus will lack an institutional structure.
>
> Without the force of law, local plans and local allocative solutions are written on the sand.
>
> The federal lands are national assets, not local fiefdoms. The public natural resources are not just local storehouses to be looted in the deregulation riots. Local collaborative solutions could impede or contravene national laws or policies.
>
> . . . local collaborators will have little if any incentive to insure adherence to positive, substantive legal rules.
>
> New collaborative mechanisms and structures and methods are unnecessary. All of these efforts try to work outside legally ordained decision-making processes.
>
> Current federal law does not recognize collaboration/devolution as an appropriate means of public land decisionmaking.

Whether you agree partially, whole-heartedly, or not at all with Coggins' list of indictments, with his skepticism about finding common ground about public lands, or with his facility for turning collaboration into collaboration-*ism,* for our purposes here, he does outline the many governance questions of CBEM. That is, Coggins' intensive, legalistic evaluation of community-based ecosystem management underscores the need to ask some tough, formal, long-term, socio-political and public interest questions. How can community-based ecosystem management processes be governed in the public interest through a stable and just social institutional process? What institutional measures need to be established or affirmed in order to insure that the common good is protected as new community ecosystem stewardship processes are attempted, demonstrated, monitored, and evaluated.

Indeed, we must raise the most important question about appropriately governing and about establishing safeguards for the public with regard to their lands. Think back over the points about social disunity, polarization, and the lack of common ground raised thus far, and then ask, is it a matter of formulating new, formal, *legal-rational governance institutions*, or is it necessary to develop a new civic conversation that would inspire a collaborative method of *community governance?* And if more formal governance institutions are needed in time, can they not more sensitively reflect the civic conversations that most CBEM proponents envision as this movement unfolds.

If community-based stewardship or ecosystem management is to proceed toward successful implementation, resting upon the four foundational elements noted above–process, stewardship, reinvestment, and monitoring, a fifth undergirding principle needs to be worked out as well. For sake of brevity, let us refer to this feature as *civic governance*, but let this shorten form stand for a set of socially and communally recognized methods of enhancing and protecting *public interests* in *public land* resources, while preserving the critical tenets of community-based, collaborative stewardship.

COMMUNITY GROUNDED CIVIC GOVERNANCE

In the early days of the community-based ecosystem management movement, Margaret Shannon, a member of the 1998 Committee of Scientists charged with producing a new set of planning regulation for the US Forest Service, presented a paper entitled "Community Governance: An Enduring Institution of Democracy." The paper was presented at a workshop entitled Multiple Use and Sustained Yield: Changing Philosophies for Federal Land Management? (March 5-6, 1992, Committee on Interior and Insular Affairs, US House of Representatives). Shannon began her talk with a crucial goal statement:

> We must reorganize public land management and administration so as to improve the quality of debate, to enhance the conditions of discussions, and to persuade those unmoved by the imperative need to pay attention to the quality and condition of the ecosocial community of a natural forest. (p. 219)

Shannon's statement helps us frame several important and much needed actions: *reorganization of public land management and administration, improving the quality of debate, enhancing the conditions of discussion, and the imperative need to pay attention to the ecosocial community of a national forest*. These foci are mutually supportive and interdependent, and establish important objectives for a governance structure. If to these themes are added values elsewhere being put forth, such as sustainability of the ecosystem, sustainability of communities, increased stewardship, etc., CBEM proponents have coherently crafted a new vision of public land governance.

Shannon goes on to argue that "the lack of a civic republican model of governance is because of our shackling of the public as *discrete interests* who simply seek to influence expertise to achieve narrow ends" (p. 220) (emphasis added). This is another way of saying, with regard to both public decision-making processes and outcomes, that the public land manager has invariably been caught in the middle between discrete, segmented interests, with no acceptable means of listening to and gathering the *public* interest. A little further on she expresses concern that there is a "lack of institutions for effective community governance, and thus the easy domination of experts over the domain of public life" (p. 221).

Shannon continues,

> . . . The solution to the problem of *the Public* and the problem of public participation in administrative decision making, lies in the necessity to create, empower, and use political institutions which are based on and require *citizenship*. (p. 228)
>
> . . . Rather, it is that the current organizational structure and current policy and programs are supported by the questions framed as *private claims* to public resources. (p. 231) (emphasis added)

Shannon concludes by saying, "recognition that governance requires institutions which are based upon civic literacy . . . means that current political institutions which reward interest driven behavior . . . must be redesigned so as to require civic conversation when the civic good is at stake" (pp. 235-236). Here most deeply she gets to the heart of the matter–to the need for "civic conversation" which must in a very fundamental and abiding sense be based upon a "civic literacy," which will then make less likely the use of a vocabulary of disunity. If *civic* conversations about *public* land resources are to be fairly held, they must be constructed with a high degree of caring awareness about eco-social communities and natural resource systems. In other words, if civic conversations can hope to achieve publicly acceptable and beneficial outcomes, i.e., the civic good, they must rest upon a more solid foundation of community and ecosystem literacy. W*hat then are the fundamental and essential civic governance components of a "ecosocial" community ecosystem stewardship process, or what we have been calling CBEM?*

CIVIC CONVERSATION AND ECOSOCIAL STEWARDSHIP

Over the last six years in Southwest Colorado, involving the San Juan National Forest, a number of public land planning and stewardship activities have taken place that have brought ecosystem literacy and civic conversation into a community governance process capable of supporting community-based ecosystem management. Details about these processes have been described in other places and center around four themes:

- A local county (Montezuma County, Colorado) federal land program to improve dialogue and working relationships with public land managers (P. Larimer);
- A forest restoration project in a 140,000 acre ponderosa pine ecosystem, removing small diameter materials in overstocked landscapes, followed by fire reintroduction, known as the Ponderosa Pine Forest Partnership (T. Richard);
- The use of three US Forest Service district level community-based study groups and six forestwide working groups to increase citizen participation in resource management planning for the San Juan National Forest (T. Richard and S. Burns); and
- Emerging from these concrete projects and actions, the evolution of a conversational and institutional network based on a set of community ecosystem stewardship principles, and commitments to work cooperatively and collaborative for sustainable communities and sustainable ecosystems, called the Community Public Land Partnership (CPLP brochure and staff papers).

While these initiatives are on-going, they have already demonstrated an improvement in civic literacy, civil conversation, forest stewardship, and regional and communally accepted frameworks for cooperation between the US Forest Service, communities of place, commodity production industries, local environmental representatives, county government, higher education institutions, and others.

The lessons and themes that have emerged from these initiatives suggest a set of acceptable practices that will foster a civic governance structure for community-based ecosystem management. The primary orientation of the emergent organizational and community practices is a community development process designed to construct an acceptable, common ground, problem-solving network. The essential and foundational elements of these initiatives that pertain most centrally to community governance are described in the following sections.

Cooperative and Collaborative Relationship Building

The planning, decisionmaking, and restoration activities proposed by community-based ecosystem management cannot proceed beneficially without the creation of new working relationships between all parties. These parties are diverse. They consist of citizens, businesses, public organizations, traditional users of public lands, ecological interests, among others.

The work of the members of the CPLP in Southwest Colorado have demonstrated that these new working relationships do not require that any partner entity give up its separate identity or perspective when it enters into the collaborative process. Common goals and values for ecosystem and community sustainability have been quite adequate to sustain a civic conversation, to establish formal and informal networks, and to develop collaborative stewardship activities that are leading to real land resource improvement.

With regard to building a community governance structure, the major advance has been the integration of existing institutional authorities. Rather than asking all parties to come to the table, and set aside their own agenda or needs in order to join a "coalition" model, inter-institutional networking has occurred, which over time has become the preferred method of community governance and civic conversation. At times the networking relationships are between a college and county government, at times between the Forest Service and the state timber association, at times among citizens, a college, and the Forest Service. The participants contribute and receive, adapt and change, and challenge and cooperate to the degree they are able and willing, but all become oriented towards common goals of community and ecosystem sustainability and stewardship.

Have the formal, institutional social and legal systems changed? No, not really. The federal laws that govern public land management, public participation, water resources, and endangered species have not been altered or abandoned. Throughout the process of restoring a ponderosa pine ecosystem through thinning and fire reintroduction, monitoring for habitat impacts, assessing economic feasibility, and studying national forest conditions and needs for stewardship, there remained in place the National Forest Management Act, the Federal Advisory Committee Act, the Clean Water Act, the Endangered Species Act, and the Federal Lands Policy and Management Act among numerous others affecting public lands. What has changed is the way concerned people and interests relate to each other. The ways they share, converse, and act in a civic manner.

The political philosopher Sheldon Wolin, in a recorded conversation with Bill Moyers, about nurturing public life, noted that ". . . what is fundamental to a civic language . . . is, why should I contribute, sacrifice, and cooperate in a particular way that's not going to advance my interests and may even ask me to sacrifice some of those interests . . . for the common good or for others whom I may not know personally . . ." (Flowers, Betty Sue, 1989, p. 109). So then are we going to expect that a civic conversation might lead to not just contribution or cooperation, but sacrifice? Are we to anticipate that a county government would take the legal risk of purchasing a timber sale to promote forest restoration, that several small family loggers would be willing to loose money by thinning five ponderosa pine demonstration sites, that local environmentalists would be willing to stay the course long enough to monitor the ecological outcomes of returning an overstocked stand to a range of natural variability, that dozens upon dozens of citizens would meet monthly for nearly three years to develop their own ecological literacy and share their traditional knowledge and vision for a healthy future for the San Juan National Forest? Can we expect strangers or neighbors to do this? Maybe we can't expect these activities to always take place, but they have occurred in properly organized situations and they can continue. But not without a civic conversation at the core of the relationships. What's changed? Nothing much, or a whole lot? Just the way people talk and listen, just the way they relate to each other, just what they now believe is possible. Now there is a civic conversation in which participants can trust. Now there is a civic literacy about communities, economies, ecology, silviculture that is grounded, not in ideological vocabulary, but in authentic improvement.

Collaborative Learning and Civic Literacy

Crucial to the collaborative decisionmaking structures has been a highly interactive learning process. In the Ponderosa Pine Forest Partnership these have been conducted through informal, field oriented discussions and concentrated workshops that transfer and disseminate knowledge. In the San Juan National Forest Plan Revision process, learning has occurred in study groups and topic working groups in on-going monthly meetings that resemble the democratic forums called for by Margaret Shannon. In both cases, civic literacy about the "ecosocial community of a natural forest" has improved considerably among the participants. Often times the same words that would have previously created social disunity are used, but now there is a common ground, an understandable land based context upon which to hear each other's civic and communal voices. The Community Public Land Partnership, a loosely affiliated institutional network, sustains the learning process through ongoing communication and dissemination of community ecosystem stewardship concepts, initiatives, and lessons learned, and by providing supportive resources for facilitation, documentation, and fundraising.

Critical to collaborative learning is the combining of scientific and community knowledge. Topic area specialist from federal and state natural resource management areas share with college and university based scientists and local community members who have a wide range of local knowledge, values, and interests. At a "neutral table," a safe place is established to share, learn, and debate. What occurs at this table is what Margaret Shannon describes as a "democratic forum."

> Within a democratic forum model . . . interest is simply what brings you there. The connection establishes individual and groups as necessary participants at the table. Once at the table, however, migration across social worlds allows for political discussion which creates from empathy, a greater understanding of the issues and alternative policy re-

sponses. From a policy design perspective, the presence of multiple social worlds leads to questioning and opens new options for policies. Thus, while interest helps structure the membership of policy communities, once at the table it is the multiple social worlds which support a democratic forum for policy deliberation. (p. 243)

In the paragraph above, Shannon further emphasizes features that all collaborative learning processes must possess in order to establish community-based problem solving, or "policy communities," underscoring the movement from mere interests, to information, fairness, values, permeability, and interdependence. Especially important to the present discussion of community governance is the feature of "permeability" which emphasizes the inclusion of "new voices" and "interdependence." Interdependency is key to collaborative processes because it addresses how issues and their consequences are linked, not separated. As Shannon again notes, "No policy community is an island unto itself" (p. 244), whether the context is an economic region, a bio-region or the nation.

Organizational Capacity Building

Central to civic governance of community-based or collaborative stewardship processes is the capacity of all the member organizations. Each entity must possess or develop capacities to participate in and accomplish the other two foundational components of the collaborative process–namely building new relationships and participating in open learning and participatory planning situations. Some capacities involve working to build staff teams and value systems that support collaboration. Other capacities address community development skills to develop community conversations, and partnerships leading to stewardship activities on the ground.

While every organization, agency, and group needs to develop some new capacities, public land management agencies need especially to learn how best to share and disseminate information in more socially and communally understandable modes. Line officers and program specialist need to present technical and regulatory information and decision spaces in accessible and participatory ways. Convenor skills and hands on community learning methods need to be acquired by scientists, planners, and policymakers at local, regional and national levels of all public land management agencies that both encourage and facilitate civic conversations about public land resources in terms of long term stewardship.

Citizens, and community leaders at local, state, and bioregional levels need to equip themselves with skills and knowledge to integrate their values and perspectives into on going resource planning, management, and stewardship initiatives. Sometimes this will mean establishing a public lands program as was done by Montezuma County. Sometimes it will mean joining or participating in various forms of an interagency or inter-community, public land partnership as has been demonstrated by the Community Public Land Partnership in Southwest Colorado. These will need to be "permeable" so as to include the national perspectives as well as the community-based values and knowledge.

Conversely, it should not be overlooked that communities, as they enter into collaborative ecosystem stewardship activities, are contributing to fulfilling the national interests in CBEM through their knowledge, stewardship actions, and monitoring. To envision this point most clearly we may have to imagine an unsustainable community, either one growing too rapidly because of tourism or one dying because of economic decline in timbering, existing right next door to an unsustainable, unhealthy ecosystem. That declining forest is about to be impacted by a catastrophic fire or disease outside the range of natural variability. From where and whom will the human and communal stewardship values, commitments, and resources be obtained, in this case to fight the fire or thin the tree stands? The dying community does not have the capacity to implement a restoration stewardship initiative. Industry has left, the Forest Service has closed its office and reduced its staff because of downsizing the federal government. Who is left to care for the land? No one. Indeed no one if a civic conversation cannot begin to renew the process of community-based ecosystem management.

Implementation: Systemic or Communal?

The implementation of CBEM remains problematic and open to further developments of its principles and methods. It is not yet clear that new legislation or formal regulatory authorities are needed to support the institutions and governance concerns related to community-based ecosystem management. That is, it could well be the case that existing federal acts (NFMA, NEPA, and even FACA) and related authorities (RCA–Rural Community Assistance, economic recovery program) are adequate for the time being to implement the *relationship building, collaborative learning, and organizational capacity development* needed to revive a civic governance framework, to produce a civic conversation as envisioned in this analysis. While we can not say that the legal foundations of CBEM make no difference, surely for many activities at this time, it is probably more a matter of learning how to talk to each other and listen cooperatively.

If We Do Not Start with the Premise That New Authority Is Required, What Steps Can Be Taken Now?

First, existing government to government structures, from county, municipal, and tribal governments, to federal levels could be utilized more fully.

Partnership, or collaborative stewardship or community-based groups could more broadly utilize the powers of local governments to work with, coordinate, and advise the federal public land management agencies. FACA does not even prohibit this activity. See 41 CFR-101-6.1004, paragraph f, which states:

Examples of advisory meetings or groups *not covered* (emphasis added) by the Act or this subpart.

> . . . (f) Any local civic group whose primary function is that of rendering a public service with respect to a Federal program, or any State or local commission, or similar group established to advise or make recommendations to State or local officials or agencies.

If a county or multi-county area establishes a community-based ecosystem management council, it could in turn advise, coordinate, serve as a learning or study group, and become a "democratic forum" as Margaret Shannon has suggested. Such an entity would be capable of establishing a permeable civic governance process for further relationship building, learning, and capacity building, for the purpose of "rendering a public service," or of being civic.

Current Rural Community Assistance (RCA) Programs operated by Cooperative Forestry–USFS (established by subtitle D of the 1990 Farm Bill) to improve economic diversity and mitigate problems in forest dependent communities is mandated to establish action teams for strategic planning and development. Recent research conducted by Voth, Burns, Frentz, and Sperry has shown that while the RCA program is being utilized by the US Forest Service to build economic and social capacities in communities, little attention is being given to how these capacities could also be utilized to enhance planning, management, and stewardship activities on forest lands. Part of the problem is with the US Forest Service organizational structure, where benefits of the RCA program in Cooperative Forestry, a division of the US Forest Service, are not being utilized fully by the Forest System side of the larger agency. Nevertheless, there are reasonable administrative policy remedies to this lack of collaboration that would more clearly and effectively link programs and resources of RCA to the Forest Service as a whole, in order to utilize the RCA's capabilities to make a contribution to community ecosystem stewardship.

The National Environmental Policy Agency has developed a similar program for community-based stewardship activities within communities. While the Bureau of Land management does not have the supplemental funding to assist local communities or the authority to form community action teams, it has been utilizing range stewardship councils to improve range conditions. While the authority or governance structure of the BLM model might not be totally transferable directly to other community-based situations in ecosystem management, considerable learning about civic governance processes, improved management and resource management and resource standards, and stewardship responsibility should be applicable over time.

With regard to land management agency capacity building, there are several activities that require further attention and support. In 1997, Chief Dombeck of the USFS established the Collaborative Stewardship Team under the leadership of then Regional Forester, Elizabeth Estill, Region 2–USFS. The Team has prepared a report that outlines numerous action steps that would move the process of community collaboration forward. The report speaks of encouraging collaboration among field units, creating a network of mentors, and designating a collaborative support Team made up of agency and non-agency personnel, among other proposals.

As of the time of this writing, the Collaborative Stewardship Team Report and Action Plan has not been disseminated in official form by the leadership of the USFS, however, it would appear that this will soon occur. The challenge obviously will be how best to implement the proposed actions. There are some objectives that by their nature call out for a joint effort between the USFS, other land management agencies, local government, stewardship councils, and community partnerships. Through pragmatic, and adaptive implementation strategies, the USFS Collaborative Stewardship Team Action Plan could address many of the civic governance and conversation issues in community-based ecosystem management, including the evolution and strongly positive acceptance of various collaborative forms of relationship building, learning, and institutional capacity building.

Training and education in community-based, collaborative methods is seriously needed within all sectors, but especially within agencies. Willing community oriented staff within the USFS have commented that existing community collaboration training provided by the agency only introduces the values or need for collaboration, but does provide a means of hands on learning. The Pinchot Institute has for sometime been providing workshops primarily for agency leadership in "open decisionmaking" under the leadership of Jeff Sirmon and Terry Tipple. The National Training Center, administered by the Bureau of Land Management has been of late offering a week-long course entitled "Community-Based Partnerships and Ecosystems for a Healthy Environment." There is discussion currently of developing a shorter version of this course to make it more accessible for staff working in the field at a forest or resource area level. Greater awareness of and more resources for these training opportunities are needed in order to build community collaborative capacities with federal land agencies, local governments, community associations, and other constituencies who now attend this course. Some of these opportunities could be provided jointly for citizens, governmental leaders, conservation groups, and non-governmental organizations to extend the opportunities for more "permeable" civic conversations.

On a national level it would be appropriate to establish a series of CBEM pilot projects. Through a thoughtful process of project development, the pilots could address the major questions within the broader community-based, collaborative stewardship arena. Demonstration site selection should take place in such a way as to provide a range of ecosystem management goals, stewardship approaches, and community governance models. If the pilots were conducted within a systematic and organized analytical framework, many questions could be posed for analysis, and many civic governance and institutional arrangements could be properly evaluated. Findings from the pilots could then be proposed for codification in law and policy.

A recent draft of the Forest Options Group report, entitled "Second Century: Options for the Forest Service," proposes five budgeting and governance systems–namely entrepreneurial budgeting, collaborative management, collaborative planning, forest trust, and a rate board. This report also calls for a pilot project for each mode. The central purpose of a series of pilots, whether of the type proposed by the Forest Options Group or others, is to create a set of demonstration, community-based ecosystem management models that establish a set of "best stewardship practices" with a context of civic governance.

This piloting approach for developing community-based ecosystem management is believed to be more beneficial than large, formal system changes, in order to explore, demonstrate, and disseminate the practical improvements of community ecosystem stewardship, within real community development situations. Indeed we must ask, is it possible to develop the needed civic conversations? If they cannot be held, cannot be democratic, cannot be open and inclusive, will any law make them so? Can we better learn to talk and listen, to converse because we are required to, or would we arrive there more assuredly because we now see the need to cooperate, indeed sacrifice for each other and the land?

CONCLUSION

Central to the adaptations proposed by community-based ecosystem management is the need to change the ways people connect to each other and the public lands. This shift, as clearly described by Margaret Shannon, needs to be based on civic governance, with related foundations in "civic literacy" about sustainable communities and ecosystem. It would be a civic literacy that is capable of deeply informing a new community-based ecosystem stewardship.

A new community governance process must be civic in character to reflect the collective rather than private nature of public land resources. Similarly, it must be civic if it is to authentically mirror the central principle of community-based ecosystem management, which is to reconnect public lands to communities. The processes of community governance, in this context, should be conversational, democratic, inclusive, and grounded in face to face problem solving.

Essentially, the land management and governance institutions, both formal and informal, must adapt and move away from the centralization of resource management, and an over-reliance upon top down, scientific analysis as a singular, autonomous method of inquiry. Existing, adapted, and new forms of civic governance are needed to facilitate the building of relationships between multiple parties, create on going collaborative learning, and form and encourage organizational and individual capacities within land management agencies, communities, citizenship organization, and governments at all levels.

Pilot community-based ecosystem management projects should be initiated that would cumulatively promote a systematic analysis of the civic governance needs and capabilities of a variety of community-based stewardship activities–forest restoration, urban interface fire risk reduction, integrated recreation planning, and watershed management. Such a systematic and purposive demonstration and analysis strategy would utilize the grass-roots, community-based processes upon which community-based ecosystem management most depends. A highly formal, centralized, and "one shoe fits all" approach to developing civic governance processes and patterns would threaten the inherently collaborative, bottoms up nature of community-based ecosystem management. However, to ignore the fact that community-based ecosystem management needs to be based on trustworthy civic governance structures, protecting the public interest, and affirming principles of ecosystem sustainability is to proceed naively towards a large-scale, societal train wreck. Healing the public lands of decades of private claims through evolving a truly civic conversation, attending carefully to how they will require community cooperation and the sacrifice of personal interests, not the loss of individual rights, will require a level of public dedication to be neighborly of the highest order. At this time, such dedication seems possible out of deepening civic sense of stewardship and public responsibility. To legally or authoritatively mandate a civic conversation is contradictory to the spirit of authentically connecting people with their public lands. Conversely, a literate civic conversation that produces a believable community governance process is at the heart of community-based ecosystem management. There it can renew citizen responsibility, stewardship, and commitments in the most fertile place possible–not in arbitrary mandate and requirements of law, but in the inescapable, yet challenging relationships of people to the land.

REFERENCES

Community Public Lands Partnership (1997). Informational brochure and case studies. Durango, Colorado: Office of Community Services, Fort Lewis College.

Coggins, G. C. (1998). Of californicators, quislings and crazies: Some perils of devolved collaboration. *Chronicle of Community* 2 (2): 27-33.

Daniels, S. E., Walker, G. B. (1996). Collaborative learning: Improving public deliberation in ecosystem-based management. *Environmental Impact Assessment Review.* 16: 72-102.

Flowers, B. S., editor. (1989). *Bill Moyers: A World of Ideas*, a conversation with Sheldon Wolin, 96-100. New York: Doubleday.

Forest Options Group. (1998). *Second Century Options for the Forest Service.* The draft report of the Forest Options Group, Oak Grove, Oregon.

Frentz, I. C., Voth, D. E., Anderson, S., and Pell, W. (1997). A talk in the woods: An impartial advisory committee eases conflict on the Ouachita. *Journal of Forestry* 95(1): 26-31.

Frentz, I. C., Voth, D. E., and Anderson, S. and Sperry, C. (1998 in process). Linking Community Strategic Planning and Forest Planning (funded by the USDA National Research Institute). Fayetteville, Arkansas: University of Arkansas.

Gray, G. and Kusel, J. (1998). Changing the rules. *American Forests.* 103 (4): 27-31.

Grumbine, R. E. (1994). What is ecosystem management? *Conservation Biology* 8 (1): 27-38.

Larmer, P. (1996). A Colorado county tries a novel approach: Work the system. *High Country News* 28(9): 11-13.

McCoy, K. L., Krumpe, E., Cowles, P. D. (1994). The principles and processes of public involvement: A state-of-the-art synthesis for agencies venturing into ecosystem management. Unpublished paper. Moscow, Idaho: Department of Resource Recreation & Tourism, College of Forestry, Wildlife & Range Sciences.

Office of Community Services. (1997). San Juan National Forest community study group summary report. Unpublished report. Durango, Colorado: Office of Community Services, Fort Lewis College.

Office of Community Services. (1998). San Juan National Forest community working groups summary report. Unpublished report. Durango, Colorado: Office of Community Services, Fort Lewis College.

Richard, T. (1995). Timber on a tightrope: A demonstration project to reshape the future. *Inner Voice* August: 16-17.

Richard, T. (1998). Ponderosa Pine Forest Partnership: Forging new relationships to restore a forest. (unpublished case study in progress). Durango, Colorado: Office of Community Services, Fort Lewis College.

Richard, T. and Burns, S. (1998). Beyond 'scoping': Citizens and San Juan National Forest managers, learning together. *Journal of Forestry* 96(4): 39-43.

Shannon, M. A. (1992). Community governance: An enduring institution of democracy. (Printed for the use of the Committee on Interior and Insular Affairs, U.S. House of Representatives), Proceedings and summary of the workshop entitled Multiple Use and Sustained Yield: Changing Philosophies for Federal Land Management? pp. 219-250. Washington, DC.

USDA Forest Service Collaborative Stewardship Team. (1997). Report of the Collaborative Stewardship Team. Washington, DC: USDA Forest Service.

Integrating Forest Labor Participation into Community-Based Ecosystem Management Processes

Beverly A. Brown

SUMMARY. Community-based ecosystem management (CBEM) premises its effectiveness on participation of "community." In this paper, the role of forest and watershed laborers as part of rural community, and the importance of this multicultural and often mobile group in the success of CBEM is analyzed, especially in relationship to dominant business interests.

Mainstream CBEM is currently conducted through educated middle class networks and often highlights local small business development, including various forms of local business preferences for access to local watershed resources. There are few checks and balances inherent in current CBEM initiatives that would structurally prevent business from becoming dominant in local forest decisions over time.

Blue collar forest workers are integral members of social, cultural, economic, and environmental networks in rural localities. Because forest business depends on forest labor, and the sustainability of most forest jobs depend on ecosystems integrity, groups of workers who are free to speak out can act as structurally integral monitors and brakes on business excesses. Although participation of this central constituency in community efforts is currently limited because weakened U.S. labor law precludes the realistic right of labor to speak out publicly without retaliation by business, and a long history of exclusion results in lowered capacity, labor participation in CBEM processes may be indispens-

---

Beverly A. Brown is Coordinator, Jefferson Center for Education and Research, P.O. Box 279, Wolf Creek, OR 97497.

[Haworth co-indexing entry note]: "Integrating Forest Labor Participation into Community-Based Ecosystem Management Processes." Brown, Beverly A. Co-published simultaneously in *Journal of Sustainable Forestry* (Food Products Press, an imprint of The Haworth Press, Inc.) Vol. 13, No. 1/2, 2001, pp. 291-303; and: *Understanding Community-Based Forest Ecosystem Management* (ed: Gerald J. Gray, Maia J. Enzer, and Jonathan Kusel) Food Products Press, an imprint of The Haworth Press, Inc., 2001, pp. 291-303. Single or multiple copies of this article are available for a fee from The Haworth Document Delivery Service [1-800-342-9678, 9:00 a.m. - 5:00 p.m. (EST). E-mail address: getinfo@haworthpressinc.com].

able to the structural effectiveness of CBEM as an enduring institution. *[Article copies available for a fee from The Haworth Document Delivery Service: 1-800-342-9678. E-mail address: <getinfo@haworthpressinc.com> Website: <http://www.HaworthPress.com>]*

KEYWORDS. Community-based forestry, forest labor participation, multi-cultural, mobile workforce

The commitment to "community" in community-based initiatives presents an opportunity and obligation to reflect on who are considered legitimate community representatives for the many low-income natural resource workers and rural residents who are intimately concerned about natural resource affairs. Among these groups are ethnically diverse people, including low-income European-American, Latino, Native American, African American, and Southeast Asian constituencies. In this paper, the role of non-timber forest workers is examined as an example of a multicultural occupational community with (1) a direct stake in participating in natural resource dialogues, and (2) a pivotal, indispensable role in the success of community-based ecosystem management (CBEM) processes.[1]

The current natural resources debate in the U.S. accepts as the defining norm a middle-class-dominated discussion from which blue/pink collar workers are absent, and in which the participants are almost entirely European-American. The foremost presence of middle-class European-Americans is so thoroughly established in rural environmental, industry, and community activist affairs, that few people question how or why this constituency has come to represent natural resource issues not only in general, but also in the "community-based forestry movement."

The middle-class white folks involved in these natural resource debates correctly note that there is a fractious diversity of cultural approaches even *within* middle class European-American participants. This internal diversity makes it even more difficult to see who and what questions are not "at the table" in the discussions and negotiations that will define and implement CBEM. Pressing needs to act quickly on controversial initiatives tend to keep participants focused on already-involved actors, who have been described as "winning their place at the table."[2] Thus while heated controversy takes place within a territory defined by current mainstream participants, the limitations of current strategies are obscured.

CBEM STRUCTURES

CBEM currently focuses on three pragmatic approaches stressing: (1) a strong concept of ecosystem management, partnered with community development programs that have a local business locus; (2) challenging public forests to be more response to local business interests; and (3) stressing pragmatic cooperation with private industrial forestry. As discussed below, these approaches developed out of their own historical contexts and with participants that moved easily from recent historical precedents.

Workers are one constituency that has been a weak or nonexistent player in forest and water issues nationwide over the last few decades, and they are poorly if at all represented in CBEM processes. Can the issues with which current participants are occupied be resolved without expanding the dialogue to include blue/pink collar working people and a diversity of ethnic participation? In its real-world application, resource management takes place through the day-to-day labor, diverse specialized knowledge, and hundreds of small decisions made every day by on-the-ground working people. Without the people who are so intimately involved, who is the "community" in community-based ecosystem management? This question can be addressed by stepping back from the debate to look at how the current discussion evolved.

In the last three decades, the history of community-based activism within an ecosystem management framework emerged from three principal concepts: (1) embracing a hybrid of (a) conservative, sometimes libertarian, devolution of governance and (b) liberal/radical small-scale and direct-democracy-is-beautiful management paradigms, both resulting in a focus on "local" community; (2) developing local community-linked business strategies (associated with resources from either public or private lands); and (3) creating a venue for action by setting legal and regulatory precedents in public land and water policy (often initiated by the environmentalist community or reactions to precedents set by the environmentalist community).

These three approaches were pragmatic, politically astute ways of proceeding and have produced some important advances. In the first instance above, the focus on "local community" elicits sympathy from many people alarmed by either big government or big business or both, and from many others who long for an intimate sense of community perceived as lost in the contemporary world. The definition of "local" and "community" have been consistently fuzzy, with little recognition that individuals in a locality have radically different interests and levels of access to power. In the second instance, the idea of developing small businesses and connecting larger natural resource businesses with local community development finds resonance in both dominant conservative ideologies and more liberal or "progressive" strategies. Third, establishing an arena in which success might be realistically attainable has led many environmental and now CBEM initiatives toward federal and state lands, because public land and water agencies are required to respond to appeals and legislation in a way that private industrial lands are not.

Two other perspectives that are logically entwined with public land resources and business development have been underused or downplayed:

(1) Large business interests, from local to global, are pivotal players and can overwhelm most local initiatives–with the backing of powerful regional, national, and global allies–whenever it is strategic for them to do so.

Until CBEM significantly challenges the power of business, large natural resources industries are likely to ignore or work quietly against CBEM, or position themselves to coopt the movement to its own ends. This is not necessarily with any ill-intent, but follows the routine self-interest that is part of maximizing one's position in the business world.

Of course, large business can accomplish many positive outcomes, but it operates with minimal public accountability and dense informal networks of power. The choice to leverage a positive or negative business-related outcome is, except for regulatory conditions, outside the democratic process–that is, within the decision making space of business managers rather than public bodies. In the current situation, with an emphasis on community-based natural resources business development, CBEM may pivot on "local" business development leaders as the principal definition of "local community" or people may speak of a diffused sense of "community" or "capacity" development. These "development" and "capacity" approaches dodge a central question of who can muster sufficient power to create a counterbalance to business. They also beg the question of who in a rural area can realistically afford to participate publicly in controversial business-related issues, given that many worker's jobs and other community connections (for instance, funding for sports fields) depend on patronage relationships requiring smooth relations with business owners and managers.

Considerable risk is required to challenge the power of large business and bring a publicly accountable democratic system of checks-and-balances into place. The power of the business community to retaliate overtly or covertly against people or organizations that challenge them can be broad, well-financed, ill-monitored, and often not particularly concerned about ethics. *This is crucial to CBEM, as the mechanics of decisions made at the regional or national level ultimately will be made through business interests and carried out by employees or others directly or indirectly dependent upon "local" business structures.*

(2) Without a definition of "community" that takes into account conflicting interests and access to power within and beyond a locality, a discussion of "community" is elusive. The groups most effective at leveraging local power are actively organizing and lobbying at larger-than-local levels. A paradox is present here: The fewer regional and national allies (business, social, political, interest group) a *local* interest group has, the less likely it is to have any effect on the local power structure. Alternately, the more a group has regional

and national allies, the more likely it is to have the strength and credibility not only to challenge local power structures, but also to enter into reciprocal obligations with other diverse "local" CBEM interests focusing on specific local issues. These reciprocal relationships among autonomous but well-connected local groups increase the likelihood that a broad network of diverse individuals within a locality will be involved in CBEM. Within these parameters of participation, the concept of "community" describes a dynamic process rather than a static aggregate entity or a descriptive similarity of views within a bounded geographical area.

Thus a practical definition of "community" may be less useful as a description of place, or even of common interests, but as recognition of *reciprocal* bonds of responsibility and accountability among a socially, economically, and culturally diverse set of peoples with local bases and nonlocal affiliations. If "community" describes a place but leaves wholesale groups of individuals out of democratic participation, then the term "community" could be considered a ruse to cover nonreciprocal, nonaccountable power (usually business-related, but could be based on status, gender, ethnicity, etc.) over others, with little or no chance of a challenge from less-powerful groups.

The most prominent CBEM model for business development on public land mimics private industrial forestry. It is loosely based on a *concept* of private property in land (one private party is vested a controlling bundle of decision making rights) and expressed in various proposals for long-term restricted access by one party. Examples include leases, licences, concessions, local preference, some stewardship contracts, and similar mechanisms that have quasi-private property implications on public land.[3]

The positive outcome of CBEM efforts to date is the involvement of local interests in discussions and initiatives for ecosystem management of forests in a locality. Many exciting experiments are taking place in the community-based movement. As cited in elsewhere in this volume, models are being developed that expand our concept of scale, products, monitoring, and democratic participation by educated middle-class European-Americans. Such models are likely to have substantial positive effects in improved ecosystem management, and potentially may offer employment opportunities for other residents in the locality.

Among the drawbacks of current CBEM efforts is the lack of attention paid to the structure of social relations in natural resource affairs. Without developing broad democratic participation by less powerful social, economic, and ethnic stakeholders, power in CBEM will inevitably concentrate in large business (including small businesses grown large). The likelihood of this concentration rests on normal desires to increase business success and size within a system of community development that gives primacy to entrepreneurship. If, in the absence of a significant shift in CBEM's players, a focus on some version of geographical local-preference contracts is instituted (especially if local preference is vested in business deals), then CBEM could well contribute to a concentration of business power over social, political, and economic democracy within a locality. With the growth of that nonreciprocal relationship, the influence of business will increase. Industry will possess a concentrated power to exert leverage against activities that maintain ecological as well as social well being. By centering "community" interests in business, which can control disproportionate local access to resources and markets, CBEM may contribute to an increase of non-democratic concentration of power. This is the kind of concentration that took place in the innovative cooperative natural foods businesses of the 1970s in the Pacific Northwest, as successful, highly participatory cooperative networks were bought or pressured out of business by corporate interests (including some that emerged from the movement).[4]

Environmental groups have quickly seen the potential for business to dominate ecological interests (see esp. McCloskey, 1995), and would probably be right in resisting CBEM if that were the only consideration. However, environmentalists have been as slow as business or most CBEM activists to consider the importance of reciprocal systems of checks-and-balances at all levels of social/community interactions with natural resources. CBEM networks have placed too much faith on "local" development led by business development and ignored less powerful interests that can easily be bowled over by business. Environmentalists, meanwhile, have placed most of their chips on a legalistic, defensive, top-down federal strategy that eliminates the potential for building creative, adaptive, reciprocal systems built on participation by diverse interests (including business) in a multi-level checks-and-balances model. The environmentalist vision is sometimes difficult to distinguish from other rigid command-and-control systems that have a single fixed mission supported by a particular constituency. Thus both CBEM and the environmental movement may be based on what may turn out in the long run to be anti-democratic social processes.

One could say that CBEM is susceptible to encouraging within it self-serving interests analogous to, say, the boll weevil burrowing into and destroying a good stand of cotton. The environmentalists, meanwhile, could be said to be using a strategy analogous not with integrated pest management which could bring things back into balance, but with DDT, which would kill anything–good, bad, benign–that might develop legs to stand on. Neither circumstance is likely to result in long-term ecological and social health. Meanwhile, the positive checks-and-balances contributions of interactions among diverse social interests–analogous in a natural system to ecological diversity–are suppressed or ignored, treated as either irrelevant or dangerous.

THE ROLE OF WORKERS

One of the largest groups of less-powerful interests in natural resource affairs is also the backbone of forest-related populations. Both residents in rural localities and migratory people who labor in those same localities regularly perform work that has a direct ecological impact on forests and watersheds. These timber-related employees, non-timber contract labor in the woods, and the independent non-timber forest products (NTFP) gatherers are individuals who live, work, and recreate in the forests during part or all of the year. A core group that melds community and ecological interests, they are broadly connected to the public at large through kinship and friendship links across many constituencies. Like recreational users, these workers represent the multiple dimensions of the forest-community interaction (see for instance, Brown, 1995). They are perhaps one of the most pivotal groups in this relationship, because, among non-business owner stakeholders, their livelihoods are the most directly tied to the forest. The sustainable health of forest ecosystems inevitably affects them as workers, as individuals, and as members of communities.

Unlike recreational users, workers/harvesters (hereafter referred to as "labor") play an essential role *within* business, as well as affecting business generation or decline. Business depends on employees or their equivalent (such as independent NTFP harvesters selling to a company). As individuals and community members, forest contract laborers and NTFP harvesters depend upon the health and integrity of the ecosystem. The relationship would seem to be intrinsically reciprocal among businesses, labor, community, and environment. Most middle-class recreational or environmentalist groups/individuals make their living outside the patronage relations that exist between forest-related employer and employee. Forest/watershed laborers, though, must learn to exist within social environments that are strongly affected by the power that business-related individuals wield over them.

Rather than a reciprocal relationship of responsibility and accountability, the employer-employee dynamic is more typically a dominant-subordinate relationship, especially in labor-based businesses. In rural communities this relationship extends beyond the workplace, for instance into which political cause one supports, or whether a logger is opposed to, say, the institution of strip mining in the neighborhood. This is true for both resident and migratory forest workers; it is the *relationships with the business, not the specific place,* that determine in which public actions a worker can engage without risking overt or covert retaliation.

Thus labor has a pivotal role in an inclusive concept of CBEM process, because workers are integral to "community" in two respects: (1) Laborers fulfill multiple aspects of community membership–as residents, as individuals engaged in civic activity, and as a workers embedded in economic relations; and, (2) especially for under-the-canopy forest workers, laborers have an occupational stake in the maintenance of ecosystem integrity. But labor cannot speak out effectively for one pillar of their livelihood–the environment–because if the other pillar of their livelihood–business–is offended, workers can be terminated or their ntfp products rejected. Labor-based rural work usually exists in a labor-surplus situation–a worker who is cut off usually sustains serious consequences. Again, employer threats against individuals or their family members, as a consequence of speaking out, are well known in rural areas, including situations known personally to this writer. The numerous forest workers with whom I have spoken are incredulous that some middle-class people do not believe that latent or overt threats by employers are routine both within and outside the workplace. In the United States, resource workers cannot assume that they have a *realistic* democratic right to speak out publicly on ecosystem management affairs.

In forest labor, the situation is complicated by the presence of multiple ethnic groups in forest floor work. These generally low-income workers/gatherers, predominantly from Southeast Asian, Latino, Native American, and European-American backgrounds, have low social status, often scant formal education, frequent language difficulties, and/or little facility with public speaking or writing documents. They are subject to persistent cultural stereotyping and prejudice. Ironically, this group often has a sophisticated knowledge of what actually goes on in the woods, with information that can significantly add to or even correct the conclusions of scientists and activists (Brown, 1998).[5]

Labor in the United States has a theoretical right to organize and to negotiate a contract that outlines of reciprocal responsibilities and accountability between the workforce and the employer. Over the last sixty years, the promise of the Wagner Act (1936)–which declared it to be the policy of the United States to encourage collective bargaining–was undercut by the employer-friendly revisions of the Taft-Hartley Act in 1947. During the last twenty-five years, conservative political appointments to the national bodies that decide the practical details of labor law have further weakened worker rights. The result is that the once realistic right of labor to negotiate limits to employer power, establish limited democratic rights inside the workplace, and create democratic space for workers to act publicly outside the workplace without fear of retaliation, no longer exists realistically for the overwhelming majority of workers (Gross, 1996). Among low-income non-timber forest workers, the lack of this effective power is virtually absolute. Among loggers, this right effectively evaporated when timber work was transformed into contract labor (West, 1987). Among mill workers, labor leaders in the United States have followed a top-heavy leadership and industry-accommodating business unionism (Lembcke and Tattam, 1984).

While workers are effectively powerless, they are also strategically positioned in a joined-at-the-hip relationship with business. Natural resource businesses depend on natural resource laborers. This intrinsic dependency suggests possibilities in which CBEM processes could make explicit the importance of the rights and participation of workers in maintaining social, economic, and environmental well being.

At the beginning of this paper, I suggested that CBEM has pragmatically avoided challenging the power of business. I also suggest that grappling with this challenge is a necessary step on the way to truly "community-based" ecosystem management, with a definition of "community" grounded in diverse reciprocal and accountable relationships with local and nonlocal allies and stakeholders.

Forest labor's joined-at-the-hip relationship with business means that labor has the structural potential of moderating business-generated environmental excesses, while also promoting the interests of the business community in maintaining long term economic health through ecosystem management. No other non-business-owner group in the public ecosystems dialogue occupies this structural position vis-a-vis business. Labor's role as pivotal to environmental as well as community health means that the freedom of labor to participate in CBEM without retaliation from employers becomes pivotal to the success of CBEM processes. Labor's right to participate is not only an essential democratic right and an expression of fundamental justice, it plays a pivotal role creating a reciprocal and accountable system of checks-and-balances in a genuine "community"-participatory ecosystem management system.

HOW DO WE GET FROM HERE TO THERE?

The history of labor relations in the United States has been so ideologically skewed that it is difficult to conduct a rational discussion about workers as citizens within a democracy without descending into partisan world views that carry one far from the fundamental question of people's right to participate in public affairs that may or may not please their employers.

Why aren't forest workers not only represented "at the table," but taken seriously and perhaps even courted by other constituencies, such as environmentalists or CBEM advocates? The barriers are many, aggravated by a long history of conservative ideologues denigrating the *concept* of workers rights through the guise of denouncing certain corrupt labor leaders (while quietly undermining labor rights at the legislative and regulatory level). Middle-class community development people simply ignore the importance of workers.

Participatory capacity is not built under the thumb of business, or "helped" by community development scenarios that include training programs designed to transfer skills in labor-based tasks, but not in participatory decision making. After decades of exclusion from natural resources decision-making discussions, it is unlikely that labor would take a seat "at the table" merely on invitation to the CBEM process. Throw in a history of being used or slandered, plus deeply ingrained mainstream prejudice against ethnic minorities, tribal groups, and low income people, and the result is a forest workforce with a deep cynicism and active *avoidance* of being "at the table."

Thus, although labor's right to exercise social and political democracy may be essential to the best hoped-for goals of CBEM, making up for years sidelined by politics and business opposition is a formidable challenge. It is also entirely possible.

One of the first requirements is the *will* of those involved in CBEM and other allies to incorporate labor's diverse voice in ecosystem management affairs. Although creating this conceptual space in their thinking is a task called for by all CBEM participants, actual efforts to support capacity building and willingness to take a risk to become involved will be accomplished by several sets of activities:

1. Create opportunities for forest labor to speak without retaliation by business and business allies. This involves labor law activism, and an awareness of business-supported mechanisms to undercut natural resource labor. Examples include not only generic retaliation against employees in an individual workplace, but also endeavors such as the H2B "guestworker" program, which undercuts forest worker efforts to improve working or environmental conditions in the woods.
2. Encourage forest laborers' willingness to become involved by offering practical support, information, etc., including covering costs associated with participation, and providing bilingual interpreters if necessary. It is essential to cooperate with a broad and expanding base of forest workers from each culture involved. This will avoid a small group of "gatekeepers" becoming vested with disproportionate leverage within their home community or advanced as tokens in the mainstream community.
3. Among all CBEM stakeholders, build conceptual as well as literal space "at the table" for integration of diverse forest labor perspectives. As Ranachith "Ronnie" Yimsut is fond of noting, "They sometimes invite us to the table, but when we get there, there are no chairs for us to sit in."

Several constituencies are well-placed to assist in including low-income and diverse cultural perspectives in CBEM:

1. Foundations have a special role to play in bringing together social and environmental justice concerns in natural resource affairs. At this time,

with the exception of toxic pollution issues, most foundations clearly separate social justice and their environmental program grants. It is time to integrate the social and environmental challenges of low-income and multicultural workers–as part of "community"–into a natural resource based concept of environmental justice.

2. Nongovernmental organizations (NGOs) involved in social, economic, and environmental justice, especially those committed to cross-cultural work, have a natural role in helping move forward programs that build networking, peer analysis, information contextualization, and in other ways support the self-strengthening of forest labor.
3. Environmental-labor coalitions could strengthen the position of their respective agendas, while mitigating against the increasingly rigid, nonnegotiation stance that is turning what used to be "progressive" environmental organizations into undemocratic, nonparticipatory wielders of power.
4. Public agency and community-development groups could make a conscious commitment to understanding that while inclusion of labor may slow and significantly complicate matters, it results in a superior process in the end. (This is the price we have to pay for marginalizing pivotal stakeholders for so long.)
5. Labor's social, economic, and environmental concerns must be integrated into all dialogues and documents. This means addressing worker/harvester, low-income, multicultural issues in all natural resource topics, not downgrading them to a special session of "diversity concerns" at the end of the day.
6. Perhaps most fundamentally, the constituency of labor itself will need to seize the opportunity to become involved, and to make a place for itself even when others do not realize labor's importance. University of California professor Jeff Romm notes, CBEM "has created probably the first real opening for labor participation in resource management decision processes. That is an historic gain. But the opening will close by default if it is not used" (Romm, personal communication, 1998).

Both CBEM and the environmental movement have revolutionized dialogue on ecosystem management in the United States. Their creative momentum in partnership with forward-looking agency personnel and businesses will radically change the design of forest and watershed management in the next century.

Establishing mechanisms in which large businesses must build reciprocal relations of responsibility and accountability within a dynamic process of "community" is key to building a sustainable relationship between communities and the environment. This goal cannot be accomplished if we pioneer 21st century ecosystem management programs that rely on 19th century worker-employer relations. Such antiquated relations make civic democracy impossible for workers who constitute a group that by its nature has pivotal leverage in business, community, and environment. This has been well recognized in many of our most successful international competitors in western Europe, and to some extent in Japan, where the interests and participation of labor is built into the system (Adams, 1994; Sexton, 1991).

Some of the marginalization of forest labor in current CBEM process is borne of decades-reinforced ignorance of labor's importance. It is perhaps because labor is so pivotally positioned in forest and water ecosystems affairs that some other interests–especially business, but other established status groups as well–have been reluctant, even resistant, to encouraging and building forest workers/gatherer's participation. On the other hand, workers' pivotal role is why it is essential to pursue the processes that lead to labor being a part of the group that not only comes to the table, but decides on the shape of it.

NOTES

1. Note on Sources: This paper is intended to be a conceptual discussion of the importance of the role of labor in community-based ecosystem management (CBEM). For a more comprehensive discussion of the topics in this paper, see Brown (1997). The background framework for the paper was accumulated through participation in numerous CBEM-related meetings and informal group discussions among CBEM participants between 1993 and 1998, and from general sources on the environmentalist and collaborative/partnership movement in the U.S. between 1993 and 1998, references to and discussions of which are contained throughout this volume. For more information on the discussions, correspondence, ephemera, and minutes of CBEM-related meetings, and contacts with national CBEM initiatives in natural resources, readers are invited to contact the Lead Partnership Group (Taylorsville, CA), National Network of Forest Practitioners (Boston, MA), the Communities Committee of the Seventh American Forest Congress (via Watershed Center, Hayfork, CA), Applegate Partnership (Applegate, OR).

2. Comments made by environmentalist, community-based forestry, and industry participants in Lead Partnership Group meetings, 1996.

3. This is in contrast to concepts of, say, common property in land, trees, or other attributes; or, another way of phrasing, "common-interest" property, in which,–viable under various kinds of land ownership–certain bundles of decision-making rights are mutually agreed to be shared by or distributed co-terminously with several or "all" parties in a way that maximizes both use and protection of a resource (see esp. Ostrom, 1990). For instance, access to fish in a large lake; or the different, but overlapping interests of seasonal mushroom harvesters and floral greens harvesters on the same piece of industrial land that a company manages only for timber.

4. Based on direct experience of the author in the natural foods business in Portland and the Willamette Valley of Oregon, 1973-1980.

5. Based on 1994-1998 testimony and analysis by non-timber forest workers from SE Asian, Latino, Native American, and low-income European-American constituencies participating in a three-state network catalyzed by the Jefferson Center, a popular education organization based in Wolf Creek, Oregon. The author is coordinator of the Jefferson Center.

REFERENCES

Adams, R. J. 1994. Union certification as an instrument of labor policy: a comparative perspective. Pp. 303-13 in *Restoring the promise of American labor law*, eds. Sheldon Friedman, Richard Hurd, Rudolph Oswald, and Ronald Seeber. Ithaca, NY: ILR Press.

Brown, B. A. 1995. In *Timber country: Working people's stories of environmental conflict and urban flight*. Philadelphia: Temple University.

Brown, B. A. 1997. *Contingent labor on public land: a worker-centered perspective of the structures of public forest management*. M.A. thesis, Antioch University, Yellow Springs, Ohio.

Brown, B. A. 1998. Voices from the woods. *American Forests*. 103(4):35-7.

Gross, J. A. 1996. *Broken promise: The subversion of the U.S. labor relations policy*, 1947-1994. Philadelphia: Temple University Press.

Lembcke, J. and W. M. Tattam. 1984. *One union in wood*. NY: International Publishers.

McCloskey, M. 1995. Report of the Chairman of the Sierra Club to the Board of Directors. Unpublished report.

Ostrom, E. 1990. *Governing the commons: The evolution of institutions for collective action*. New York: Cambridge University Press.

Sexton, P. C. 1991. The war on labor and the left: understanding America's unique conservatism. Boulder, CO: Westview.

West, T. 1987. Timber fallers: a dying breed. *Siskiyou Journal* 27(December/January 1986/1987):14-7.

Testing Stewardship Concepts on Federal Land

Mary Mitsos
Paul C. Ringgold

SUMMARY. The goal of this paper is to describe recent attempts to define the stewardship concept and to translate it into action on the ground. The focus is on national forest lands and the unique challenges federal land managers face in working with the public to implement a concept that has different meanings to a variety of stakeholders. Stewardship requires more flexibility than is often found under existing administrative authorities. Large, bureaucratic agencies face many barriers in applying creative and innovative solutions to existing problems, not the least of which are internal and external reward structures. While there exist some authorities that allow for the creative testing of stewardship contracting projects on the ground, a number of challenges remain. Two significant ones are: (1) the need for a clearer, consistent understanding (by both Forest Service employees and constituents) of the authorities that can be used to test stewardship concepts; and (2) a more broad-based agreement on the desired outcome of stewardship projects. Given the current lack of understanding and agreement on how to implement stewardship concepts, the most effective means of moving the issue forward will be to apply innovative tests of contracting methods under current or new temporary authorities. These tests should be designed, implemented, monitored, and evaluated using broad-based, inclusive, on-the-ground coalitions that create effective ties and linkages between local and national levels. *[Article copies available for a fee from The Haworth Document Delivery Service: 1-800-342-9678. E-mail address: <getinfo@haworthpressinc.com> Website: <http://www.HaworthPress.com>]*

Mary Mitsos and Paul C. Ringgold are Research Associates, Pinchot Institute for Conservation, 1616 P Street NW, Washington, DC 20036.

[Haworth co-indexing entry note]: "Testing Stewardship Concepts on Federal Land." Mitsos, Mary, and Paul C. Ringgold. Co-published simultaneously in *Journal of Sustainable Forestry* (Food Products Press, an imprint of The Haworth Press, Inc.) Vol. 13, No. 1/2, 2001, pp. 305-320; and: *Understanding Community-Based Forest Ecosystem Management* (ed: Gerald J. Gray, Maia J. Enzer, and Jonathan Kusel) Food Products Press, an imprint of The Haworth Press, Inc., 2001, pp. 305-320. Single or multiple copies of this article are available for a fee from The Haworth Document Delivery Service [1-800-342-9678, 9:00 a.m. - 5:00 p.m. (EST). E-mail address: getinfo@haworthpressinc.com].

KEYWORDS. Community-based forestry, stewardship contracting, national forest management, monitoring

The term "stewardship" has been used with increasing regularity in the land management arena, usually to describe management activities that have either a neutral or restorative impact on the environment. It is often difficult, however, to agree on what stewardship means and how to implement it. According to Webster, stewardship is "the individual's responsibility to manage his life and property with proper regard to the rights of others" (Merriam-Webster 1995). Here, stewardship is described in an outwardly active context, relating to the way individuals should manage their environment. Many have described the concept in a more introspective manner, relating it to a principle that should guide the way we think about our place within the environment. In his 1949 book, *A Sand County Almanac*, Aldo Leopold describes a land ethic, which "reflects the existence of an ecological conscience, and . . . in turn reflects a conviction of individual responsibility for the health of the land" (Leopold 1949). This ethic has become central to the way many of us think about stewardship. A Forest Service field forester in Oregon recently described the stewardship concept as "a shift away from contractors providing one management service across a wide variety of locations, and toward contractors providing a variety of services in one location." Such an approach fosters a more thorough knowledge of local conditions, thereby enabling the contractor to more effectively meet site-specific management objectives. All these definitions are related–the first is a kind of activity, the second, a way of thinking about one's relationship with the environment, and the third an example of a practitioner applying the concept to his or her thinking about specific contracting processes on the ground.

Over the past decade there have been numerous efforts to foster new approaches to land stewardship on the national forests. These efforts have been initiated both by community groups and by Forest Service personnel. Community group efforts have been driven by the desire to promote local involvement in national forest management and to strengthen local economies. Forest Service efforts have been driven by the desire to increase collaboration, address forest health concerns in areas with low-value material, and to improve contracting flexibility and efficiency in times of shrinking personnel and budgets. A goal common to both groups has been to develop ecosystem management strategies that are sustainable for the long-term and flexible enough to meet local conditions.

With these desires and goals in mind, a variety of interested stakeholders have been exploring innovative contracting mechanisms that might help to more effectively achieve management goals on federal lands, while promoting long-term public involvement. Mechanisms used in stewardship approaches include: bundling a variety of management tasks within a single contract, bundling contracts across ownerships, multiple-year contracts, long-term cooperative agreements, and contract performance based upon descriptive end-results. These approaches involve both testing new contracting mechanisms, and applying traditional contracting mechanisms in new ways.

The goal of this paper is to describe recent attempts to translate the stewardship concept into action on the ground. Our focus will be on national forest lands and the unique challenges federal land managers face in working with the public to implement a concept that has different meanings to different stakeholders and which requires more flexibility than is often found under existing administrative authorities.

A BRIEF HISTORY OF STEWARDSHIP CONTRACTING

In the past, much of the resource work accomplished on the National Forest System was tied to, and dependent upon, the timber sale program. The timber sale contract was specifically designed to achieve the harvest and disposal of federal property–in this case, timber. However, as more diverse resource demands were placed on public lands, the timber sale contract was increasingly used as a means to undertake additional management activities (Sample and DiNicola 1996). These activities are related less to the actual disposal of federal property and more to meeting ecosystem health objectives. Such activities include riparian and watershed restoration, road decommissioning, wildlife habitat improvements, and improved recreational trails and access.

While some of these activities can be achieved using traditional timber sales (such as a commercial thinning to improve certain types of wildlife habitat), many do not require any commercial harvest of timber. Including such projects as part of the timber sale contract is difficult because of the narrow and often inflexible requirements of the timber sale process. These requirements have been developed to ensure fairness, efficiency, and resource protection during timber sales but do not easily adapt to nontimber objectives. At the same time, using traditional service contracts to accomplish these activities is hampered by a lack of appropriated funds. Neither service nor timber sale contracting instruments have traditionally been used for projects that attempt to integrate the social, economic, and ecological objectives desired in community-based ecosystem management. Stewardship contracting has been proposed as one alternative that combines existing elements of timber sales and service contracting instruments, along with several new mechanisms (such as longer-term contracting and enhanced community involvement), with the goal of more efficiently and effectively accomplishing a broader range of ecosystem restoration and maintenance tasks.

Many of the mechanisms being explored for use in stewardship contracting had their origins in land management services contracting, which was first tested on national forests during the 1980s. These contracts were primarily a response to shrinking federal budgets, reduced personnel, and demands for a broader range of outputs from public lands. They were meant to create significant savings of public funds through improved efficiency in contract administration by consolidating multiple stand-improvement activities under a single contract, and by using end-results specifications[1]. Furthermore, the flexibility of the contracts allowed for the execution of unique, site-specific prescriptions (Lynn et al. 1994).

Although initially developed as a means of facilitating traditional timber management activities (such as timber sale layout, site preparation, reforestation, timber stand improvement, and tree marking), these contracts evolved into more comprehensive approaches to ecosystem management with the inclusion of such objectives as wildlife habitat improvement, development and maintenance of dispersed recreation facilities, improvement of sustainable economic opportunities for local communities, soil and water conservation practices, and aesthetic improvement. These later contracting approaches, which involved both timber and nontimber objectives, are more commonly referred to as stewardship contracts. The variety of purposes for which stewardship contracting mechanisms might be used include:

- improving the health of federal forest resources through implementation of scientifically based ecosystem management programs;
- encouraging the creation or retention of jobs in and through ecosystem management activities, stimulating economic diversification, and contributing to the economic health and viability of rural communities near federal forest lands;
- encouraging a greater involvement by a diverse group of community members in the design, implementation, and monitoring of land management projects on federal lands;
- accomplishing projects in which the harvest and sale of forest products is secondary to creating specific resource conditions; and
- improving the use of local workers' knowledge and skills.

Some of the basic goals of stewardship contracting can be more clearly defined through a brief review of past pilot projects and federal legislation on this issue. Prior to 1992, land management service contracts had been discussed in congressional hearings but had not been legislatively authorized. The FY 1992 Interior appropriations legislation called for the initiation of "stewardship end results contract" pilot projects on the Kaibab National Forest in Arizona and the Dixie National Forest in Utah. This legislation authorized using all or part of the revenues received for timber removed under these contracts to offset the cost of stewardship services (i.e., "goods for services"). The identified services included site preparation, replanting, silviculture programs, recreation, wildlife habitat enhancement, and other multiple-use enhancements (Allan et al. 1994).

The FY 1993 Interior appropriations continued authorization for projects on the Kaibab and Dixie and expanded the program to include projects on Idaho Panhandle National Forest, Coconino National Forest in Arizona, and Lake Tahoe Basin Management Unit in California. The appropriations language also called for a study of how to include payments to local governments in such a program if it were adopted nationwide on national forest lands. Since FY 1993 there has been no further congressional authorization for such projects. However, a number of separate bills were introduced in an attempt to continue existing pilot projects, create new projects nationwide, and authorize a permanent land stewardship contracting program within the Forest Service.[2] In the mid-1990s, the land stewardship concept broadened to include several key components. These included local small business participation, alternative commodity sales strategies, and site specific planning by local stewardship councils. None of these bills were adopted into law, for reasons that included budgetary concerns and conflicts with other existing laws. However, a key reason seems to have been simply the lack of information necessary to address questions on the proposed strategies' effectiveness in meeting stewardship goals. As a result of the pilot projects, a number of recommendations were developed by the Forest Service to guide future similar efforts. Among the recommendations were:

- continue with the land management services philosophy and develop a new contract that better reflects the purpose and intent for the concept;
- make twenty-five percent fund payments for each project based on the appraised price for the timber;
- apply existing Small Business Set-Aside requirements; and
- handle each contract as a separate account. Revenues generated on the project and not used will be returned to the US Treasury (Lynn et al. 1994).

A number of years have passed since these recommendations were developed. Since that time, interest in stewardship contracts has evolved to include broader issues of sustainability.

CHANGES IN THINKING ABOUT STEWARDSHIP

Much of the thinking about stewardship has developed in the 1990s as sustainability issues have become more of a national topic. Many may agree that sustainable development means development "that meets the needs of the present generation without compromising the needs of future generations to meet their needs" (Bruntland 1987). However, the application of this concept is still subject to great debate. If one agrees with the statement that sustainable management is ecologically sound, economically feasible, and socially acceptable (Aplet et al. 1993), translating that agreement into action on a project still raises additional difficulties. As issues of sustainability have risen to the forefront of American public thought, there has been an increase in the demand for greater community involvement in the process of designing, implementing, and monitoring stewardship work on public and private land. For many communities, especially those surrounded by and traditionally dependent upon federal lands, greater involvement is seen as necessary to gain a better understanding of, and to help shape, what a sustainable future may look like on the ground. Those communities directly faced with the need for a system that integrates economic, environmental, and social issues soon began to get involved in implementing these concepts on the ground.

Economic growth and environmental quality have often been posed as either/or possibilities. Often, economic growth gets the highest priority and the local environment suffers as a result (Power 1996). Forest practitioners concerned with this rift see stewardship contracts as one possible mechanism to help mend the perceived divide between jobs and the environment (Carey et al. 1994). From our observations in working with rural, forest-based communities, as local groups wrestle with implementing concepts of sustainability, many are coalescing around issues such as attractiveness of the local environment, maintenance of healthy ecosystems, commitment to place, and quality of life. Many of these community groups see stewardship contracting as a means to increase local civic awareness and responsibilities, while helping to stem the tide of corporate consolidation and globalization that often results in the transfer of natural resource profits to distant areas (Brown 1995).

The federal government, too, is beginning to incorporate community-based and collaborative concepts into its agenda. Part of this effort is an attempt to make the federal government, in general, more service-oriented and less bureaucratic (National Performance Review 1997). An additional reason for looking to community-based collaboration is the desire to change federal agencies from entities merely charged with solving other people's problems and regulating activities to entities that are part of the solution and can facilitate improvements (President's Council on Sustainable Development 1996). Finally, as federal agencies such as the Forest Service are downsized, the need to work in partnerships to complete necessary activities is becoming more apparent.

BARRIERS TO INNOVATION

Although federal agencies are slowly developing innovative stewardship contracting mechanisms, a number of barriers significantly constrain any consequential move forward. These barriers have evolved over a number of decades and many exist for historical reasons. For example, there is a myriad of laws and regulations that govern Forest Service activity, each created for a specific set of concerns, such as multiple-use resource management, clean water, endangered species, or public participation. Arguably, each law was necessary to improve natural resource management and address public concerns and demands regarding the use of federal lands. However, each successive law passed has added to an increasingly complex mosaic of federal rules and regulations. Most of these rules and regulations, in turn, have a direct influence on agency operating policies.

These new laws have come about, in large part, because of public demands for consistent, accountable management of public resources (Plater et al. 1992). By nature, the laws prescribe and standardize management actions to be undertaken on public lands. For instance, the National Forest Management Act (NFMA) established a specific set of standards and procedures for the sale of timber on national forest lands. As a result, timber sale contracting procedures have become completely separated from the Federal Acquisition Regulations (FARs), which govern the procurement of services throughout the federal government. This distinction presents at least two barriers to implementing the stewardship concept. First, NFMA states that timber sales must be awarded to the highest qualified bidder, while service contracts can be awarded on the basis of "best value"[3] purchasing. Best value allows the government to consider factors in addition to price (such as past performance, quality, delivery, and experience) when making decisions on the award of procurement contracts. The ability to include "best value" considerations in the award of timber sale contracts would likely be of great benefit. These contracts do not merely involve the sale of federal property–they also offer a service in the form of "on-the-ground" forest stand manipulation. If quality considerations cannot be integrated into the decision to award a timber sale contract, the quality of the work on the ground may be compromised.

Second, existing policy makes it difficult for the Forest Service to require a timber sale purchaser to perform services not directly related to the removal and sale of timber. The sale of goods and the procurement of services are very different processes and a number of accountability problems arise when trying to combine a sale and a purchase. However, in attempting to implement community-based ecosystem management at a watershed scale, there is increasing interest in combining some merchantable goods (generally low-value, small-diameter trees) with a service contract dealing with restoration services. The agency has not yet developed such an instrument due to the extreme difficulty in creating an instrument that meets the legal requirements and follows the processes for both timber sales and service contracts while still providing clear accountability for taxpayers' dollars.

Citizens of the United States are afforded two unique privileges. First, there are approximately 550 million acres of federal land in this country (US Department of Interior 1998). Second, each and every citizen or group of citizens can sue the federal government for certain violations of the law. This privilege was greatly expanded in the late 1960s when the courts broadened the definition of standing to include environmental objectives as it pertained to land management decisions (Dana and Fairfax 1980). The rise in litigation over national forest lands during the past thirty years has forced the courts to increase their role in the outcome of land management decisions. Lawsuits, often with conflicting judgments from different districts that ultimately must be resolved by the Supreme Court, are adding to the complexity of the situation and subsequently constraining innovation [e.g., Ohio Forestry Association, Inc. v. Sierra Club, 140 L.Ed.2d 921 (1998); Babbitt v. Sweet Home Chapt. Comms. for Ore., 515 U.S. 687 (1995)].

For the Forest Service, the internal constraints on innovation often fall into two different categories: litigation and internal rewards. It seems that those charged with interpreting policies and procedures often do so in an extremely narrow manner, possibly to keep the agency free from litigation–or at least to lessen the number of lawsuits engaged. This narrow interpretation, while certainly limiting innovation, has not been able to stem the tide of litigation. Second, while Americans' expectations and desires for public land resources have shifted dramatically in the last couple of decades (Inglehardt 1997), the internal reward structure has not kept pace (Sample 1990; Hirt 1994). Although the current leadership extols the virtues of collaborative stewardship and working within the limits of the land, many employees feel their evaluations remain based on whether they meet specified targets, not on how they interact with their constituents or on the change in the condition of the land (pers. comm., USDA Forest Service Collaborative Stewardship Team, 1997). Yaffee (1994) provides an excellent analysis of the difficulty an agency such as the Forest Service has with addressing new issues and applying new management techniques. Externally, disincentives for creativity often come from either congressional directives or national interest groups, both of which are protecting and supporting the demands of their constituents. While few may be truly happy with current conditions, both national interest groups and Congress seem, as yet, unwilling to talk about significant change. Due to the contentious nature of current debates over the future intent of public lands, now may not be the time to legislatively mandate significant change.

EXISTING AUTHORITIES THAT ALLOW FOR INNOVATION

The sharp regulatory distinction between timber sale and service contract authorities presents a number of barriers to implementing the stewardship concept. Much work remains to be done in developing a clear and consistent framework for a more integrated approach to contracting. However, the regulations governing each type of contract do contain some existing authorities that allow the development of innovative contracting approaches on federal land. The following examples provide a sense of the range of existing authority to test a variety of approaches.

One area that has recently become more flexible is the procurement of services. As mentioned above, the "best value" purchasing concept allows federal agencies to use other factors in addition to price when awarding procurement contracts. Other factors include past performance, quality, delivery, and experience. In making award decisions, the agency may, among other techniques, compare offers and hold discussions, and may choose more qualified firms at a higher price. As a result, the competitive advantage belongs to those vendors who have performed well in the past, provided quality work, complied with wage requirements, and have high standards of workmanship.

For service contracts, the authority to negotiate prices and cooperatively develop work specifications with potential contractors is much broader than with timber sale contracts. With the goals of flexibility, cost-efficiency, and product value in mind, recent statutes have broadened this authority even further, making it possible to use a negotiated submittal process on procurements with expected costs as high as $100,000.

One tool that increases the efficiency of service contracts and adds to the effectiveness of management outcomes is the Request for Proposal (or RFP). This is a request for interested bidders to submit proposals for how they would carry out a set of activities defined by the agency and how much they would charge to do the work. The RFP may use specifications or a statement of work and can range from broad to very detailed in its expectations. It allows field personnel and contracting officers to review a range of options for having a management service performed, based on the skills and equipment of interested bidders. This is not a process guided by potential contractors; the agency will have already determined its management objectives through the public planning process and has final authority in awarding the contract. The RFP can be especially helpful in cases in which the agency wants to test a new management technique (such as utilization of very small diameter wood from thinning as a means of reducing fuel loads). In such cases, an RFP might be developed that describes the overall goals of the management activity, as well as preferred techniques. Following receipt of proposals, the agency could review options and engage in discussions with the contractors who submitted them. A contract conforming to the selected proposal can then be awarded.

The current authority to negotiate timber sale contracts is limited in comparison to similar authorities for the negotiation of service contracts. As required by NFMA, all timber sales appraised at $10,000 or greater must be advertised, and current regulations require that these sales be competitively bid and awarded to the highest bidder. However, sales appraised at less than $10,000 must be advertised for competitive bid only when sufficient local interest in such sales is likely. In many cases involving low-value material with little or no regional market, this authority allows the Forest Service to negotiate with individuals who may have a unique ability to derive value from the purchase of these products.

In similar cases involving the harvest of low-value materials, administrative use authority[4] has been used to provide the material free of charge to contractors that perform a desired service in removing that material (e.g., fuels reduction and thinnings). Administrative use sales are allowed only under limited conditions, generally in relation to research or when disposal protects the forest from injury or improves growing conditions. Furthermore, limitations are placed on the total value and volume Forest Service line officers can authorize under the administrative use authority. While not limited to research and demonstration settings, this authority has often been used where there has been interest in testing the development of potential future markets for low-value materials.

The Forest Service generally has broad authority to work with interested local groups to design, implement, and monitor projects on national forest land. However, the Forest Service retains final decisionmaking authority on all projects. There are also limitations on the structure of formal groups that provide advice or recommendations to the agency. These limitations are governed primarily by the Federal Advisory Committee Act (FACA), which is meant to ensure openness in public processes and balanced representation on formal advisory committees. However, FACA should not be viewed as a barrier to the collaborative decisionmaking approach. Perhaps the most important thing to remember about FACA is that it relates only to formal advisory committees and generally does not preclude the formation of informal, ad hoc citizens' committees (Natural Resources Law Center 1997).

One mechanism that allows the Forest Service to accomplish stewardship projects in cooperation with outside organizations is the cooperative agreement. Such agreements may be entered into under a variety of statutory authorities. Typically, cooperative agreements involve a joint accomplishment of work and directly benefit both the cooperator and the agency. These agreements have been extremely useful as tools for promoting community involvement in land stewardship projects (such as watershed restoration work) on federal forest land (Watershed Research and Training Center 1998). The benefits of such projects accrue to the agency in the performance of desired land management services and to the cooperators in skills training and, in some cases, the use of low-value products removed. While cooperative agreements have proven useful for high capacity communities undertaking community-based land stewardship activities, some limitations remain. These agreements often require the cooperator to provide matching resources and the Forest Service may not pay fees for work to the contractor. Some communities find that these limitations make it difficult for them to use the agreements successfully.

CURRENT STRUGGLES WITH IMPLEMENTING STEWARDSHIP IDEAS ON THE GROUND

A challenge that remains is developing a consistent understanding of existing authorities for stewardship contracting among interested stakeholders. Work is under way both within and outside the Forest Service to clarify authorities that will allow for the development and testing of innovative contracting mechanisms in nontraditional situations. This ongoing work includes reviewing and analyzing existing authorities, as well as developing training sessions on allowable applications of contracting authorities for Forest Service personnel and the interested public.

Another significant obstacle to promoting various stewardship mechanisms is the lack of a common understanding among stakeholders as to the ultimate goal of the land stewardship process. To some, stewardship contracting is primarily a means to an end. This view is demonstrated in outcome-oriented mechanisms that focus primarily on facilitating a desired management activity.[5] Others view stewardship contracting through a broader lens and are interested in more process-oriented mechanisms that focus primarily on deriving corollary benefits from stewardship contracts.[6]

In addition to the lack of consistent understanding of goals, there is a lack of agreement on the desired outcomes of stewardship projects. In other words, which ecosystem functions and values are desired on federal forest lands? Stakeholders from around the country continue to debate the physical and symbolic role these public forests should play in the American landscape. Until this role is more clearly established, there will continue to be differences of opinion over which stewardship mechanisms are most appropriate for implementation on the national forests. However, this debate does not preclude continuing to explore a set of broadly defined mechanisms (such as those that facilitate collaborative decisionmaking) that can be applied equally to a range of management scenarios. Among the additional issues and questions that remain to be addressed are:

- *Sustainability:* What constitutes community sustainability? How do we compare the stability of existing community against long-term sustainability? (e.g., The jobs are where, for whom, and for how many?) Where and to whom do benefits flow?
- *Accountability:* What criteria are used to determine maximization of value? (i.e., Conflicts between the valuation of public assets based on goods sold versus quality of services provided.)
- *Flexibility:* How might adaptive, collaborative, and flexible stewardship-based approaches be fit into the existing rigid operating framework?

ACTION STRATEGIES FOR FUTURE ADVANCES

As the concept of stewardship continues to be developed and tested through implementation on federal lands, a number of strategies can help move the issue forward. Perhaps the most important step is to ensure that, at the ground level, there is broad-based, open, and inclusive community participation in designing and implementing the contract. Collaborating with area universities and researchers will help communities secure the information needed to implement the best available scientific techniques on the ground. Creating ties to national interest groups is also necessary to integrate national-level concerns and to ensure that national public land goals are implemented.

The best available science is a necessary component of sustainable management; continuous testing of that implementation will be essential for adaptive, flexible management. While it is important to involve multiple stakeholders in designing and implementing stewardship projects, it is equally important to integrate local and national interests in monitoring and evaluating on-the-ground management. As it is still unclear how sustainability will work in practice, monitoring and evaluating the on-the ground results will be to inform future direction.

Because stewardship projects have such a variety of goals, it can be extremely difficult for forest managers to implement them using traditional timber sale or service contracts. One solution may be new contracting strategies, which can be effectively and efficiently meet stewardship goals. Any tests of these strategies should attempt to use existing timber sale and service contracting authorities in innovative ways and should be designed to answer questions about the effectiveness of various contracting strategies. The results would then form the basis for an open discussion of which strategies are the most useful for achieving stewardship goals, and whether those strategies could be enhanced by developing new authorities, policies, and contract instruments. The outcome of new projects conducted under existing authority could be combined with information from past and present pilot projects to determine the direction of future legislative and agency action.

The testing of new applications for timber sales and service contracts is one example of how these projects might help define future action. As the variety of on-the-ground ecosystem management objectives increases, along with contractors' ability to use nontraditional or low-value forest products, the distinction between services and sales becomes less clear. Projects that integrate a variety of ecosystem management tasks help build a local workforce that is more resilient to economic changes. This creates a need to consider strategies that combine elements of services and sales within single contracting instruments to more effectively implement ecosystem management goals.

There may be opportunities through these types of contracts to make use of the value of materials that are not targeted for sale, but must be removed to achieve stewardship objectives. Examples include forest stands in need of thinning for small fuels reduction, certain prescriptions for fisheries or wildlife habitat improvement, or prescriptions to restore forested areas to desired stand structures or species composition. Each of these projects may involve the removal of materials that have some commercial value. This is especially true as new markets are created for previously unmarketable material (such as small-diameter wood and special forest products). These by-products of ecosystem-based sustainable management can serve multiple goals and using them can help offset ecosystem-based management costs while providing material to new markets for specialty products. The creation of a high-skilled workforce trained to perform ecosystem-based management activities will help to diversify local economies. Furthermore, the innovative use of material previously deemed unmerchantable can provide additional resources to meet consumer demands in a more efficient manner.

Given the current debate about the goals of stewardship contracting and the question of how best to implement these goals sustainably, the best course of action appears to be exploring concepts through pilot projects. While this may be frustrating and time consuming, testing concepts in well-designed pilot projects will help answer questions about the effectiveness of new instruments.

Due to the complexity of existing laws and regulations and the slow pace of change in bureaucratic organizations, enacting new laws that provide broader authority is often seen as a quick and necessary fix to existing problems. This approach, while potentially effective, should be taken with caution. Developing legislative language that allows broader, more flexible authority will not guarantee the resulting implementation of new laws will be more flexible once agencies put additional regulations in place. Furthermore, putting legislation through Congress allows for language changes that meet the specific needs of congressional delegates but not necessarily the needs of those initially proposing the legislation. Rather than waiting for this type of broader permanent authority, federal agencies should continue to test innovative methods for achieving stewardship goals. Temporary authority may be needed to test some new mechanisms. Included within this authority should be language and appropriations that allow for a significant amount of monitoring and evaluation. Sustaining this kind of incremental approach over the long-term will require leadership and initiative from all levels within federal agencies, from field offices to the Washington, DC, headquarters. Furthermore, the encouragement and commitment of community-based groups, nongovernmental organizations, and others in the testing and monitoring of new tools will be essential to ensure these new mechanisms are effective. In the words of one federal employee, incorporating the elements of stewardship into agency culture is a matter of "heart, guts and soul" (pers. comm., USDA Forest Service Collaborative Stewardship Team, 1997).

CONCLUSION

Implementing land stewardship contracts on federal land has received increasing interest in the last quarter century. While the original thinking behind stewardship contracts was increased efficiency in contract administration on public land, coalitions formed to carry out community-based ecosystem management tasks have significantly broadened the concept to include environmental, economic, and social considerations. For many of those involved, implementing these new concepts on the ground has often seemed frustratingly slow. While some authorities can be used creatively to accomplish land stewardship objectives, innovation within federal agencies is often a time-consuming task.

A challenge that remains is the lack of a consistent understanding of goals and the desired outcome of stewardship projects. While some see the primary benefits of stewardship contracting as its use of process-oriented mechanisms (e.g. collaborative decisionmaking), others view stewardship contracting more as a tool to increase efficiency of management activities (e.g., goods for services). This lack of shared understanding should not preclude the testing of new mechanisms designed to facilitate community-based ecosystem management. Well-structured pilot tests that include a significant amount of monitoring and evaluation will allow for the long-term development of instruments designed to meet sustainable management goals. Pilot testing and subsequent adaptation of new tools will help to ensure clear accountability for taxpayers' dollars spent on public land.

NOTES

1. An end-results specification describes the desired outcome of a given project. Rather than being required to conform to a specification that prescribes a given technique or process, contractors are able to use their preferred technique for achieving the desired outcome, as long as the technique is approved by the Forest Service.

2. For a more detailed summary of this legislation, see Ringgold and Mitsos, 1996, "Land Management Stewardship Contracts: Background and Legislative History."

3. The concept of "best value" was incorporated into the FARs through legislation enacted in 1994 in response to concerns that awarding a contract on bid price alone will not always result in quality work.

4. Administrative use authority allows for the disposal of timber or other forest products by permit as opposed to commercial timber sale.

5. Examples include: (1) the use of end-results contract specifications to better achieve desired conditions, and (2) providing salvage rights to the contractor for certain low-value products to help reduce the costs of a service contract.

6. Examples include: (1) expanded authority to involve the public in project development and monitoring, and (2) the use of bundled contracts to develop a workforce with diverse ecosystem management skills.

REFERENCES

Allen, D.K. et al. 1994. Implementing Land Management Service Contracts. Report to Congress by the Dixie National Forest, Intermountain Region, USDA Forest Service.

Aplet, G.H. et al. 1993. *Defining Sustainable Forestry.* Washington, DC: The Wilderness Society.

Babbitt v. Sweet Home Chapt. Comms. for Ore., 515 U.S. 687 (1995).

Brown, B.A. 1995. *In Timber Country: Working People's Stories of Environmental Conflict and Urban Flight.* Philadelphia: Temple University Press.

Bruntland, Gro Harlem. 1987. What is Sustainable Development?. In *Towards Sustainable Development.* Nottingham, UK: Russell Press

Carey, H.H. et al. 1994. Developing Partnerships for Forest Stewardship and Rural Economic Development. Proceedings of the National Network of Forest Practitioners. Washington, DC.

Dana, S.T. and S.K. Fairfax. 1980. *Forest and Range Policy: Its Development in the United States (2nd ed.).* New York: McGraw-Hill, Inc.

Hirt, P.W. 1994. *A Conspiracy of Optimism: Management of the National Forests Since World War Two.* Lincoln, NE: University of Nebraska Press.

Inglehardt, R. 1977. *The Silent Revolution: Changing Values and Political Styles Among Western Publics.* Princeton, NJ: Princeton University Press.

Lynn, R. et al. 1994. Land Management Services Contracting. Report to Congress by the Washington Office, USDA Forest Service.

Leopold, A. 1949. *A Sand County Almanac, and Sketches Here and There.* New York: Oxford University Press.

Merriam-Webster. 1995. *Merriam-Webster's Collegiate Dictionary*. Springfield, MA: Merriam-Webster, Inc.

National Performance Review. 1997. The Blair House Papers. Washington, DC.

Natural Resources Law Center. 1997. Innovations in Forestry: Public Participation in Forest Planning. Boulder, CO: University of Colorado Natural Resources Law Center.

Ohio Forestry Association, Inc. v. Sierra Club, 140 L.Ed.2d 921 (1998).

Plater, Z.J.B., R.H. Adams and W. Goldfarb. 1992. *Environmental Law and Policy: Nature, Law, and Society*. St. Paul, MN: West Publishing Company.

Power, T.M. 1996. *Lost Landscapes and Failed Economies: The Search for a Value of Place*. Washington, DC: Island Press.

President's Council on Sustainable Development. 1996. Sustainable America: A New Consensus for the Future. Washington, DC.

Ringgold, P.C. and M.K. Mitsos. 1996. Land Management Stewardship Contracts: Background and Legislative History. Washington, DC: Pinchot Institute for Conservation.

Sample, V.A. 1990. *The Impact of the Federal Budget Process on National Forest Planning*. Westport, CT: Greenwood Press.

Sample, V.A. and A.A. DiNicola. 1996. Land Stewardship Contracts: Issues and Opportunities. Discussion paper based on a symposium held September 29-30, 1994. Washington, DC: Forest Policy Center.

USDA Forest Service Collaborative Stewardship Team. 1997. Meeting held April 22, 1997, Missoula, MT.

US Department of Interior. 1998. Public Lands Statistics–1997. Vol. 182, BLM/BC/ST–98/001 + 1165. Washington, DC: USDI Bureau of Land Management.

Watershed Research and Training Center. 1998. Chopsticks Small Diameter Demonstration Project (Grant #R5-14-97-30). Report prepared for USDA Forest Service.

Yaffee, S.L. 1994. *The Wisdom of the Spotted Owl: Policy Lessons for a New Century*. Washington, DC: Island Press.

Surviving the Transition: Issues for Maturing Community Forestry Groups

Carol Daly

SUMMARY. Community-based stewardship forestry (CBSF) on public lands in the United States is still a fledgling concept. Neither the number of projects completed nor the length of time that has passed since their initial implementation has been sufficient to draw definitive conclusions about the movement's success in improving economic, social, and environmental conditions in forest-dependent communities. What is of most concern now is whether the CBSF fledgling will have a chance to reach its full potential or will instead fall victim to organizational exhaustion, economic malnutrition, or political sniping. Experiences with stewardship collaboratives in northwest Montana have helped define 10 key factors CBSF groups must effectively address if they are to survive: (1) organizational burnout; (2) membership composition, expansion, diversification, and utilization; (3) change in organizational mission or focus; (4) financial viability; (5) responsiveness of federal decision makers to CBSF groups' legislative concerns; (6) continued skepticism among both industry and environmental communities; (7) movement from small demonstration projects to implementation on a larger scale; (8) need for procedural simplification; (9) outreach to the broader community; and (10) institutionalization of a meaningful role for communities in the natural resource policy making and management process. There are no simple solutions to these challenges, but the collective record of CBSF groups so far suggests a level of commit-

---

Carol Daly is President of the Flathead Economic Policy Center, 15 Depot Park, Kalispell, MT 59901.

[Haworth co-indexing entry note]: "Surviving the Transition: Issues for Maturing Community Forestry Groups." Daly, Carol. Co-published simultaneously in *Journal of Sustainable Forestry* (Food Products Press, an imprint of The Haworth Press, Inc.) Vol. 13, No. 1/2, 2001, pp. 321-334; and: *Understanding Community-Based Forest Ecosystem Management* (ed: Gerald J. Gray, Maia J. Enzer, and Jonathan Kusel) Food Products Press, an imprint of The Haworth Press, Inc., 2001, pp. 321-334. Single or multiple copies of this article are available for a fee from The Haworth Document Delivery Service [1-800-342-9678, 9:00 a.m. - 5:00 p.m. (EST). E-mail address: getinfo@haworthpressinc.com].

ment and staying power sufficient to propel the fledgling through the transition and into full flight. *[Article copies available for a fee from The Haworth Document Delivery Service: 1-800-342-9678. E-mail address: <getinfo@haworthpressinc.com> Website: <http://www.HaworthPress. com>]*

KEYWORDS. Community-based forestry, stewardship forestry, natural resource management, community collaboratives

Community-based stewardship forestry (CBSF) on public lands in the United States is still a fledgling concept. Neither the number of projects completed nor the length of time that has passed since their initial implementation has been sufficient to draw definitive conclusions about the movement's success in improving economic, social, and environmental conditions in forest-dependent communities. Still, preliminary results are encouraging, and the findings from those early projects are being used by practitioners to refine the questions they ask and the processes and techniques they use in carrying out and evaluating additional demonstration projects. What is of most concern now is whether the CBSF fledgling will have a chance to reach its full potential or will instead fall victim to organizational exhaustion, economic malnutrition, or political sniping. Experiences with CBSF collaboratives in northwest Montana illustrate the difficulties.

Flathead County, Montana, covers 5,257 square miles, roughly the size of the state of Connecticut. Approximately 77 percent of its land area is in government ownership, including large portions of the Flathead National Forest and Glacier National Park. Of the remaining 23 percent in private ownership, 62 percent is forested. The forest products industry is the county's largest manufacturing activity, directly employing more than 1,800 persons and supporting another 3,000 derivative jobs, together totaling nearly 18 percent of the county work force (Flathead Economic Policy Center 1997). The tourism industry is also heavily dependent upon forest resources (water, wildlife, scenery) for its viability. Roughly 8 percent of all private sector wages paid in the county (and 30 percent of the jobs) are estimated to be derived from visitor expenditures (Flathead Economic Policy Center 1997).

Frequent and impassioned arguments ensue over how the area's public forest resources should be managed, and all too often the failure to achieve acceptable resolution of those disagreements has led to cancellations, appeals, and/or litigation of proposed federal or state forest projects. In fiscal year 1987 the Flathead National Forest offered 114.7 million board-feet of timber for sale. Since then, the numbers have declined steadily, with 31.9 million sold in fiscal year 1997. Most critically for the future of the industry, volume under contract fell from 308 million board feet in 1987 to 44 million in 1997 (McKinney 1998). Other equally contentious ecosystem management issues include wildlife habitat protection, road building and road closures, noxious weed control, and the effects of human-wildlife interactions on threatened or endangered species such as grizzly bears, grey wolves, and bull trout.

In early 1994, two independent logging contractors optimistically began exploring ways to turn down the verbal heat and engage people instead in positive approaches to finding acceptable solutions. A series of small meetings was organized, and participation grew steadily. By late spring the group (now called the Flathead Forestry Project) was meeting every other week, and anyone with an interest in collaboratively addressing public forest management issues was welcome. The biweekly meeting schedule has been maintained ever since, with approximately 25 regular attendees and more than 130 receiving the minutes of those meetings. People with a broad range of concerns–logging, mill operation, environmental protection, recreation, education, public policy, etc.–participate, but as individuals, not as representatives of specific organizations or interests.

COMMON CONCERNS, UNIFYING GOALS

The mission statement the group crafted is ambitious:

Recognizing the need for pro-active, community-based solutions to mutually dependent social, ecological, and economic concerns, the Flathead Forestry Project (FFP) has as its foundation three unifying goals:

1. to promote community trust and collaborative processes,
2. to ensure forest and ecosystem health, and
3. to provide for a sustainable resource-based economy within the region.

FFP's projects are based on a commitment to stewardship forestry, to managing for the long-term health and sustainability of the entire forest ecosystem, including soil, water, vegetation, wildlife, and humans. The group's strategy has been to start small, carrying out a number of projects deliberately limited in acreage and impact but with maximum public participation in project planning, design, contracting, monitoring, and evaluation. A key component is the use of steward loggers–loggers trained and/or experienced in the philosophy and practice of stewardship forestry. One way contractors can demonstrate their qualifications is by successfully completing the Montana Logging Association's Accredited Logging Professional (ALP) training program, which includes a 40-hour Montana Forest Stewardship Workshop.

Another concept FFP is exploring is the "separation of the logger from the logs;" that is, paying a contractor to conduct all the stewardship work on a particular project, then selling any resulting merchantable products in a totally separate transaction, either "roadside" or from a public sorting yard. If the contractor's task is to achieve a desired future condition on the land, instead of to deliver logs of a certain type or size to a mill, he should be motivated to make cutting decisions based primarily on ecologic, not economic, considerations.

Demonstration Sites

The group's first on-the-ground effort was a 30-acre ponderosa pine restoration project in Condon, in the southern part of the Flathead National Forest. Planned in 1995 and carried out in 1996 and 1997, it was a cooperative effort with another community collaborative, the Swan Citizens *ad hoc* Committee. The Swan *ad hoc* group was formed prior to FFP and provides a forum for discussion and action on a broad range of community issues.

The Condon project was done as a regular Forest Service timber sale, with the contract award made solely on the basis of price. However, it involved considerably more local participation in planning, implementation, and monitoring than is usual in federal sales. In the course of putting together that first project, FFP members learned that without authorizing federal legislation, they would not be able to test many of their key concepts on federal land. That led to the drafting of a proposed five-year national stewardship demonstration program. The bill was introduced in 1995 by two of Montana's three congressional delegates but, lacking the endorsement of the third, was never afforded a hearing by the committees to which it was assigned. The effort, however, brought FFP to the attention of other CBSF groups around the country, and spawned many valuable working and learning relationships.

In 1996, a 20-acre project was completed on private land, enabling FFP to test the value of using contractor qualifications as well as price in choosing among bidders. It packaged in a single contract a number of activities, including stand thinning and improvement, wildlife habitat enhancement, and development of limited recreational facilities (trails and a cabin access). The family-owned F. H. Stoltze Land and Lumber Company assisted the project by sharing in the construction costs for a bridge and road across Stoltze land to reach the project area.

FFP's third effort, a 106-acre package divided into five commercial thinning sales, was designed to evaluate different treatment designs and contractor approaches to stand improvement. It was appealed, and as a result a 15-acre unit containing some old growth was dropped from the project. Work on the remaining units began in 1997 and continued in the 1998 season. A nearby environmental education center, the Glacier Institute, is conducting long-term monitoring of a portion of the project, and area high school students did several stewardship plot inventories.

Frustrated by the limitations of the Forest Service's timber sale process,

FFP chose as its fourth demonstration a 20-acre multi-task project on state school trust land next to a large rural subdivision on the Flathead Valley floor. The Montana Department of Natural Resources and Conservation (DNRC) was eager to try a collaborative stewardship project. At FFP's urging, the agency offered the sale with two key changes from its normal operating procedures. First, rather than just agreeing to follow DNRC "cut" or "leave" tree markings, bidders instead were asked to submit a detailed technical proposal explaining how they would achieve a DNRC/FFP-described desired future condition for the area. DNRC agreed to give the successful proposer significant discretion in making cut/leave decisions to implement the plan. Secondly, bidder qualifications were considered in addition to price in making the final selection. A bid review committee, with both DNRC and FFP members, evaluated technical and financial proposals, interviewed bid finalists, and recommended the proposals members felt would provide the best overall value to the state school trusts. The responsible DNRC officer followed the bid review committee's recommendations in making her selection.

By late 1996, some FFP participants and outside observers were questioning the group's effectiveness. "We've been meeting and working for three years now, and we've only treated a little over 150 acres," the argument went. "All the legal barriers to fully implementing our concepts remain in place. At this rate, we aren't doing much to help either the environment or the economy."

Advocates pointed out that, due in significant part to FFP efforts, the small timber sale program on the Flathead National Forest had become much more active. New legislative proposals were being discussed in Congress, with helpful suggestions and comments not only from other community collaborative groups, but also from some national industry and environmental organizations. Finally, considerable learning had taken place in the Flathead, as FFP efforts expanded public knowledge of forest management issues and options, and the level of trust between some segments of the public and the Forest Service appeared to be improving.

Enhanced Credibility

Further, the value of collaborative community involvement in the management of public lands was gaining credibility. In 1997 DNRC hired the Flathead Economic Policy Center (FEPC), which volunteers staff services to FFP, to organize a collaborative to develop options for management of a portion of the Stillwater State Forest. The agency had proposed a project in the Beaver Lake area some years before but had withdrawn it because of widespread public opposition. Now DNRC offered to turn its usual process upside down. Rather than starting with an agency-developed proposed action, DNRC invited the public to research and design from scratch one or more options that would be economically, socially, and environmentally sound and that DNRC could use as the alternative(s) to be studied in an Environmental Impact Statement for the project.

The Flathead National Forest, meanwhile, placed approximately $200,000 with the National Forest Foundation, earmarking it for cost-sharing projects with Flathead-area community collaboratives. Both the Swan *ad hoc* group and FFP developed proposals, the former for operational costs for its new Swan Ecosystem Center and a wilderness ranger program, and the latter for a roughly 100-acre forest restoration project on Cedar Flats, just north of the city of Columbia Falls. Cedar Flats, FFP participants decided, would be their flagship demonstration, finally employing all their key concepts on federal lands.

Necessary additional funding for Cedar Flats was obtained from two private foundations, plus the US Environmental Protection Agency, and the North American Fund for Environmental Cooperation. It was subsequently chosen nationally as one of 23 Forest Service Management Reinvention Pilot projects. Two years in development, it went out for bid in June 1998. Designed in a series of paired service contract units, it will test six FFP hypotheses:

a. A service contract is a better instrument than a conventional Forest Service timber sale for achieving a desired future condition for a specific area when a number of ecosystem components are being addressed and/or when the value of the timber estimated to be removed is close to or less than the cost of the work to be done. Timber sale purchasers cannot legally be required to perform services unrelated to the removal of timber and the reforestation of the area. Also, sale purchasers have no incentive to do more than the minimum work required on a job which has little profit potential.
b. Better on-the-ground work results when the logger is separated from the logs (that is, when the contractor has no financial interest in any forest products removed during the project).
c. Forest Service planning and administrative costs can be reduced by placing some of the responsibility for planning, layout, and pre-marketing on prospective quoters. One way to achieve this is to provide those making the offer with a complete description of a desired future condition for an area (but not with specific prescriptions) and then have the quoters decide and explain how they would do the work to best achieve that condition.
d. The use of qualitative factors (such as prior stewardship training and field experience) in addition to price, when selecting a contractor will increase the likelihood of a better, more complete job being done.

e. When a qualified stewardship contractor is used, and when the desired future condition for an area is clearly described, it should be unnecessary for the Forest Service to either mark or designate by description the trees to be removed (or left).
f. Using qualified contractors and allowing them to set their own cutting guidelines will not result in a loss of product sale value.

WHAT'S NEXT

With Cedar Flats nearly underway, the pressing question for FFP participants and observers is, "What's next?" The group must decide whether doing more demonstration projects will further the collaborative's aims, or whether having done all it can under existing legislative authorities, it now should simply go out of business. In assessing its future, FFP confronts issues that face many CBSF groups which are maturing (and perhaps tiring) while their fledgling stewardship progeny are still unlikely to survive without active organizational support.

Ten Factors for Survival

Organizational burnout–In addition to attending every-other-week regular meetings, active FFP members have participated in special planning sessions, field tours, stewardship plot inventories, site photography, plant and bird identification and counting, monitoring work, community orientation/education programs, contractor workshops, Forest Service meetings, research interviews, media interviews, and more. The toll has been particularly hard on members who are forestry workers. Already working schedules that require them to be on the road or in the field from 5:00 a.m. until late at night, they have to weigh the value of FFP participation against the compelling needs of family and business. Frustrated by the federal system's slowness in responding to local stewardship concerns, and feeling ever more threatened economically, many are opting to limit or end their active FFP work, spending their time instead in pursuits that produce more immediate, satisfying results. Most still keep track of FFP's progress through the minutes and occasional conversations with other members, but their absence from day-to-day activities is sharply felt.

Other participants also are wearied by the long, slow process and find the diminished forest worker presence at the table troubling. Without the people who would be the most directly affected (at least economically) by changes in federal forestry management practices they say, is it really worthwhile for the rest of us to carry on? To date, the answer has been mostly "yes." The continued attendance of some independent loggers and local mill employees buoys the group and provides it with the necessary diversity of interests, experience, and opinions. There is a strong sense of commitment to a job not yet completed, to a responsibility not yet fulfilled. The challenge is difficult to walk away from, but the going gets tougher every day. Some tangible, positive response from federal policymakers to the CBSF movement is badly needed.

Membership expansion, diversification, and utilization–FFP welcomes new participants and encourages everyone interested in the group's work to come to meetings, field tours, and other activities, even if they do not have the time to be regular attendees. Members recognize the need for continued growth and diversification, not just to involve more people in CBSF projects, but also to make sure FFP does not become too organizationally comfortable or self-satisfied. A steady infusion of fresh ideas and opinions should keep the group from falling into the trap of doing things "because that's the way we've always done it." Public, legislative, and scientific views on forest management are continually evolving, changing, or expanding, and those views need to be incorporated into the FFP process. Two recruitment targets have been teenagers (to involve, educate, and learn from tomorrow's leaders) and people from the more confrontational and litigious environmental groups (to ensure their opinions are clearly heard and understood in FFPs deliberations).

A few high school science students have attended one or two FFP meetings, but no on-going participation has resulted. Some regular participants seemed unsure how to relate to the students, and well-meant efforts to draw them into group discussions may have been perceived as patronizing. Also, students have many claims on their after-school time, and FFP's admittedly not very exciting meetings may not compete well with clubs, sports, studies, and other youth activities. Students do seem interested in hands-on projects, and quite a few have participated enthusiastically not only in doing plot inventories on FFP sites, but also in carrying out extended environmental research projects under the tutelage of mentors from federal and state natural resource agencies. While the research is not directly tied with FFP's work, it may in future provide a pool of motivated young people for FFP's monitoring and evaluation teams.

Sadly, none of the leaders of the targeted environmental groups has attended an FFP meeting, although invited repeatedly. The groups do, however, receive meeting minutes and carefully track FFP activities. Occasionally they raise issues with FFP by letter, and they have appealed two of its four public land projects (sometimes quoting passages from FFP's minutes in their appeals). While not at the table, these groups do have a presence in FFP's work.

In addition to trying to be responsive to the concerns of people who don't come to meetings, FFP also struggles to integrate sporadic participants into its process. These members receive all meeting minutes and notices of special activities, but tend to come only when the agenda items are of particular interest to them. This leads to meetings where regular attendees are frustrated by the need to take time to bring others up to date on current activities and, often, to revisit issues they thought had been settled in previous sessions. Sometimes the infrequent attendees do persuasively raise new arguments, leading the entire group to change or modify previous decisions. Such reversals or unexpected turns in the collaborative process have proven particularly frustrating for Forest Service officials trying to be responsive to FFP concerns. "When does a collaborative decision become final?" they ask.

Most groups use a facilitator to try to keep their deliberations civil and on track. FFP, a generally very civil group, rotates facilitation duties among the participants, and the technique seems to work well. The Swan *ad hoc* group (also quite civil) uses a local professional facilitator who volunteers his time, but the group rotates meeting planning responsibilities among its members. The Beaver Lake Collaborative had an overall process facilitator paid for by DNRC, but rotated meeting facilitation responsibilities among the participants.

Changing mission or focus–Over time, there has been a gradual shift in the tone and substance of FFP discussions, with less emphasis now on small sale stewardship logging *per se* and more on overall ecosystem management issues. This may be both a cause and a result of diminished forest worker presence at meetings.

Because of the successful implementation of its own demonstration programs and the increased agency attention FFP brought to the issue of small sales, the group achieved in large measure its goal of restarting the small sale program on the Flathead National Forest. While the number and volume of such sales is still not all that the logging community would like to see, small sale opportunities have improved significantly. FFP participants for whom that was the major goal may feel their work is done.

Others are more concerned with the broader goals of restoring ecosystem health and re-building trust between federal and state land managers and the general public. For them, stewardship forestry projects are a means, not an end. That outlook creates some tension among participants who rely on wood products for their economic well-being.

While they agree with the need to maintain or restore ecosystem health, they believe it is best achieved with conventional harvesting and thinning practices. Participants without forest industry ties, on the other hand, are often equally interested in fostering the increased use of other tools, including the aggressive reintroduction of fire as both a management technique and an ecosystem process. But, in addition to viewing large-scale burning as an unwarranted waste of timber, many loggers and mill employees are wary of taking a further step down the slippery slope toward less harvesting on public lands. As one industry-related participant frequently points out, "Coming to the table in groups like this to try to work out solutions always results in [the timber industry] having to give up something. We can't win; we just keep losing more."

Members with ties to the environmental community respond that they may not be at financial risk, but their participation is still hotly criticized by some of their peers, who accuse them of collaborating with the enemy. They are risking their credibility and their standing in their community of interest. To date, FFP has held together in spite of such concerns, but the idea that not everyone at the table bears the same degree of risk in collaborating is a bone of contention that must acknowledged and addressed.

Financial viability–FFP has been both aggressive and creative in finding ways to experiment on-the-ground with its ideas. The Flathead National Forest and the state DNRC have pushed their legislative authorities to the limit to facilitate the process but cannot be expected to exceed those limits. Local private contributions of money, time, and skills have been outstanding. Land has been provided for demonstrations. Foresters and loggers have given their time to instruct interested amateurs in stewardship site inventorying and have loaned their expensive equipment for the workshops. Thousands of dollars in postage, printing, and administrative costs have been donated by individuals and by FEPC, which also has beaten the foundation and agency bushes to secure grants and cost-share agreements for current FFP projects. Funding each new demonstration, however, is increasingly difficult, and this "forestry by bake sale" approach frustrates participants. They see it as an unwarranted failure by state and federal governments to provide adequately for lands under their management.

FFP's concern is shared nationwide by private foundation directors, environmental interests, and many of the other nonprofits that compete for scarce private grant funds. "If [federal agencies] are successful in [attracting significant private grants], it will give Congress even more reason to cut their funding on the premise that they can go to the private sector to get this money," complained one critic quoted in *The Chronicle of Philanthropy.* "We don't have sufficient funding as it is. To have to compete with the federal government puts us into a whole other realm" (Able, Jr. quoted in Demko 1996).

Slow or no response by federal decisionmakers to CBSF groups' legislative concerns–If the local politics of CBSF collaboration are difficult, conditions are much worse at the national level. FFP and other community-based groups must have the approval of Congress and the support of Forest Service headquarters in Washington, DC to move local demonstrations onto truly innovative paths. Possible implementation strategies such as service contracts that exchange goods for services, the creation of agency revolving funds to pay stewardship work, and the use of qualitative factors in selecting stewardship timber sale purchasers all require changes in existing laws. Requests from local groups for modifications, even on a pilot basis, have routinely been ignored, quashed by factions within the Congress or Administration, mauled and misrepresented by hostile lobbyists, or left to languish, unheard, in Congressional committee processes.

Forest- and district-level Forest Service employees generally have been very supportive of FFP and Swan *ad hoc* efforts, but the lack of positive, tangible administrative or congressional response to stewardship forestry initiatives sends a troubling message. Are the supportive employees out on a limb that could be cut off behind them? Speeches from the Washington Office are encouraging, but to date there has been no sign of rewards for participation in collaborative efforts.

While agency employees wrestle with the implications this has for their careers, other FFP members wonder if they are wasting their time or (worse) being duped. Is verbal encouragement of collaboration that is unaccompanied by either financial or legislative support simply a ploy to justify prolonged inaction? It is almost preferable to consider the federal nonresponse as deliberate, rather than as evidence that the current government is unable to break its gridlock over public land management.

Continued skepticism among both the industry and environmental communities–With some demonstration projects completed and more under development, FFP and other local collaboratives like to think they have achieved a level of success, but criticism is still strong. Many forest workers complain about spending time in meetings and doing small-scale demonstrations, while mills are closing and the future of the federal timber sale program hangs in limbo.

Some environmental activists take the opposite tack: five years they, is far too short a time for anyone to talk about success or even to draw meaningful conclusions about the value of CBSF. Discussing his group's appeal of a Forest Service plan to place interpretative signs in the ponderosa pine restoration project area where FFP and the Swan *ad hoc* group conducted their first demonstration, a local environmentalist said, "The main thing we're concerned about is that they will use that trail to promote logging in old-growth ponderosa forests. It's a subject that the scientific community is greatly divided on." Another group's spokeswoman added, "Biologically speaking, there is no basis in fact that the Condon project was a success (Hammer & Yanishevsky quoted in Mann 1998).

Are opponents' qualms about CBSF a matter of science, trust, or philosophy? Probably all three. A letter to the editor of the *Daily Missoulian* charged, "The Forest Service is more than willing to collaborate with those who agree with it, and it is more than willing to label as extremists and illegally cold-shoulder those with whom it does not agree. . . . [It] is currently attempting to change [the] laws so it can push aside the democratic process in favor of a collaborative process it can buy off, control and orchestrate to meet its own ends (Hammer 1998)."

FFP participants, meanwhile, are frustrated by the naysayers' unwillingness to actually participate in the collaborative process. Most recently, FFP invited (unsuccessfully) two of its most strenuous critics to join the all-party monitoring team for the Cedar Flats project. While other, more "middle of the road" environmentalists are serving on the team, the absence of the project's most severe critics is a sore spot. Can true community collaboration occur if some parties are not at the table, even if that absence is of their own choosing?

Those who have doubts about the CBSF movement do raise valid issues that need to be satisfactorily addressed as community groups press for greater flexibility in planning and conducting stewardship projects. Three of the most pressing are (1) creating safeguards to prevent any demonstration authority from being misused, (2) ensuring adequate accountability to the public for any actions taken, and (3) providing a means for including the concerns of non-local constituencies in the process.

Questions of scale–FFP has chosen to pursue its goals through a series of very small projects for a number of reasons. Small projects are less likely to become as controversial as larger ones. If a technique tried in a small demonstration is less successful than hoped, less potential environmental harm is done. Smaller projects are also easier to finance.

There are environmental and socio-economic downsides to the small-scale approach. From an ecosystem management perspective, small projects produce what a consulting fire ecologist in the Flathead calls the "Swiss cheese" effect. Rather than facilitating ecosystem restoration at an appropriate landscape scale, the small demonstration areas create a series of "holes" in the forest landscape, and treatments, however well done, have a minimal impact on the overall needs of the system.

In addition, the planning, analysis, and administrative time and costs for small projects are often as great as if a much larger area had been addressed. With long planning times, high front end costs, and limited timber production and/or forest restoration accomplished, FFP is characterized by some critics as producing "micro solutions to macro problems." But participants maintain that small projects are better than no projects at all, and some members feel it is not necessary or even appropriate to increase the scope of activities. Others, however, would like to experiment with FFP's concepts on a larger scale and see that as a logical and necessary progression in their work.

Need for simplification–As FFP delves ever more deeply into the complexities of the federal timber sale and service contracting processes, it becomes evident that in their present form these processes are far too complicated and administratively demanding for small, largely volunteer CBSF groups to implement effectively. Further, the existing system's reliance on penalties for poor performance, rather than incentives for top quality work, runs contrary to the stewardship philosophy.

If federal and state agencies were to continue to administer all management activities on public lands, that might not be a problem. Budget cutting and downsizing, however, have become more the rule than the exception in government, and agencies increasingly look to the private sector to pick up the slack. The Swan *ad hoc* group, for instance, now substantially supports a wilderness ranger program that used to be totally Forest Service funded. It also fields a backcountry education program that maintains trails and facilities the Forest Service no longer has adequate staff to handle.

FFP considered offering and administering (through FEPC) the service contracts for the work on Cedar Flats, but the administrative requirements and liability issues were simply too much for FEPC's board and one-person staff to assume. If the trend of transferring responsibility for funding and conducting certain on-the-ground programs from the public to the private sector continues, agencies and CBSF groups will have to find a way to simplify administration and limit a CBSF organization's financial exposure without sacrificing program quality or accountability.

Outreach to the broader community–One of the valued outcomes of CBSF programs is the increase in public awareness of and involvement in natural resource issues, both local and national. In addition to expanding its active membership, FFP tries to reach a broader public through media coverage, field tours, exhibits at various public events, talks to civic groups, and other means. The Swan *ad hoc* group has established an ecosystem learning center, which offers to Montanans and visitors alike easily accessible wildlife and forestry exhibits, interpretative walks, workshops, and training programs. FEPC, the Forest Service, and local hospitality businesses jointly fund a series of informative and entertaining evening programs at motels and campgrounds, trying to reach some of the 2 million visitors who flock to the Flathead Valley each summer.

Both FFP and the Swan *ad hoc* group are active in organizations such as the National Network of Forest Practitioners and the Communities Committee of the Seventh American Forest Congress, which help inform national policy makers and the broader public about community forestry issues and options. Public support will be critical to achieving and sustaining the changes in laws, policies, and attitudes needed for CBSF to reach its full potential.

What is enough?–In community collaboratives a major questions is always, "When do we stop talking and start doing something?" The Beaver Lake Collaborative approached the problem by placing a time limit on its work; its participants agreed that within six months they would come up with a series of management alternatives for the Beaver Lake project. They also agreed to meet as often and work as hard as necessary to get the job done within that time. Neither FFP nor the Swan *ad hoc* imposes similar time limits on its work.

None of the three groups insists on consensus before action. Questions and concerns are studied and debated extensively. Somehow there seems to come a time when even those who may not agree completely with a proposed action are at least willing to give it a chance. This unspoken agreement, arrived at without votes, seems to be a product of improved trust among participants. Everyone is vested enough in the collaborative process that he or she can be sure no other member will deliberately or willingly undermine it.

Aside from the somewhat amorphous process through which decisions happen and collaboratives move from talk to action, there is another, larger question. Is the community-based collaborative a new part of the natural resource management system, one that will develop its own permanent institutional status? Or is it a temporary creation, necessary during the period when new concepts are shaped and trust levels are rebuilt between the public and government land managers?

Some FFP participants clearly see their role as a transient one, to be relinquished at some future time when conditions and practices are such that "we can just let the professionals do their jobs." Others envision community interests having a continuing and necessary place at the natural resource management table. How to assure a meaningful long-term community role in the process will be a pivotal issue in the next few years for maturing CBSF groups, agencies, policy makers, and other natural resource interest groups.

REFERENCES

Able Jr., Edward H. 1996. Controversy Brews Over Aggressive Fundraising for Federal Programs. *The Chronicle of Philanthropy.*

Daly, Carol. 1998. Conversation with Dennis McKinney, Flathead National Forest, June 12.

Flathead Economic Policy Center. 1997. *Economic Analysis of Flathead Country and Northwest Montana*. Kalispell, MT.

Hammer, Keith J. 1998. Forest Service redefining 'collaborative.' *Daily Missoulian.* Missoula, MT. Letter to the editor.

Mann, Jim. 1998. Signs for nature trail challenged. *The Daily InterLake*. Kalispell, MT. June 12, 1998 p. 1

Participatory Research for Adaptive Ecosystem Management: A Case of Nontimber Forest Products

Yvonne Everett

SUMMARY. Participatory research involves members of interested communities who work with researchers to apply their joint skills and experience to explore issues of mutual concern. There is an increasingly explicit federal mandate for participatory research efforts involving stakeholder groups in public land management. This paper presents a case study of one effort ongoing in the Hayfork Adaptive Management Area (AMA) on the Six Rivers and Shasta-Trinity national forests of northern California. There, a network of U.S. Forest Service staff, scientists, local nongovernmental organizations (NGOs) and nontimber forest product harvesters (wildcrafters) has been working to address ecological, economic, and social aspects of harvesting nontimber forest products (NTFP) from public land. This paper discusses the challenges of NTFP management on public lands and analyzes participatory research as an approach for addressing some of these challenges. *[Article copies available for a fee from The Haworth Document Delivery Service: 1-800-342-9678. E-mail address: <getinfo@haworthpressinc.com> Website: <http://www.HaworthPress.com>]*

KEYWORDS. Community-based forestry, participatory research, adaptive management, nontimber forest products

---

Yvonne Everett is Assistant Professor of Natural Resources, Humboldt State University, Harpst Street, Arcata, CA 95521-8299.

[Haworth co-indexing entry note]: "Participatory Research for Adaptive Ecosystem Management: A Case of Nontimber Forest Products." Everett, Yvonne. Co-published simultaneously in *Journal of Sustainable Forestry* (Food Products Press, an imprint of The Haworth Press, Inc.) Vol. 13, No. 1/2, 2001, pp. 335-357; and: *Understanding Community-Based Forest Ecosystem Management* (ed: Gerald J. Gray, Maia J. Enzer, and Jonathan Kusel) Food Products Press, an imprint of The Haworth Press, Inc., 2001, pp. 335-357. Single or multiple copies of this article are available for a fee from The Haworth Document Delivery Service [1-800-342-9678, 9:00 a.m. - 5:00 p.m. (EST). E-mail address: getinfo@haworthpressinc.com].

Participatory research and management–in which academic researchers or scientists join with nonacademics to apply their knowledge, skills, and experience to "real world" problem solving–has successfully addressed difficult problems in rural development and natural resources management around the world (Chambers 1991, Poffenberger 1996). In contrast with bureaucratically centralized "top down" or narrowly disciplinary scientific research approaches to problem solving, participatory research in natural resource management is based on the notion that members of communities of place and natural resources users have a fundamental right to play a role in decisions about access to resources and environmental quality. Further, participatory research draws its strength from the recognition that community members' supportive involvement and site-specific knowledge are crucial in crafting and implementing sustainable solutions to local resource problems. Clearly, participatory research cannot replace basic scientific research. Rather, it informs scientific inquiry, raising basic and applied research questions and strengthening the link between research and applying research results in resolving problems.

Federal land management agencies in the United States, such as the U.S. Forest Service and the Bureau of Land Management (BLM), increasingly are applying participatory processes. One example is the Northwest Forest Plan to implement ecosystem management on public forests in Washington, Oregon, and northern California. Under the plan, forest management policy has shifted from a primary emphasis on resource extraction to one on maintaining forest ecosystem functions (USDA 1994). As a result, commodity production on public land is declining and federal land management agencies are charged with helping to diversify rural forest-community economies that historically have depended on resource extraction, especially timber harvest from public lands (USDA 1994). The agencies have been directed to involve staff, researchers, and communities in their efforts to develop approaches to implementing these goals. In recognition of the necessary exploratory and iterative learning, the implementation approach being promoted is called adaptive management. Adaptive management efforts include iterative planning, implementation, monitoring, and evaluation phases, with the intent that lessons learned in one cycle of activities will be explicitly used to modify the next (USDA 1994).

In 1994, 10 so-called "Adaptive Management Areas" (AMAs) were designated on public forest lands. The AMAs are places to "develop and test new management approaches to integrate and achieve ecological, economic and other social and community objectives" (USDA 1994 p. 6).

The Hayfork AMA on the Six Rivers and Shasta-Trinity national forests in northern California is the largest of the 10 AMAs and extends over more than 400,000 acres. Here and on surrounding private lands, Forest Service management staff and researchers, wildcrafters (nontimber forest product harvesters), small farmers, local nongovernmental organization (NGO) staff (including the author), and interested citizens have been working with nontimber forest products (NTFP).[1] Their ultimate goal is to develop an ecologically and economically sustainable NTFP industry in the region.

The following discussion has two parts. The first section includes a brief overview of NTFPs and the particular challenges associated with their management on public lands. It concludes with a short discussion of institutional frameworks that might meet these challenges. The second section presents a case study of ongoing work with NTFPs in the Hayfork AMA as an example of efforts to develop an institutional capacity for research and management on public land in collaboration with local communities and resource users.

NONTIMBER FOREST PRODUCTS

The Demand for NTFPs

Nontimber Forest Products (NTFPs), or Special Forest Products (SFPs), are items gathered from the forest that have spiritual, subsistence, or market value (Molina et al. 1997; DeBeer and McDermott 1996; Economic Botany 1993). While a very broad range of products can be included, for the purpose of this discussion the term NTFP applies to products from plants, fungi, and lichens gleaned from the forest.

NTFPs have been harvested for as long as people have lived in and near forests; many–from honey to mushrooms to Brazil nuts–have significant market value. While NTFPs usually bring less income on a product-by-product basis than timber, these products can provide important interim income while trees are growing (e.g., Pilz et al. 1998). In the Pacific Northwest, commercial exploitation of NTFPs began early this century and ranges from mushrooms, medicines, and floral greens and boughs to berries, nuts and Christmas trees. In the last two decades, interest and market demand surged (Freed 1997); the market for medicinal herbs, many of which are wildcrafted, is skyrocketing as people seek alternatives to conventional western health care (Mater 1995).

This demand may provide economic diversification opportunities for rural forest-dependent communities. At the same time, there may be less desirable consequences associated with a rapid increase in NTFP exploitation.

Overexploitation of NTFP resources is a common problem globally (Homma 1992; Hall and Bawa 1993) and demand and a high price for NTFPs may portend over-harvesting and resource depletion. Land managers and landowners often perceive NTFPs as more easily renewable than timber, failing to recognize potentially destructive harvest activities (Godoy and Bawa

1993). Where resource depletion occurs, a range of associated environmental, social, and/or cultural impacts are likely to follow (Homma 1992). In the South and midwestern United States, highly marketable species such as American ginseng *(Panax quinquefolium*) and goldenseal (*Hydrastis canadensis*) are in serious decline in the wild (United Plant Savers 1998; American Herbalists Guild 1998). In the Pacific Northwest, a run on Pacific yew (*Taxus brevifolia*) bark for cancer treatment in the early 1990s required passage of the Pacific Yew Act of 1992 to limit the felling of this slow-growing late-successional conifer (Molina et al. 1997). On the Olympic Peninsula, salal (*Gaultheria shallon*), a mainstay of the half-century-old floral greens industry (Schlosser and Blatner, 1994) is now beginning to show signs of overharvesting amid reports of confrontations over harvest sites (Collaborative Learning Circle 1998). Numerous forest and woodland herbs, well known to herbalists and long-term users of alternative medicines, could become the next popular discovery akin to the recent "St. John's Wort is the new Prozac" rush (Bloomfield et al. 1996). For most of these products, there as yet has been no assessment of the impact of harvesting on other plant or mushroom species, animals or soils.

Capturing Benefits for Rural Communities

Further, if economic diversification is a goal, it is unclear whether much of the economic benefit from large-scale NTFP harvesting stays in communities near the forests where products are gathered. Over the last decade, for example, demand for the matsutake mushroom (*Tricholoma magnivelare*) has fueled a seasonal "gold rush" of thousands of mushroom hunters into the forests (Lipske, 1994, Schlosser and Blatner 1995; Redhead 1997). Matsutake mushrooms from national forests in Oregon and California are commercially harvested largely by short-term visitors, sold to out of town buyers, exported from the region in bulk, processed in Washington state or British Columbia, and flown to Japan. The real profits go to a few exporters in the Seattle and Vancouver areas (Hosford et al. 1997).

Managing NTFPs on Public Lands

Currently, NTFPs also fail to provide adequate revenues to cover the cost of their management on public land. Managing the matsutake phenomenon, for example, has been a complex and challenging task for the Forest Service. When the mushroom market emerged in the late 1980s and early 1990s, large numbers of people began to arrive in the forest. Each Ranger District had to devise a mechanism for handling the demands for a range of services with little warning, no extra budget, and limited cross-cultural skills. At first, with people lining up at the door, even simple communication was often difficult; many of the harvesters were recent immigrants from South-East Asia with limited English language skills, and few of the Ranger Districts had access to interpreters.[2]

Institutional Frameworks for NTFP Management

With numerous NTFPs just waiting to be discovered, land management agencies will need to stop managing the NTFP harvesting by reaction and develop a more coordinated approach. Agency staff have had little opportunity or funding to develop a planned approach that maximizes long-term ecological and economic sustainability. While there are efforts to control the harvest's impact, no one knows the long-term ecological implications of current harvest volumes and techniques. The literature on common property resources provides some insights on how institutional approaches to managing NTFPs might be developed.

NTFPs harvested from national forests are an example of what Ostrom has called "common pool resources," those drawn from a system large enough to make it costly to exclude potential beneficiaries from its use (1990, p. 30). Matsutake mushrooms, for example, are small, lightweight and valuable and therefore more subject to theft than timber. Without costly investment in access control and resource management on public lands, uncontrolled exploitation and resource depletion may result. In the case of matsutake mushrooms, for example, individual harvesters have little concern for the future. There is little incentive to leave a small mushroom behind to grow into a larger one, when the harvester cannot be sure that someone else will not come along and take it.[3] NTFPs in the national forests are comparable to resource commons that–without concerted intervention, defined, for example, by Hardin, as either privatization of land or more intensive (and expensive) government control–are likely to be overexploited until they are depleted (Hardin, 1968). If Hardin was correct, and assuming that national forests are unlikely to be privatized in the near future, it is possible that the challenges of managing the harvesting of NTFPs from public lands are just beginning and will become increasingly costly for the federal agencies in charge.

However, Ostrom and other researchers have pointed out an alternative to privatization and increasingly strict government control. They argue that the institutional capacity for resource management and control may emerge through coordination with the resource users. Under the right circumstances, users may be able to develop frameworks of control that counter the vagaries of boom and bust resource exploitation. Numerous case studies show that common pool resources and associated human communities can indeed be sustained over time (Ostrom, 1990; Ostrom and Schlager 1996). These researchers suggest successful institutions of this kind have common underlying factors including having the beneficiaries of resource harvesting involved in both resource management and the development of guidelines for access and monitoring. When these factors are applied to NTFP harvest from public lands, the assumption is that when NTFP harvesters help make the rules, they will be more likely to agree to follow them, particularly when there is a mechanism in place through which the resource can be monitored and infractions against harvest rules are made public and sanctioned. In contrast to the current reactive approach, collaborative institution building would be proactive–allowing land managers and harvesters to plan for a more sustainable industry in the future.

While common property researchers have identified numerous underlying principles affecting institutional development for natural resources management, three factors make NTFP management particularly challenging for public lands in the United States. First, wildcrafters harvesting NTFPs from public lands in the Northwest are extraordinarily mobile. As David Aurora, an internationally experienced ethnomycologist has noted, nowhere else in the world do wildcrafters have the same degree of access to private vehicles and good roads across large areas of forest with relatively few local inhabitants (Aurora 1995). For booming NTFPs, such as matsutake harvesting, communicating with thousands of transient harvesters from the West and Canada is extremely challenging. Second, while most illustrations here have been drawn from the matsutake mushroom industry, each NTFP has its associated harvest and market structure. Most harvesters specialize. The third factor critical for NTFP management in the United States is reliable information about the specified NTFPs' abundance, distribution and impacts of harvesting. Without such knowledge, users and managers have no basis for making decisions about sustainable harvest levels (Ostrom 1990; Ostrom and Schlager 1996; Hanna et al. 1996). Keeping these constraints in mind, one method of initiating participatory NTFP management is to work with individually identifiable wildcrafters and agency staff from a localized geographic region to develop reliable information and management approaches for NTFPs not yet subject to the strong market forces of a resource boom.

PARTICIPATORY RESEARCH ON NTFPs IN THE HAYFORK AMA

NTFPs on the Hayfork AMA

When the Hayfork AMA was designated under the Northwest Forest Plan in 1994, local interest in NTFPs was emerging. Yet the local Ranger Districts had little information about NTFP frequency, abundance, and harvest impacts. Wildcrafters and interested citizens had valuable knowledge, but there was no mechanism for tapping into their expertise or applying it to NTFP management. At the same time, the agency had a new mandate to involve communities in AMA research. Thus the opportunity presented itself to address NTFP harvest management with participatory research.

The region surrounding the AMA is mountainous with diverse microclimates and a wide variation in plant distribution and abundance. The dominant vegetation is Klamath mixed conifer forest but includes oak woodland and alpine ecosystems as well. The area is sparsely settled, with an average of only four people per square mile living clustered in small communities throughout the bioregion. Close to 80 percent of the land base is managed by federal agencies. Until recently, beside the public sector (government agencies and schools), timber harvesting and processing were the mainstays of the local economy. Nearly every settlement had at least one lumber mill; most had several. Today, there is only one mill left in the entire county, and many people have left in search of employment elsewhere. The people who remain, often families who have lived there for generations, are seeking to diversify their income opportunities and some see NTFPs as a possibility.

When the Hayfork AMA was set aside, several NTFPs were already being harvested from the six area Ranger Districts. The matsutake mushroom, which in the area grows in tan oak forest, was the most sought-after NTFP on the two northern Ranger Districts. Managing the harvest was problematic; there was considerable concern among Native Americans, who traditionally harvest matsutake for food, about competition from commercial harvesters. On another district, NTFP harvesting focused on bear grass (*Xerophyllum tenax*), a commercial floral green and traditional basketry material. Again, Native American basketweavers expressed concern about competition with commercial bear grass harvesters. In addition to these already somewhat controversial NTFPs, a range of species of medicinal plants were being harvested throughout the AMA. Most of the medicinal plant harvesters lived in the region and many processed the plant parts they harvested (e.g., by drying the plants and making tablets, tinctures, or salves) to add value to sale price. Others gathered for personal use only. In the AMA districts there was as yet no gold rush for any of these plants; however the number NTFP permits requested increased from 1989 to 1997, and market forecasts (Mater, 1995) predict that the trend will continue.

Participatory Research

Forest Service activities in the Hayfork AMA began in 1995 with efforts to involve the community in developing an implementation plan. The plan's direction was defined in a series of public meetings led by two Forest Service AMA coordinators. Participating were agency implementation staff and specialists, community members, resources management and community development NGO staff and researchers, visitors to the area, and scientists from the U.S. Forest Service Pacific Southwest Research Station (PSW) participated (USDA Forest Service 1997). Two key issues were how ecosystem management activities would affect forest health and which resources would still be available to the local economy. Participants wondered what NTFP harvest levels might be sustainable and asked about these products' potential for contributing to local economic diversification. In response, the Watershed Research and Training Center (WRTC), a local organization, began coordinating participatory research on NTFPs.[4] The participatory research process progressed through several stages during its first several years: from early identification of participants and definitions of research priorities to joint research and initiation of participatory decision-making on NTFP management.

Coordinating a Participatory Process: Participants

The first step in the WRTC's coordination effort was to identify interested wildcrafters, agency staff, and other community members (e.g., those with concerns about the ecological sustainability of commercial NTFP harvest). The lack of trust among potential participants was a major challenge. In general, wildcrafters and agency staff had previously had minimal or sometimes even confrontational interaction over resource access. Wildcrafters were reluctant to divulge information about their activities, fearing competition or lost access to resources. In the region, NTFP harvesting is often a supplementary economic strategy for people with very marginal incomes. Any change in the existing situation risks wildcrafters' economic well-being. Nevertheless, a few members from two local herb marketing cooperatives showed interest in working together and with WRTC researchers. Few Forest Service staff were interested in either NTFPs or in activities to encourage increased public involvement. Among those who were, only a handful had any time to allocate to this minimally funded program. Several who did had to negotiate for or donate some of their time. In general, knowledge about NTFPs and wildcrafting among agency staff was very limited. However, the AMA coordinators were solidly behind the effort, and two District Rangers who were ranking Forest Service line officers lent their support. Some community members from the small towns closest to the AMA opposed any commercial NTFP harvesting, fearing potential overharvesting and negative ecological impacts. Others were interested in NTFPs as a source of income but had no experience. Still others had worked together a few years earlier to promote NTFPs as an economic enterprise. calling their effort the Trinity Project.

Previous Experience

The Trinity Project, led by a county supervisor, showed through bitter experience that for bulk extraction of floral greens or mushrooms, most of the comparatively remote AMA region cannot compete with coastal and moister climates in the Pacific Northwest. Nor could people from the AMA area make much income from bulk harvesting of any one item. This group of small landowners, wildcrafters, county representatives, and Forest Service staff found they could profit and be the most competitive by taking advantage of the comparatively dry climate and specializing in growing and wildcrafting high quality medicinal herbs. Further, the highest returns would come from value-added processing of herbs into ointments, pills, tinctures, and other consumer-ready products (Special Forest Products Committee 1992; The Trinity Project 1992; Trinity County Board of Supervisors 1991). Each lesson was difficult to learn. The group struggled with a lack of business management skills and received poor advice from some of its consultants. After three years, management and communication problems had so strained personal relationships that the project collapsed; by 1995 only a very few of these now-experienced people were willing try again. Those who were, including a well-connected product buyer, were key to getting the participatory research effort going.

Initiating a Participatory Process

To initiate a discussion about NTFPs, the Watershed Center and the U.S. Forest Service in 1995 co-hosted a two-day public workshop attended by wildcrafters, buyers, agency staff, researchers, and a range of citizens. The 70 participants listened to speakers from outside the region and began to network. Some wildcrafters met their local Forest Service representatives for the first time and were able to directly raise issues of concern. Other participants made initial business contacts that have since developed into mutually profitable relationships. This was the first of three such summer workshops to date. As a communication mechanism among a fairly large group of people in this sparsely settled area, the workshops are central to the participatory research effort. Participants then work on issues raised at these meetings during the subsequent year.

In addition to the broadly attended workshops, working meetings drew anywhere from five to 25 attendees. These meetings were facilitated by WRTC, agency, or NTFP cooperative staff. Workshops and meetings were jointly planned, either at the end of previous meetings or through communications between WRTC staff and the 15 or 20 most active participants from the wildcrafter cooperatives, Forest Service, local tribes, NGOs, and interested individuals. Agendas could be modified by participants at the meeting. Minutes were kept and sent, along with the participant address list, to everyone who attended. Announcements were made by telephone tree, e-mail, in the local newspaper, and through bulletin board postings in communities around the AMA. In general, meetings were designed to be inclusive and open to participation from newcomers. This type of participatory process can be slow. At times wildcrafters, who pay the highest cost for participating, because they are not paid for time spent attending meetings, expressed frustration at the time needed to bring new participants "up to speed" during meetings. For this reason, not all meetings were broadly publicized. In general, the participatory process would benefit from increased funding to cover participants' travel, which in this AMA area can quickly cover 100 miles or more.

The first funding for this process was a $1,000 grant from the Shasta-Trinity National Forest to rent the facility for the 1995 summer workshop. That fall, the Forest Service Pacific Southwest Research Station gave the WRTC another grant to pursue NTFP research. Thus the WRTC now had the financial capacity and infrastructure to coordinate activities. By 1998, as communication capacity and additional funding became available, other participating groups began to take the lead in organizing meetings.

Identifying Problems, Developing Goals, and Defining Research Questions

Key issues raised at the first summer workshop included the need for more information on NTFP marketing, the ecological sustainability of wildcrafting, the potential for certification of sustainably harvested NTFPs, and Native American concerns about commercial harvesting. The concerns expressed by Hupa, Karok, and Wintun people were of particular importance since much of the wildcrafting takes place on their aboriginal territory, now managed as national forests. The question of sustainability lay at the heart of their concerns. Several attendees, including most of the Native American participants, opposed any commercial harvesting. To some, the idea of focusing any attention on NTFPs was equivalent to opening Pandora's box. After the second summer meeting, several participants did not return. Others believed there must be a way to harvest some quantity of NTFPs sustainably. Still others focused on market values, showing little regard for the sustainability issue, which further strengthened the first group's concern.

A significant meeting was held in November 1995. The goals were to address some of the key concerns raised in the summer workshop and to define research questions. About 25 wildcrafters, Forest Service managers and scientists, NGO staff, and interested citizens attended and discussed the lack of adequate information about NTFP species' distribution and ecology and the impact of NTFP harvesting.

One issue discussed was the need for NTFP inventory data. Representatives from the Rogue Institute for Ecology and Economy from Ashland, OR and Trinity Alps Botanicals, two groups that had developed NTFP inventory methods, presented their ideas. The lead ecologist and lead botanist from the

Shasta-Trinity National Forest explained existing vegetation inventory methods applied on the forest, which did not include an explicit inventory of NTFPs. Forest Service representatives warned however, that existing inventory protocols would have to be adjusted to include NTFPs, since it was unlikely the financially strapped agency would introduce radically new inventory approaches. It became clear that none of the area's ongoing NTFP inventory approaches could meet these constraints. The Trinity Alps Botanicals protocol, though well-designed and useful for the NTFP cooperative's needs, was too expensive for AMA-wide application. It focused on identifying commercial-sized populations of NTFPs based on intensive (and therefore costly) field inventory. The Rogue Institute system (Rogue Institute, 1995) was too comprehensive and complex to be implemented without a major shift in current Forest Service inventory practices. At the meeting, the group discussed an alternative approach: using existing Forest Service Geographic Information Systems (GIS) data layers to model NTFP distribution according to habitat need. If models could be fine-tuned enough after several iterations of field testing, they might be useful to the Forest Service and for efforts to develop and monitor for NTFP harvest guidelines. The WRTC had in-house research capacity in GIS and agreed to take on this effort with data-sharing support from Shasta-Trinity National Forest and PSW research funds (Everett 1997a).

Other questions were more contentious; the process of addressing these issues illustrates the strength of participatory research. In this case, resolving of conflicting concerns led to the development of new ideas about NTFP management, crystallized research questions, and increased mutual respect and understanding among participants. The central issue was which NTFP species wildcrafters should be allowed to harvest. Some participants initially opposed any commercial harvest fearing overharvesting. Others pointed to the abundance of many NTFPs and the fact that some were exotic weedy species botanists and ecologists would like to see eliminated. Still others pointed out that unenforceable rules would not protect valuable species and argued that it would be better to communicate with wildcrafters and work toward joint harvest management.

During the discussion the group began to differentiate between economically valuable and abundant species and those that are valuable and comparatively scarce. For the area's top 20 to 30 commercially interesting species, participants worked out a dichotomous key of their relative sensitivity to harvest as a basis for guiding research and management priorities. Developing the key was a joint research effort no one had all the information, including the Forest Service botanist who guided this portion of the discussion. Confusion over variation in frequency and abundance of species in different microhabitats within the AMA underscored the need for site-specific determination of permitted harvest volumes.

With the key, the group was able to agree on several general principles for future action. Commercial activities, especially for bulk exploitation, would focus on exotic, weedy, and abundant medicinal species first. Field research trials would begin for species with good market value, but believed to be sensitive to harvest. Research trials would focus on learning what, if any, amounts might be harvested and how often from different areas in the AMA and which harvest methods were both low-impact and economical. Rare and endangered species were clearly off limits, as was commercial harvest of species spiritually significant to local Native American tribes. While agreeing to an initial focus on wildcrafting, the group recognized that, in the long run, wildcrafting would need to be supplemented with native plant propagation as well as agroforestry and silviculture techniques that enhance NTFP species growth.

Participatory Research and Monitoring

With research support from the Forest Service Pacific Southwest research station, GIS-based inventory modeling at the WRTC began in spring 1996 and field research trials were initiated that summer. Based on the dichotomous key, prince's pine (*Chimaphila umbellata*) was the top priority species for field research trials. WRTC research staff took the lead in developing a research protocol for regeneration trials through literature reviews and discussing prince's pine with Forest Service researchers, botanists, district rangers, special forest products coordinators, and wildcrafters. The protocol mimiced different intensities of wildcrafting's impact on plant populations and measured the populations' rate of recovery (Everett 1997b). Forest Service staff, wildcrafters, and WRTC researchers cooperated to identify suitable field study sites which they then treated and monitored for successive years. Nonagency people who worked on the project were paid out of Forest Service Pacific Southwest research station funds. In the following years, similar trials were initiated for yarrow (*Achillea millefolium*), yerba santa (*Eriodictyon californicum*) (Everett, 1996a), and Oregon grape (*Berberis nervosa*) (Lonner, 1998).

While for the most part these regeneration trials progressed well, there were a few set-backs. In the case of the prince's pine study, one of three study areas was inadvertently logged. Some of the yarrow plots were trampled by cattle.

Many additional species regeneration trials, as well as a range of ecological studies, will be needed to determine what levels of NTFP harvest might be sustainable. The initial results, however, should help agency staff work with wildcrafters to define conservative NTFP harvest guidelines. By 1998, monitoring results from the first regeneration trials began to trickle in. After the first two years of the study, prince's pine (*Chimaphila umbellata*) was just beginning to sprout back from treatments that mimicked typical harvest practices in the region, and there was general agreement that it should not be open for commercial harvest under current conditions in the AMA. More definitive discussions were expected after completion of the study. Results from the yarrow (*Achillea millefolium*) study, on the other hand, indicated that it was little affected by moderate harvesting (clipping 25 percent of the flower heads in a patch). Yarrow harvesting permits continued to be available.

There are many definitions of participatory research. In this instance, research questions were defined by a group of self-selected participants from the management agency and the community of resource users. Meetings were open and broadly publicized, although it is important to note that there will always be people who are interested in participatory research but are difficult to reach. Over time, these individuals and groups came to partcipants' attention and efforts were made to bring their opinions to the table. The community members' role was to define key research questions to be pursued. The researcher's role was to work with participants to develop a research design that would scientifically address those questions. The researcher led the field and analysis portions of the study in close cooperation with community members so that everyone understood the methods and ultimately the results of the study. Community members participated in the field implementation and monitoring. In the prince's pine study, for example, a family of wildcrafters who lived near one of the field sites occasionally monitored the plots during the year. Then, once a year, they worked with other participants who revisited the area to measure how the plants were responding to treatment. This type of participatory research approach is one way to work with a small group to develop information that can then be shared with a broader audience.

Information Documentation and Dissemination

At the second annual NTFP workshop in 1996, the lack of information about species and sustainable harvest was a continuing concern. While research trials would provide some answers, it would take years to achieve even basic results. Further, while participants were learning through the process, many of those requesting NTFP permits from the Forest Service were not participating and unaware of the potential impact of poor harvest methods. Meeting participants decided to work together to pool and publish existing information on NTFPs for the area, agreeing to leave out potentially sensitive species to avoid encouraging their harvest. The emphasis would be on explaining sustainable harvest methods. Participants' information about NTFPs was collated, verified, and cross-checked with the literature. Two drafts of a "Guide to NTFPs for the Hayfork AMA" were completed and circulated with editorial comments drawn from all interested participants from the annual workshops and from several peer reviewers (herbalists and botanists). The resulting manual, published by the Forest Service PSW Research Station, is now given out at the AMA Ranger Districts to all who seek permits to harvest NTFPs from public land (Everett 1997c).

One year later, the key concern at the 1997 annual summer meeting was training wildcrafters for sustainable harvest. Reports were increasing of people appearing at marketing cooperatives with loads of improperly harvested or processed plant materials, so several participants produced a generalized training approach for harvesters. The training was given a trial run by 1997 workshop participants, who offered further advice. Additional training days adapted for locally relevant NTFPs were held at the WRTC in Hayfork. Trinity Alps Botanicals compiled a standards and training manual focused on herbs of interest to its members (Klein and Johnson 1997).

Learning by Example

Another method applied on the Hayfork AMA was to draw upon the experiences of groups managing NTFPs elsewhere. A key concern for the AMA effort, for example, was to understand how a sudden rush on one or another product would affect the region. Wildcrafter cooperatives were particularly interested in ideas for "scaling up." As participants became more organized, they sought to move from production of bulk exports with very limited processing to more economically viable levels of value-added herb processing.

Several participants in the AMA research effort–including the WRTC, Trinity Alps Botanicals, and the Hoopa tribe–were also involved in the Collaborative Learning Circle (CLC), a regional peer-training network for forest practitioner organizations. These participants decided to work through the CLC to organize field trips that would include wildcrafters as well as NGO and agency staff. In October 1997 a group of 14 traveled together for three days (CLC 1997), then followed that up with another trip in April 1998 (CLC 1998).[5] Participants visited the Goosenest Ranger District on California's Klamath National Forest and spoke with staff about their experiences with and responses to the strong demand for matsutake mushrooms. They visited and spoke with floral greens wildcrafters and land managers in Shelton, Washington, where salal (*Gaultharia shallon*), a once highly abundant shrub, is being overharvested. They met with wildcrafters, NGOs staff, and Forest Service and Bureau of Land Management representatives who are working with NTFPs in the Ashland, Oregon, area. The group was impressed by successful herb production facilities in Williams, Oregon, and Trout Lake, Washington, and some began to identify the steps needed to move toward larger-scale processing. Not only did participants bring back insights, but the time spent traveling together further deepened the level of trust and communication among agency staff and wildcrafters.

Industry Development and Communication Among Wildcrafters

While the participatory research process focused on joint research and increasing communication across the AMA, participants worked more independently in cooperatives or specific communities to develop fledgling NTFP harvesting and processing capacity.

In 1995-1996, WRTC associates in the Hayfork area carried out field trials on mullein (*Verbascum thapsus*) and yarrow (*Achillea millefolium*), two species with comparatively low sensitivity to harvest, to see if they could gather enough to make a profit. They began with an initial order for several hundred pounds dried. A commissioned merchant from a neighboring community provided the market. By 1996 they moved to more industrial-scale processing of around 1,000 pounds of dried mullein, doubling that in 1997. That summer more than 30 people from the area found part-time work harvesting and drying mullein. The effort stalled in 1998, when the harvest's crucial coordinator became ill. By mid-summer a replacement team took over the effort, and though momentum had been lost, the harvest continued.

Cooperative members from another AMA community worked to build their herbal tincture processing capacity. They developed a business plan and a marketing label and increased inventory and distribution. A third community has a well-equipped nursery facility from the days when clearcutting created a large demand for seedlings for reforestation. Now nursery staff are working to grow native medicinal and ornamental plants in the underused facility. In a fourth community, the local NTFP and herb-growing cooperative completed construction on an industrial-scale herb processing facility with financial support from the Forest Service and Trinity County. However, in a major blow to the NTFP effort there, the financing on this investment failed and the facility was sold to a private venture from outside the area. As is so often the case in small, rural communities, business management skills were lacking.

The groups working on these industry development efforts cooperated on some economic development aspects. At times, they shared orders with one another within the region and beyond. They discussed joint marketing strategies. Some explicitly sought to avoid competition with others from the area; others were less collaborative. All struggled with low levels of community capacity, in this case particularly the skills and capital needed to develop and run the business of NTFP processing and marketing. The groups found numerous people ready to harvest products to sell locally; few however, would take on the more responsible leadership and entrepreneurial roles needed to make the effort work for everyone. A few individuals carried most of the burden. They began to network with each other and to learn from other herb marketers, sometimes from outside the region.

Participatory Management: Communication Between Wildcrafters and Forest Service Staff

After months of effort, participatory management activities began to emerge out of the trust building, joint research, and NTFP industry development activities.

In 1996 wildcrafters and Forest Service staff began to address concerns about NTFP harvest administration. In ongoing meetings they discussed, for example, the general Forest Service practice of charging wildcrafters 10 percent of the wet weight value of a given product. Wildcrafters found this problematic. For one thing, prices can vary dramatically by product quality and by day. For another, the practice made harvesting lucrative in some areas and economically prohibitive in others. As wildcrafters pointed out, some proposed units of measure (e.g., bushels vs. pounds) were outmoded or impractical. If Forest Service staff set permit prices too high, wildcrafters warned, it was likely that some gatherers would continue to harvest illegally. That could sever the Forest Service's only point of communication with wildcrafters and handicap NTFP management and monitoring. From the other vantage point, Forest Service staff expressed frustration with fluctuating prices and the difficulty in setting fair permit rates. In response, members of two AMA marketing cooperatives volunteered to share production and marketing information with their local Ranger Districts to assist in district- and forest-level NTFP monitoring and permit price development.

Forest Service staff increasingly communicated with each other across administrative districts, forests, and regions to learn how these issues were handled elsewhere. They sought to standardize permits where possible, to keep administrative costs low, and reduce harvesters' confusion over varying rules and standards.

Wildcrafters recommended using a mechanism for rewarding good stewardship among harvesters. Assuming that permitted harvest levels are linked to production conditions in the forest, monitoring information is important. Wildcrafters proposed harvesters in the woods help Forest Service managers by providing that monitoring information from year to year. Agency staff could randomly spot-check harvester performance. Wildcrafters who complied with permit guidelines and contributed reliable monitoring information would be allowed the privilege of harvesting. People found breaking the rules would be refused permits in the future. No formal action has been taken on the proposal, however in several districts, Forest Service staff now use this mixture of trust in past performance and random spot checks to assure compliance with permit guidelines.

While progress was made in the first several years of the participatory research effort, wildcrafters and NGO researchers encountered a certain level of frustration in working with the Forest Service bureaucracy. Staff turnover was a significant issue. The fourth Forest Service AMA coordinator in as many years arrived in the fall of 1998. As individuals, each of these staff people contributed a great deal but beginning all over again with each one was a challenge. Further, the effort's visibility inside the Forest Service always depended on carefully cultivated contacts.

ISSUES FOR THE FUTURE

Next steps in the research process are likely to focus on the need to move beyond pure wildcrafting to more active management of NTFPs through silvicultural manipulation. Certain species, for example, are associated with areas recently disturbed by fire. The WRTC has developed a simple NTFP identification and inventory method for woodland valley landowners and hopes to work with those wanting to practice agroforestry or silvicultural methods to enhance NTFPs on their land (Everett 1997d). In cooperation with the Collaborative Learning Circle, Trinity Alps Botanicals and staff from the Hoopa Tribe's Tsemata Nursery took the lead in organizing a native plant propagation workshop in spring, 1998.

One issue for the future is how to reach the many wildcrafters who come from outside the region, to incorporate their concerns in developing NTFP management practices and to promote sustainable harvest ethics among them. Many are repeat visitors who should be judged by the same criteria for permit compliance and monitoring as local harvesters. One approach could be to work with those who organize forest workers and hold meetings when non-local harvesters are in the area. This could also benefit NTFP management in higher demand or "gold rush" situations.

Industry development support, including extending access to capital and business management skills, will be needed if a core of local community members are to persist in making the dream of a sustainable NTFP industry a reality. As leaders in their communities, they in turn, are fundamentally important to the long-term process of building institutions for common pool resources management.

CONCLUSIONS

The NTFP management case presented here provides one example of a participatory research and institution-building process. While too soon for an ultimate analysis, strengths and weaknesses can be identified. Given the region's low population density, an impressive number of people from the AMA area (40 to more than 100 in public workshops and between five and 25 in ongoing working meetings) have been involved in the effort since 1995. At the same time, not all potential interests were represented. While there were a few very active wildcrafter participants, they spoke for a far larger group that did not participate. Over time, the number of participants in working meetings declined. This may be due to a failure by the research process or its participants to respond to individuals' priorities, which reduced their interest in the effort. It may also be due to a declining local economy, the cost of attending meetings when measured in time and gas money, and other demands on participants. It may also be that, with research projects underway, participants shifted into a cycle of more intensive focus on their own progress, which may be followed by renewed energy for joint processes in future.

There were incremental changes made in NTFP permit language, and good relationships developed between community members and Forest Service staff. Meetings, knowledge of NTFPs shared by participants, research protocols, and initial results were documented and thus can be incorporated into the communities' and agency's future work or shared more broadly if interest arises.

As far as moving toward institution building for common pool resources management, clearly there is a long road ahead. Luckily for the Hayfork AMA, there is still time, with no NTFP gold rush or significant resource depletion in sight. The participants in this effort have begun to lay a foundation for future cooperation. There have been considerable challenges in bridging gaps in culture, geography, and lifestyle among participants even with the narrowed focus on just those residents in the region interested in medicinal plant harvesting. There has been success both in developing community capacity to participate in research and in harnessing the capacity and interest of some Forest Service staff. While ongoing NTFP regeneration and inventory research and monitoring continue, it remains to be seen whether the institution-building effort will continue. The key WRTC research coordinator has taken on other duties and in the near term at least, funding and infrastructure to support the participatory effort are correspondingly reduced. However, as noted above, other groups active in this process have already begun to take the lead in organizing workshops and meetings. The case demonstrates that participatory research and management processes are difficult to sustain but also that they can produce valuable results for more sustainable resource management in the future.

AUTHOR NOTE

As a researcher, it has been a privilege and a multifaceted challenge to be part of this process. The author thanks all the participants for the many hours they shared and for the insights gained. This was a team effort and she apologizes for not listing everyone by name. Particular thanks to Lynn Jungwirth, Jim Jungwirth, Phil Towle, Cecilia Danks, Christina Johnson, Jolie Lonner, Julia Riber, John Larson, Marcia Andre, Julie Nelson, Janet Griffin, Elton Baldy, and Christine Klein. Garland Mason from the U.S. Forest Service Pacific Southwest Research Station made the whole effort possible by placing his trust in them and providing crucial funding support. Thanks to Michelle Robbins for her excellent copy editing. Most encouragingly, from her perspective, there have been no commercial-scale shipments of prince's pine from the AMA area since 1995. At the same time, there are a growing number of wildcrafted products marketed from the region by people involved in the participatory research process, and she is confident that these products have been and will continue to be sustainably harvested.

NOTES

1. From 1994-1998, the author worked as a staff researcher for the Watershed Research and Training Center, one of the community-based nongovernmental organizations involved in this effort. She played a leading role in the organizational effort behind the participatory research process.

2. Front desk staff went from responding to the occasional visitor's inquiry to dealing with tens and sometimes hundreds of mushroom permit requests per day. Staff came up with harvest guidelines, designed and administered permits, and sought to communicate with harvesters about all aspects of the harvest. They clarified restrictions on camping, campfires, and trash; directed people to areas open to harvest; and taught sustainable mushroom harvest methods. Over the past decade of mushroom seasons, Forest Service employees and mushroom harvesters have learned what to expect and the initially rather chaotic situation has improved.

Still, local costs to the Forest Service for these and other aspects of NTFP program management, e.g., law enforcement, are high. In theory, the NTFP programs could pay for themselves. At $10 per day per matsutake permit, Oregon's Deschutes National Forest, for example, collected $308,600 in fees in the month of September, 1997 (pers. comm. Smith and Yimsut, 1997). Yet, only a small portion of the revenues from permits stay on the forest. In most cases, under the current congressionally mandated Forest Service budgetary structure, all of the NTFP permit proceeds are returned to the General Treasury in Washington D.C. and are not reinvested in NTFP harvest management on the ground. At times, Forest Service Ranger Districts confronted with high demand for NTFPs have had to draw funds away from other programs to cover costs.

3. Mushrooms are relatively unpredictable, ephemeral fruiting bodies, and (if not trampled or raked) may turn out to be more resilient than other NTFP taxa. Slow-growing perennial herbs with medicinally valuable roots (e.g., wild ginseng) are even

more vulnerable. Once a patch of herbs has been pulled up, it may take years for it to recover depending on how much is left behind and in what condition.

4. The WRTC, a registered (501) c (3) not-for-profit organization, was founded in Hayfork in 1993 by local residents in response to the shift in U.S. Forest Service policy under the Northwest Forest Plan. It promotes sustainable community economy and healthy forests through research, training, education, and economic development (WRTC 1998).

5. Funding for these field trips came from the Ford Foundation and the Weyerhaeuser Family Foundation.

REFERENCES

American Herbalists Guild. 1998. Comments submitted to the Environment and Natural Resources Sub-Committee regarding medicinal uses of plants and their protection under the Endangered Species Act. Online: *http://www.healthy.net/herbalists/position.htm* (9/27/98).

Aurora, David. 1995. Conversation with Yvonne Everett. November 5. Orleans California.

Bloomfield, Harold H., Mikael Nordfors and Peter McWilliams. 1996. *Hypericum and Depression*. Santa Monica, CA: Prelude Press.

Brown, Beverly. 1997. Conversation with Yvonne Everett. October, 7. Chemult, Oregon.

Chambers, Robert et al. 1991. *Participatory Rural Appraisal*. London: International Institute for Environment and Development, Sustainable Agriculture Project.

Collaborative Learning Circle. 1997. Report from the NTFP Theme Group field trip, October, 1997. Rogue Institute for Economy and Ecology, Ashland, Oregon.

Collaborative Learning Circle. 1998. Report from the NTFP Theme Group visit to Oregon and Washington, April 1998. Rogue Institute for Economy and Ecology, Ashland, Oregon.

DeBeer, Jenne and Melanie J. McDermott. 1996. *The Economic Value of Non-Timer Forest Products in Southeast Asia*. 2d.ed. Amsterdam: The Netherlands Committee for the International Union of the Conservation of Nature (IUCN).

Economic Botany. 1993. Special issue: Proceedings of a workshop on economic valuation and sustainable management of non-timber tropical forest products. 47(3).

Everett, Yvonne, 1997a. Modeling distribution of non-timer forest products in forest landscapes: a GIS based methodology. Paper presented at the 1997 International Association for Landscape Ecology Meetings, Durham, North Carolina.

Everett, Yvonne. 1997b. Community based collaborative research on regeneration of prince's pine (*Chimaphila umbellanta*). *Proceedings Native Plants as Minor Crops*, Washington State University, Tri-Cities.

Everett, Yvonne. 1997c. A guide to selected non-timer forest products of the Hayfork Adaptive Management Area, Shasta-Trinity and Six Rivers National Forests, California. PSW-GTR-162. Berkeley, California: *USDA Forest Service Pacific Southwest Research Station General Technical Report*.

Everett, Yvonne. 1997d. Special forest products and agroforestry. Part 5. In L. Jungwirth (ed.), Final report to USDA Forest Service and Trinity County on Grant #R5-14-96-8: Linkages Phase III. Watershed Research and Training Center, Hayfork, California.

Everett, Yvonne. 1996. Research activities focused on non-timber forest products in the Hayfork AMA. Fiscal Year 1996 Report of the Watershed Research and Training Center to the USDA Forest Service Pacific Southwest Research Station, Hayfork, California.

Everett, Yvonne. 1995a. Summary of special forest products workshop, Willow Creek. Watershed Research and Training Center, Hayfork, California.

Everett, Yvonne. 1995b Special forest products inventory workshop summary. Watershed Research and Training Center, Hayfork, California.

Freed, James. 1997. Special forest products: past, present and future. In Alan Drengson and Duncan Taylor (eds.). *Ecoforestry*, pp. 171-178. Gabriola Island, BC: New Society Publishers.

Godoy, Ricardo A. and Kamaljit S. Bawa. 1993. The economic value and sustainable harvest of plants and animals from the tropical forest: assumptions, hypotheses, and methods. *Economic Botany 47*(3): 215-219.

Hall, Pamela and Kamaljit Bawa. 1993. Methods to assess the impact of extraction of non-timber tropical forest production plant populations. *Economic Botany* 47(3): 234-247.

Hanna, Susan S., Carl Folke and Karl-Goeran Maeler. 1996. *Rights to Nature*. Covelo, California. Island Press.

Hardin, Garret. 1968. The tragedy of the commons. *Science* (13 December): 1243-1248.

Homma, A.K.O. 1992. The dynamics of extraction in Amazonia: a historical perspective. *Advances in Economic Botany* 9:21-23.

Hosford, David et al. 1997. Ecology and management of the commercially harvested American matsutake mushroom. Corvallis, Oregon. General Technical Report PNW-GTR-412. *USDA Forest Service Pacific Northwest Research Station*.

Klein, Christine and Christina Johnson. 1997. Standards and guidelines for the harvesting of selected medicinal use NTFPs. Trinity Alps Botanicals. Burnt Ranch, California.

Larson, John. 1997. Conversation with Yvonne Everett. October 20. Willow Creek, California.

Liegel, Leon. 1994. Impacts of non-timber harvests on long term productivity and sustainability in northwest human and natural resource systems. In Chris Schnepf (ed.), *Dancing with an Elephant. Proceedings: The Business and Science of Special Forest Products*. Portland, Oregon: Western Forestry and Conservation Association.

Lipske, M. 1994. A new gold rush packs the woods in central Oregon. *Smithsonian* January: 35-45.

Lonner, Jolie. 1998. Research protocol: determining the sustainable harvest of Oregon grape root. Master of Science thesis proposal. Arcata, California, Humboldt State University.

Mater, Catherine. 1995. Trends in non-timber forest products markets. Presentation to the workshop on non-timber forest products, Willow Creek, California.

McRae, M. 1993. Mushrooms, guns and money. *Outside:* (October): 64-154.

Molina, Randy et al. 1997. Special forest products: integrating social, economic and biological considerations into ecosystem management. In Kathryn A. Kohm and Jerry F. Franklin (eds.), *Creating a Forestry for the 21st Century*, pp. 315-336. Covelo, California: Island Press.

Molina, Randy. 1993. Biology, ecology and social aspects of wild edible mushrooms in the forests of the Pacific Northwest: a preface to managing commercial harvest. General Technical Report PNW-GTR-309. Corvallis, Oregon. *USDA Forest Service Pacific Northwest Research Station.*

Ostrom, Elenor. 1990. *Governing the Commons*. New York: Cambridge University Press.

Ostrom, Elenor and Edella Schlager. 1996. The formation of property rights. In S.S. Hanna et al. (eds), *Rights to Nature*. Covelo, California: Island Press.

Palm, Mary E. and Ignacio Chapela (eds). 1997. *Mycology in Sustainable Development: Expanding Concepts, Vanishing Borders*. Boone, NC: Parkway Publishers.

Peters, Charles et al. 1989. Valuation of an Amazonian rainforest. *Nature* 339:655-656.

Pilz, David et al. 1998. Relative value of chantarelles and timber as commercial forest products. *Ambio* Special Report No. 9. September, 1998.

Pilz, David and Randy Molina. 1997. American matsutake mushroom harvesting in the United States: social aspects and opportunities for sustainable development. In Mary E. Palm, and Ignacio Chapela (eds), *Mycology in Sustainable Development: Expanding Concepts, Vanishing Borders*. Boone, North Carolina: Parkway Publishers.

Poffenberger, Mark. 1996. Communities and forest management. A report of the IUCN working group on community involvement in forest management. Washington DC: IUCN.

Redhead, Scott A. 1997. The pine mushroom industry in Canada and the United States: why it exists and where it is going. In Mary E. Palm and Ignacio Chapela (eds), *Mycology in Sustainable Development: Expanding Concepts*, Vanishing Borders. Boone, North Carolina: Parkway Publishers.

Rogue Institute for Ecology and Economy. 1995. Special forest products inventory ecosystem assessment guide. Rogue Institute, Ashland, Oregon.

Schlosser, William E. and Keith A. Blatner 1994. An economic overview of the special forest products industry. In Chris Schnepf (ed), *Dancing with an Elephant: Proceedings of a Conference on the Business and Science of Special Forest Products*. Portland, Oregon: Western Forestry and Conservation Association.

Schlosser, William E. and Keith A. Blatner. 1995. The wild edible mushroom industry of Washington, Oregon and Idaho. *Journal of Forestry*, (March): 31-36.

Smith, Jerry. 1997. Conversation with Yvonne Everett. October 11, Chemult, Oregon.

Special Forest Products Committee. 1992. Special forest products inventory proposal to USDA Forest Service economic recovery implementation program. Shasta-Trinity National Forest, Redding, California.

The Trinity Project. 1992. Final Report to the USDA Forest Service economic diversification studies program, Shasta-Trinity National Forest, Redding, California.

Trinity County Board of Supervisors. 1991. The Trinity Project Proposal to the

USDA Forest Service economic diversification studies program, Shasta-Trinity National Forest, Redding, California.
United Plant Savers. 1998. Homepage. Online. *http://www.frontierherb.com/herbs/ups/vision.html#goals*
USDA Forest Service. 1994. Record of decision for amendments to Forest Service and Bureau of Land Management planning documents within the range of the northern spotted owl. standards and guidelines for management of habitat for late successional and old-growth forest related species within the range of the northern spotted owl. USDA, Washington DC.
USDA Forest Service. 1997. Hayfork Adaptive Management Area Plan. Shasta-Trinity National Forest, Redding, California.
WRTC. 1998. Homepage (read 10/6/98). Online *http://www.tcoe.trinity.k12.ca.us/~wsc/tcgis/wrtc.html*
Yimsut, Ranachith. 1997. Conversation with Yvonne Everett. October, 11, Chemult, Oregon.

Assessing Well-Being in Forest Dependent Communities

Jonathan Kusel

SUMMARY. This paper presents a new approach to the conceptualization and assessment of well-being in forest-dependent communities. Studies of well-being in agrarian communities, boomtowns (communities undergoing rapid growth), and forest-dependent communities are examined to highlight common themes in natural resource-dependent community studies. Social indicators are discussed and a summary of weaknesses presented. The county, a commonly used unit of analysis, is rejected in favor of a more socially relevant unit. This new approach to well-being in forest communities begins with definitions of the terms *community* and *forest dependence*. The work of Amartya Sen, whose conceptualization of well-being focuses on the real opportunities people have and their achievements in light of their opportunities, forms the foundation of the new approach proposed here. Sen's conceptualization is broadened by focusing on the community and acknowledging the importance of a sense of place. Methodologically, this new approach adds to the evaluation of social indicators an assessment of community capacity that consists of three components: physical, human, and social capital. It also involves evaluating how residents draw on and develop

---

Jonathan Kusel is Director, Forest Community Research, P.O. Box 11, Taylorsville, CA 95983.

The author would like to thank reviewers for their helpful comments on this and a previous version of this manuscript which appeared in the Sierra Nevada Ecosystem Project (SNEP): Final Report to Congress, vol. II, Davis: University of California, Centers for Water and Wildland Resources, 1996. Questions or comments should be directed to the author at Forest Community Research, 4352 Main Street, Suite 2, Taylorsville, CA 95983.

[Haworth co-indexing entry note]: "Assessing Well-Being in Forest Dependent Communities." Kusel, Jonahtan. Co-published simultaneously in *Journal of Sustainable Forestry* (Food Products Press, an imprint of The Haworth Press, Inc.) Vol. 13, No. 1/2, 2001, pp. 359-384; and: *Understanding Community-Based Forest Ecosystem Management* (ed: Gerald J. Gray, Maia J. Enzer, and Jonathan Kusel) Food Products Press, an imprint of The Haworth Press, Inc., 2001, pp. 359-384. Single or multiple copies of this article are available for a fee from The Haworth Document Delivery Service [1-800-342-9678, 9:00 a.m. - 5:00 p.m. (EST). E-mail address: getinfo@haworthpressinc.com].

these aspects of their community to meet local needs and create opportunities. This results in a well-being assessment that includes static as well as dynamic measures of how communities respond to and create opportunities to improve local well-being. *[Article copies available for a fee from The Haworth Document Delivery Service: 1-800-342-9678. E-mail address: <getinfo@haworthpressinc.com> Website: <http://www.HaworthPress.com>]*

KEYWORDS. Community-based forestry, community well-being, forest-dependent community, social indicators

Forest ecosystems in North America have become the focus of comprehensive and broad-scale ecosystem studies. Many have adopted an "ecosystem management" approach (Bormann et al. 1993; Forest Ecosystem Management Assessment Team [FEMAT] 1993; Ministry of Environment 1994, Quigley and Arbelbide 1997; and Sierra Nevada Ecosystem Project 1996). Ecosystem management has been defined in diverse ways, but it is generally agreed that humans and human communities are part of ecosystems and an important area of study (Grumbine 1994; Manley et al. 1995; World Commission on Environment and Development 1987).

This paper presents a new conceptual and methodological approach to assessing well-being in communities that depend on natural resources. It takes this focus primarily because of the recent emphasis on forest ecosystem studies and secondarily because the well-being of communities in forested areas has long been too narrowly discussed in the context of extractive forest management activities. Studies involving other kinds of natural resource-dependent communities and using social indicators are examined to highlight the diverse, complex approaches to understanding human well-being.

The paper is divided into two parts. The first reviews studies evaluating the well-being of individuals living in natural resource-dependent communities, concluding with a discussion of social indicators.

The second part of this paper presents a new approach to the study of well-being in forest-dependent communities defining the terms "community" and "forest dependence," and "community capacity," introducing the ability of communities to effectively respond to issues and create local opportunities. The work of Amartya Sen forms the foundation of the new approach. The paper concludes with a discussion of how this expanded focus informs a new approach to community well-being assessment in the context of ecosystem management studies.

PART ONE: STUDIES OF WELL-BEING IN NATURAL RESOURCE-DEPENDENT COMMUNITIES AND THE USE OF SOCIAL INDICATORS

Resource Dependency and Well-Being

The inclusion of humans and the study of community well-being is in its infancy in ecosystem studies. It is therefore useful to briefly examine empirical studies of resource dependency and human well-being in natural resource-dependent communities (NRDCs). This brief review offers a glimpse of the diverse ways researchers have grappled with linking resource dependency and human and community well-being. Two classic studies of well-being in NRDCs are reviewed, followed by an examination of studies of boomtowns (communities that have undergone extremely rapid growth associated with the extraction of nonrenewable resources) and forest-dependent communities.

An often implicit and underlying aspect of resource-dependent community studies is the attempt to understand the relationship (usually termed "dependence") between resources and well-being, which overlooks the more basic questions of what constitutes well-being and how that well-being might best be evaluated. Other research, such as work on social indicators, has addressed these questions more directly, though still not without difficulties.

Two Classic Studies–Two of the best-known studies linking resource use and community well-being were done in the mid-1940s in California and Montana. Walter Goldschmidt investigated agricultural communities in California, and Harold and Lois Kaufman examined forest communities in Montana. The way these researchers conceptualized the connection between resource use and well-being laid the groundwork for studies that followed, even 50 years later.

Goldschmidt (1947) evaluated the structure of agriculture and its relationship to community well-being in California. The variables he examined included the wages of owner-operators, industrial workers, and basic laborers; employment turnover; labor security; social isolation of workers; labor participation in important community decisions; and the strength and diversity of community institutions and infrastructure. Goldschmidt found that an increase in the concentration of the farm sector led to a decline in rural economic and social well-being. In contrast to a community surrounded by large farms, one surrounded by small farms had a higher percentage of self-employed and white-collar workers; a lower percentage of farm wage laborers; more business and retail trade; more schools, parks, civic and social organizations, newspapers and churches; a better developed infrastructure;

and a more local decisionmaking structure. The most affected dimensions of well-being were living conditions and income.

Subsequent studies have shown that the inverse relationship between large-scale industrialized agriculture and well-being still generally holds true for California and nearby states in which large-scale industrial agriculture is dominant (MacCannell 1988; Swanson 1988), but researchers agree that regional and historical settings must be taken into account to assess localized effects (Lobao et al. 1993). In raising the issue of land tenure's impact on the well-being of agrarian communities, Goldschmidt's study raised the possibility that concentrating the control of other resources particularly in the hands of essentially absentee owners might have similarly adverse effects in other kinds of resource-dependent communities.

A cornerstone of Goldschmidt's study is the Jeffersonian ideal of the small, agrarian rural community. This is the model against which agricultural and other resource-dependent communities have been evaluated (Bealer et al. 1965; Drielsma 1984; Swanson 1988). The community in this model is typically stable, smallscale, and offers opportunities for healthy family life, independence, and entrepreneurial activity (Drielsma 1984).

The best-known forest community study of well-being was carried out by Harold and Lois Kaufman (1946) in the Libby-Troy area of Montana. In addressing well-being, the Kaufmans used the then-popular term "stability." But because their use of stability encompasses much more than employment, the term "well-being" is substituted for it in this discussion.[1]

The Kaufmans believe that the creation of a prosperous economy is essential to well-being, but in addition to "what people do for a living," the Kaufmans evaluate "how well they live." They state, "A characteristic of the good life is that experiences in the community and of the forest are not only regarded as means but as ends in themselves–they are appreciated and enjoyed for their intrinsic worth. Also, the good life has a depth and variety of experience" (p. 23). They point out that attainment of "the highest standard of living" can only be realized by maintaining a balance between population and natural resources and link this concern to the limits of "timber supply, production costs and markets" (p. 15). Like more conventional analysts, the Kaufmans agree that maintaining community well-being involves developing a stable timber industry and a diversified economy and practicing sustained-yield forestry. They describe five additional "approaches" toward maintaining community well-being: organizing the greater community, strengthening the rural home, making religion a part of life and the church more community-centered, promoting public participation in the determination of forest policy, and creating a forest-centered tradition. These suggestions evidence both the Jeffersonian tradition and a sense that the promotion of well-being involves process as well as product.

Kaufman and Kaufman question the wisdom of the Sustained-Yield Forest Management Act passed at the time of the study. They argue that it favors timber operators with large holdings, thereby concentrating economic power in the hands of a few while being "silent concerning controls that might be needed to safeguard the public interests" (p. 71). In one of the first calls for stakeholder involvement, the Kaufmans suggest the Forest Service involve the public in formulating forest policy to insure that the concentration of economic power does not result in the abrogation of public interests and concerns. They maintain such involvement should be "extensive" (p. 85). The Kaufmans' study is rare in its attention to these issues.

Boomtown Studies–Studies of rapid resource-related growth in small communities, commonly known as boomtown studies, focus on the impact of natural resource development. Well-being is generally discussed in terms of social disruption–primarily increased crime–resulting from rapid population increases. Population growth, as the independent variable, is associated with extractive energy projects such as oil or gas or mining development. Dependent variables have included measures of income and various aspects of employment, but the most commonly used variables by far are measures of crime (Albrecht 1982; Finsterbusch 1982; Freudenburg and Jones 1991; Gold 1982; Krannich et al. 1989; Seydlitz et al. 1993; Wilkenson et al. 1982).

Some researchers have drawn broad conclusions suggesting that rapid development leads to the loss of integrative functions and is accompanied by a loss of local control. This is caused primarily by the rapid influx of outsiders overwhelming existing social services and networks (Jobes 1984a, 1984b; Gold 1985; Kennedy and Mehra 1985; Krannich et al. 1989). Gold (1985) believes social disruption is caused by contrasts in lifeways and involves the replacement of close friendships and kin networks (*gemeinschaft* characteristics) with a less integrated social organization.

Freudenburg and Jones (1991), in an exhaustive review of boomtown studies, found that crime increased (by a factor of three, on average). This is in contrast to earlier studies, which, as Freudenburg and Jones point out, overstate the benefits of development activities. Yet, while lending support to the social disruption thesis (Finsterbusch 1982), the authors take issue with those who use grand theories and draw broad conclusions. They adopt what they term a middle-range perspective and suggest this increase in crime is due to reduced density of acquaintances. Unlike Goldschmidt's findings, however, decisionmaking controlled by outside organizations was not examined or did not surface as a significant issue for researchers in these studies.

Forest-Dependence–Building on the theme of outsider control raised in both Goldschmidt's and the Kaufmans' work, Marchak (1990), studied forest-dependent towns in British Columbia. That study explored the adverse effects of uncertainty about future employment (reflected in high rates of population turnover) stemming from control of the resource base by outside firms that make decisions "without reference to the needs of workers in these communities" (p. 99). Marchak noted that high turnover rates reflect the structure of the industry rather than the personal choices of workers. She was the first researcher to note that women are particularly demoralized by the conditions in single-industry forest towns.

Kusel and Fortmann (1991) and Kusel (1991) studied forest counties and communities in California, focusing on general well-being and the capacity of forest communities to maintain and enhance local well-being. They define capacity as "what enables communities to pull through hard times" (Kusel and Fortmann 1991:84). Kusel and Fortmann determined that communities are deeply affected by forces outside of their control, including outside employers, natural-resource decisionmakers, and outside money. They found that a higher concentration of private forest land holding is associated with a lower median income, and high percentages of public timberland are associated with higher poverty rates (at the county level). They also found pockets of high poverty in low-poverty forest counties. In contrast to studies characterizing the "inevitable" culture clash between newcomers and long-standing residents (Price and Clay 1980; Schnaiberg 1986), Kusel and Fortmann note that new residents and women play crucial roles in mobilizing community action and increasing local capacity.

In an ethnographic study of several communities, Kusel found that extensive job loss in rural forest communities was devastating in both the short and long term. Economic and social turmoil led to short-term difficulties for families and communities and resulted in a long-term reduction in community capacity. Kusel also found that local, family-run mills contribute more to community well-being than mills owned by large, nonlocal owners.

Forest communities in the Pacific Northwest were the focus of a social assessment conducted by the Forest Ecosystem Management Assessment Team (FEMAT).[2] The FEMAT study is one of the first large-scale American ecosystem studies to explicitly include and assess human communities. The primary objectives of this social assessment were to identify the consequences of forest management alternatives for communities and individuals and describe how these alternatives might affect social values and constituencies (FEMAT 1993, VII-45). Though there was insufficient time available for a comprehensive assessment, this was the first large ecosystem study to use the concept of community capacity and convene panels of experts to assess it. The concept of community capacity was defined in the workshops as an independent variable that influences community response to and the consequences of land-management alternatives. Higher-capacity communities were considered more adaptable and therefore less affected by changes in forest management. The concept of capacity, advanced in the Sierra Nevada Ecosystem Project (1996), plays a key role in the new approach to well-being and is described later.

Social Indicators and Well-Being

Social indicators are used in (1) social impact assessment (SIA), which predicts and assesses the consequences of technical projects (e.g., hydroelectric projects, waste-dump siting, etc.) and policy actions (Interorganizational Committee 1994), and (2) research that evaluates well-being or general life conditions (e.g., Allardt 1993; Campbell 1981). This portion of the paper also examines the philosophical and conceptual underpinnings of well-being. These two broad areas of research offer important insights into the relationship of resource dependence to community well-being, and are relied upon for much of this discussion.

Terms such as "standard of living," "quality of life," "welfare," "happiness," "life satisfaction," and others have all been used in studies to characterize a good and healthy life or the critical components of one. But they often have different meanings to different people, leading to confusion about well-being and how it might best be measured. The numerous approaches to the study of well-being, measuring utility, income, personal satisfaction, and happiness, to mention just a few, have yielded incommensurable and, at times, contradictory results, which have only further muddied the waters of well-being assessment. Burdge (1994:3) states, "The field of impact assessment does not have a series of agreed upon concepts or list of variables around which to accumulate research knowledge." This same lack of agreement characterizes how environmental planning and social assessment might be effectively linked.

In addition to conceptual concerns, social indicator researchers have wrestled with the problem of whether to study well-being using subjective self-report measures or sociodemographic measures, what many consider to be more "objective" measures of external conditions (Allardt 1993; Erikson 1993). Implicit in the debate over appropriate measures are the questions, What variables should be evaluated? and Who should do the evaluating? Sociodemographic measures of crime, income, employment, and poverty are frequently the measures of choice because they are easily gathered (or have already been gathered, in the case of U.S. Census Bureau data). Such measures have more direct policy relevance for governments than other measures (De Neufville 1975). (See Burdge 1994 and Interorganizational Committee 1994 for discussions of categories and indices.) Yet, despite their widespread use, both sociodemographic measures and subjective self-report measures have significant limitations. Sen (1985b), in particular, and others have provided powerful critiques.

Five Limits of Social Indicators

First, social indicators, which consist of aggregate individual data, ignore the variability of structural conditions at the level of a county or region, and of such institutional arrangements as the concentration of capital, land ownership, and power that influence well-being in a community (Kennedy and Mehra 1985; Kim 1973). Communities with greater disparities in wealth often have lower community well-being than those with more equal distribution of wealth, even though average measures such as income may be the same. Goldschmidt's (1947) evaluation of the structure of agriculture and its relationship to community well-being is a good example of why this consideration of institutional arrangements is important.

Second, Sen (1985b) points out that measures of opulence such as income confuse well-being with being wealthy in terms of material possessions. Measures of real income provide an indication of what an individual can buy, or one's "commodity command," but they provide no indication of how an individual may improve his or her life with purchased commodities. Sen (1987) states that commodities provide only a means to an end, arguing that the issue is more a "matter of the life one leads rather than of the resources and means one has to lead a life" (p. 16). In their research, Kaufman and Kaufman (1946) expressed a similar concern with how well individuals live.

Third, and related to the previous point, is the issue of what constitutes well-being for whom. For instance, sociodemographic measures of opulence do not take into account the distribution of resources within a family (Sen 1985b). A male head of household, for example, may purchase luxury items for himself while other family members are inadequately clothed and fed. Similarly, the concerns of women may differ from those of men. Nussbaum and Sen (1993) question whether the quality of female life has similar constituents to the quality of male life. Feminist research was launched out of a concern that women's perspectives and their life circumstances were not recognized. Oakley (1975) pointed out that women have been reduced "to a side issue from the start." And the concerns of adolescents may differ from those of adults. Freudenburg (1984) discovered that adolescents in rapidly growing communities were more likely to be dissatisfied with locality and less satisfied with overall quality of life than adolescents in similar towns that were not rapidly growing, whereas the same sense of dissatisfaction did not hold for adults.

A fourth problem has to do with subjective measures and the distinction between ill-being and well-being. Subjective well-being is commonly measured with scales indicating satisfaction with the self as a person, personal freedom, personal happiness, and sense of personal control (Campbell 1981; Chamberlain 1985). Yet, Headey and colleagues (1985) point out that well-being may be a different dimension than ill-being. They found that more objective measures of health and material standards of living while contributing little to one's well-being significantly contributed to measures of ill-being. Bradburn and Caplovitz (1965) and Wilson (1967) found the same to be true for measures of happiness. There are positive and negative dimensions that are independent of one another. In addition to requiring the measurement of positive and negative dimensions, this suggests that people may adjust their perceptions of well-being (or happiness) to the conditions they face.

Studying the same issue from an economic and philosophical perspective, Sen (1985b) states that subjective measures of well-being, such as pleasure and desire fulfillment, are incomplete for two reasons: (1) they are fully based on the mental states of an individual, and (2) they lack a personal metric of value ("the mental activity of valuing one kind of life rather than another"). Sen terms these reasons "physical condition neglect" and "valuational neglect," respectively. An example illustrates the incompleteness. One who is poor, without the comfort of a home, out of work, and ill-fed but happy has obviously adjusted her expectations and taken solace in small pleasures. But fulfillment of limited desires, no matter how happy the person might be, is not suggestive of a high level of well-being. Moreover, this psychological state cannot be compared to that of another individual whose desires are greater. Sen (1984) states, "Quiet acceptance of deprivation and bad fate affects the scale of dissatisfaction generated, and the utilitarian calculus gives sanctity to that distortion. This is especially so in interpersonal comparisons" (p. 309).

Fifth, researchers who have examined the relationship between objective and subjective measures have shown that sociodemographic indicators have little connection to subjective measures of well-being (Barlett and Brown 1985; Campbell 1981; Gans 1962; Mastekaasa and Moum 1984; Oppong et al. 1988; Suttles 1969). Moum (1988) found that only 10 percent of the variance in quality-of-life scales is explained by sociodemographic variables. Gans (1962) reported in his study of West Enders in Boston that despite residents' high degree of satisfaction with their neighborhood, upper-middle-class professionals declared it a low-income slum, and the area was razed for redevelopment. The difference between the two groups' judgment of the area provides a warning that not only may measures differ, but they may do so because they reflect the values (and power) of those doing the measuring rather than those whose well-being is being evaluated.

Given gender, class, and ethnic differences and the importance of local salience, it is not surprising that despite numerous measures of well-being, no standard metric has emerged (Burdge 1994; Johnson 1988; Oppong et al. 1988). These arguments suggest that an assessment of complete well-being may indeed be unobtainable. They also demonstrate that considerable humility is necessary in any such assessment and interpretation, and show the importance of involving those who are the subject of evaluation.

The Social Unit of Analysis

In the debates over the use of self-report measures versus sociodemographic measures and over appropriate metrics of well-being, insufficient attention has been paid to the unit of analysis or level of data aggregation used for assessment. This oversight has led to additional confusion about whose well-being is discussed and about causal and associational relationships. The county has been the most common unit of analysis in studies of forest-dependent communities (Machlis and Force 1988), but its exclusive use is inadequate for several reasons (for a contrasting view, see Lobao 1990).

Perry (1986) has criticized the use of counties because they are not a unit with real social meaning. People do not generally identify with counties, and, indeed, numerous NRDCs are alienated from their parent county. Relationships and life take place in communities, not counties or larger regions.

Often only a small percentage of communities (and subpopulations within them) are resource-dependent in a traditional sense, a fact important for NRDC assessment, and aggregate county or regional level data often obscure these relationships. In one of the most traditionally resource-dependent counties in the Sierra Nevada, median family income is slightly more than $24,000. The county's four largest communities are more dependent on traditional natural resource industries than the county as a whole, and their median incomes all are lower than the county average: one community is almost $9,000 lower, another, $5,000 lower than the county mean. The draft Columbia Basin Ecosystem Assessment stated that resource related jobs totaled only 4 percent in the region (Reyna 1998), yet there are numerous individual communities in which this total is considerably higher. These examples are offered not to suggest that the linkage between resource dependency and well-being at the county or regional level is unimportant but to show how aggregating data beyond a community can mislead. Whatever resource dependence there really is will more often than not be misstated.

If one desires to understand community well-being, then the unit of analysis must focus on and isolate community. County or regional data alone encompass too broad and diverse an area to be used for examination of well-being for most NRDCs. Moreover, the larger the aggregation, the less likely that measures (regardless of whether they are median income, poverty, or unemployment) can be accurately linked to resource management activities. A determination of causal or associational factors influencing community well-being often requires a specificity and detail unobtainable with county-level data, particularly when counties are large and heterogeneous.

A NEW APPROACH TO FOREST COMMUNITY WELL-BEING

The concept of community has engendered considerable debate, a debate that while it will not be resolved here must nonetheless be addressed. In sharp contrast, the concept of forest dependence has been uncritically accepted as meaning dependence on employment generated from timber harvesting. Amartya Sen's novel "capabilities and functioning" approach to individual well-being can be expanded by adding an emphasis on the community. This expansion calls for the inclusion of culture and community when considering well-being which leads to the concept of capacity and how it can be used in an assessment of community well-being.

Conceptual Clarity

The Concept of Community–In this paper community is defined in terms of a place-based, shared identity. This definition is primarily based on Gusfield's (1975) discussion of community, which includes the intersection of two components: relational and territorial. The relational component involves "the quality or character of human relationships," which includes a sense of belonging. Selznick (1992) states that, among individuals, this includes shared beliefs, interests, and commitments that unite diverse groups and activities. The relational component of community is a vital part of individual well-being; it is discussed further later.

Gusfield's territorial component involves what people have in common and share at their locale. This includes diverse institutional elements: governments and law, school districts, churches, and families, among other things (Selznick 1992). Gusfield's concept of community roughly encompasses the three aspects for which Hillery (1955), in a survey of the literature, found definitional agreement: social interaction, area, and common ties.[3] Although Gusfield does not limit his discussion to place-based communities, the emphasis here is primarily on geographical place-based, forest-dependent communities.

Despite this focus on locale, it is terribly important to recognize that forest communities are part of the larger society, and, in most cases, with extensive ties to outside organizations (Warren 1978).[4] The presence and type of linkages can profoundly affect the opportunities residents have available to them or are able to develop themselves. A small rural community containing a mill owned by a multinational corporation may have additional mill-related employment and other opportunities. Yet, this same community will also be quite sensitive to the actions of the multinational, which may have no local ties beyond its mill.

Institutional and social relationships extend beyond the formal administrative and informal boundaries of a community (Selznick 1992; Strathern

1984). Individuals may hold multiple "community" identities as a result of associations at work and through organizations and institutions outside their community of residence. Small NRDCs include overlapping sets of social groups, and these groups are important to local community well-being and how local communities are influenced by resource policy. The focus on place-based communities suggested here provides a clear and critical starting point for assessing well-being.

Broadening the Concept of Resource-Dependence and Recognizing the Importance of Sense of Place–Forest-dependent communities are those immediately adjacent to forest land or with a high economic dependence on forest-based industries, including tourism as well as timber and nontimber forest products. This broader definition acknowledges that well-being involves more than a dependence on timber products.

"Forest-dependence" suggests that a community's primary relationship is to a biological forest and, as the concept is commonly used, to wood products.[5] Forest-dependent communities rely on the biological forest resource, but communities in which a number of residents work in the wood and nontimber forest product industries depend also on an economic and social structure that permits and demands particular uses of the forest resource. This structure mediates the terms of a resident's access to the economic and social benefits of this resource. Thus, in a community with many workers employed in forest product industries, the residents' ability to gain economically from the forest, as well as their ability to create new jobs, is a function not only of the biological condition of the forest but of: (1) the extent to which controllers of the forest permit and promote commercial activities, (2) the extent to which those who create forest-related jobs make them available in or near the community as well as the extent to which those who control the jobs maintain them, (3) the terms upon which these jobs become available, and (4) the extent to which the market demands the forest products.

Also important to note, the commodity production perspective ignores forest-dependent communities that do not produce timber. Communities can be economically dependent on forest products without any forest-commodity production whatsoever. Communities whose *raison d'être* is forest tourism or retirement living are economically dependent on the forest for its amenity value, and their numbers and size are increasing throughout the United States.

In a similar vein, forest dependence may occur with no economic relationship whatsoever to the resource, based on aesthetic, symbolic, and locality-based importance (Hester 1985; Hiss 1990; Tuan 1993; Walter 1988). As such, the forest is a landscape and part of a human sense of place. As a landscape, Relph (1976:34) suggests, it represents "an expression of communally held beliefs and values and of interpersonal involvements." Meinig

(1979:46) observes that "a well-cultivated sense of place is an important dimension of human well-being. Carried further, one may discover an implicit ideology that the individuality of places is a fundamental characteristic of subtle and immense importance to life on earth, that all human events *take place*, all problems are anchored in place, and ultimately can only be understood in such terms."

For example, Wendel (1987) found that a majority of residents of a forest community in California chose the response "the trees/the forest" when asked the most important place in the community. The trees and the forest represent a link with residents' past tradition of logging and a connection to their present and future economic base of tourism and to aesthetic values. Hester (1985) calls places that reinforce and help define the community living tradition "sacred" places. Kaufman and Kaufman (1946:30), using the term *stability* rather than *well-being*, state, "A meaningful tradition is always an important part of the life of a stable community. A tradition is needed . . . which magnifies the significance of the forest and portrays the relationship of forest and people." Berry (1987), in a somewhat similar vein, asserts that a lasting community is inseparable from its place, with community and place mutually supportive. A durable, resource-dependent community is only possible when the human economy is sensitive to and works with the constraints imposed by the natural economy.

As a playground, sacred place, or resource, the forest supports local residents and contributes to the definition they have of themselves and their understanding of who they are. The lifeways of community members and the landscape are intertwined. Thus, when discussing dependence, one must recognize that the forest provides not only the means of production, diversely defined, but sustenance to the local living tradition, economically, socially, and spiritually.

Individual Capabilities and Functionings: The Basis of Collective Capacity

Amartya Sen (1984, 1985a, 1985b, 1987, 1993) offers what he calls the "capabilities and functionings" approach as an alternative way of evaluating well-being. Sen's approach to well-being counters the problem of the limitation of sociodemographic measures, such as measures of opulence, by evaluating not just the goods at one's disposal or one's wealth, but how they contribute to what a person can do. For Sen, individual capabilities consist of the freedom one has or the opportunities from which one can choose. An individual's functionings consist of one's achievements, or what he or she "succeeds in doing with the commodities and characteristics at his or her command" (Sen 1985b, p. 10). Functionings vary from the more basic, which include escaping death and malnourishment, to the more complex, such as achieving self-respect (Sen 1993). Sen argues that these elements are part of an individual's being and must be part of a well-being assessment.

In the case of a forest community, a mill job that provides an adequate income may be essential to an individual's (and his or her family's) well-being. However, diminished well-being through reduced opportunity and achievement results if an individual is unable to obtain medical benefits, cannot advance in his or her job, or desires creative opportunities that are unavailable. A job that provides adequate pay contributes to one's well-being, but the pay constitutes only a portion of one's achievement. These examples highlight what Sen (1984:317) refers to as the "capability to function."

The capabilities and functionings approach addresses the subjective-indicator problem by dividing the evaluation into two parts: "(i) specification of the functioning achievements, and (ii) the valuation of the functioning achievements" (1985b:30). Specification requires identifying achievements for which a valuation is made. To return to the example of a poor, unemployed, ill-fed, homeless person, the specification of her functionings would clearly indicate a low level of well-being, while the personal valuation of her well-being is rendered somewhat unimportant.

What is unique about Sen's capabilities and functionings approach is that it requires an analysis of the opportunities or freedom individuals have (capabilities), and their achievements or successes (functionings) in light of their opportunities. For someone to have a high level of well-being, she or he not only must feel well but also must have opportunities available and be able to take advantage of them. Sen, however, restricts the analysis of well-being to the individual and avoids the sticky problem of a contextually based valuation of various capabilities and functionings, which must be included for a more complete evaluation of well-being.

The Importance of Community–A complete discussion of individual opportunity, as well as a more effective valuation of functioning achievements and well-being, requires a focus on both the individual and community. Human actions are motivated by internalized values. Benn (1982:49-50) maintains these flow from "traditions of behavior," which do not reflect "individually conceived goals, but reflect those of our culture and communities." Selznick (1987:447) offers the perspective of the "implicated self," which holds that humans depend on others for personality development and "psychological sustenance." He (p. 451) states, "A morality of the implicated self builds on the understanding that our deepest and most important obligations flow from identity and relatedness, rather than consent." Bellah et al. (1985) and MacIntyre (1984) maintain that human identity is found in community, as a collective living tradition.

Accepting of the perspectives of "traditions of behavior" and the "implicated self" requires a well-being assessment to examine how communities define success (or functionings). This, in turn, provides a referent from which individuals view and assess their own success. Native American or Latino communities may define success differently than Anglo communities. Ethnically similar communities' definitions of success may differ from one another for any number of reasons as well. Hence, beyond the most basic of functionings, proper assessment must recognize differences while recognizing that these differences must be reconciled in the context of community. A community, its traditions, and the opportunities it creates for residents, must inform well-being evaluation. To neglect the community is to neglect context and important–indeed vital–aspects of individuals.

A community of shared values does not equal a community of conformity (Lasch 1988:178), nor is it composed of separated individuals or groups. Lasch states that social solidarity is not "an identity of interests; it rests on public conversation. It rests on social and political arrangements that serve to encourage debate instead of foreclosing it." Communal relationships, with the associated responsibilities they bring, and the freedom to choose both are coveted values. As Selznick (1992:363) points out, there must be "freedom *in* associations as well as freedom *of* association." He adds that a concern for personal autonomy ". . . assume[s] that the worth of community is measured by the contribution it makes to the flourishing of unique and responsible persons. As an attribute of selfhood and of self-affirmation, autonomy requires commitment as well as choice."

The perspective of the implicated self also recognizes that taking part in the life of a community contributes to individual well-being. Humans are constituted by social relationships found in community, and there is a reciprocal and interdependent relationship between an individual and others in her or his community. This perspective implies that a collective good exists; well-being may be improved by residents working on community projects when, narrowly conceived, are of no benefit to them personally. Selznick (1992:360, 364), discussing communities as places where people grow and flourish, notes that a "flourishing community has high levels of participation: people are appropriately present, and expected to be present, on many different occasions and in many different roles and aspects."

Individual well-being increases with a heightened sense of being part of a community and making that community a better place to live. This is part of the "relatedness" component of community discussed above and involves a category of individual behaviors termed "commitments." More broadly, this behavior may be termed civic responsiveness. Sen would disagree with the extension of well-being analysis to include "commitment" behaviors. Because of the importance of his work for the approach developed here, a brief review of this disagreement and a response to it are presented in Appendix I.

Sen (1990) nonetheless recognizes the importance of community to well-being, stating, "Some functionings are very elementary . . . others may be more complex but still widely valued such as achieving self-respect, or taking part in the life of the community."

Well-Being Assessment

Amartya Sen's approach to well-being requires an assessment of individual opportunities (capacities) and achievements or successes (functionings) in light of available opportunities. For large-scale ecosystem studies, or evaluation of well-being in a forest community, it is impractical if not impossible to evaluate opportunities and successes for each individual. Diverse secondary data combined with primary data about communities (including such things as support services) can be used to develop a rudimentary understanding of conditions. Useful secondary measures to assess condition and rudimentary functioning include but are not limited to: measures of poverty that identify those who have not secured an income adequate to escape it (escaping poverty being a very basic functioning); poverty intensity (i.e., the farther below a poverty threshold, the higher the intensity) suggestive of a lower level of functioning and greater need; and higher education levels and home ownership, which are suggestive of increased opportunities and higher functioning. Equally important, the presence of individuals with high levels of education may lead to increased community capacity for reasons that are discussed below. Crime, drug dependency, and children in families receiving public assistance are additional measures of conditions that may affect functioning and are reflective of residents' well-being.

Community Capacity

Because individual opportunities and functioning are shaped by conditions individuals face within a community, a well-being analysis must focus not only on the static condition of a community's residents, but also on their dynamic capability to create local opportunities and respond to local needs. A more complete analysis of well-being, therefore, includes an assessment of residents' capability to collectively affect community opportunities. This dynamic capability is community capacity.

Community capacity encompasses residents' ability to respond to external and internal stresses; to create and take advantage of opportunities; and to meet the diverse needs of residents. It refers also to a community's ability to respond and positively adapt to a variety of circumstances. Community capacity is comprised of three broad areas: (1) physical capital, which includes physical elements and resources in a community (e.g., sewer systems, open space, business parks, housing stock, schools, etc.); (2) financial capital, which is locally accessible capital and credit; (3) human capital, which includes the skills, education, experiences, and general abilities of residents; (4) cultural capital, which is the myths, beliefs, and norms that organize groups and facilitate survival; and (5) social capital, which includes the ability and willingness of residents to work together toward community goals. Social capital includes networks and trust that facilitate coordination and cooperation for mutual benefit. Cultural and social capital overlap somewhat, but they are not the same. Social capital appears to be one of the most important determinants of capacity.

Robert Putnam's (1993a) empirical work makes clear the importance of social capital and how it is a critical determinate of economic development and community capacity.[6] Putnam (1993b), examining the modern-day rise of regional governments in Italy from the 11th century, states:

> The historical roots of the civic community are astonishingly deep. . . . communities did not become civic because they were rich. The historical record strongly suggests precisely the opposite: They have become rich because they were civic. The social capital embodied in norms and networks of civic engagement seems to be a precondition for economic development, as well as for effective government.

An example is offered to explicate the role and importance of social capital and to show the relationship between social and financial capital. A community may have a number of residents who are quite wealthy, but if they are disinterested and detached from local matters, their financial capital does little for the community beyond their narrow self-interest concerns. Conversely, in a community with little financial capital but high social capital, residents voluntarily come together to address local needs and will conduct fundraisers as well as reach outside the community to raise money if it is needed.

Assessing Capacity–Assessing community capacity is complex and involves evaluating residents' ability to meet needs and create local opportunities. It involves assessing local "commitment" behaviors and the presence of civic groups, along with physical and human capital. To assess social capital alone, Putnam (1993a) noted, requires diverse slices of data examined over time.

In their assessment of communities for the Sierra Nevada Ecosystem Project (1996) and for the Forest Service for the Klamath Region (1997), Doak and Kusel, in addition to collecting socioeconomic data, assessed community capacity through workshops with experts knowledgeable about local communities. Individually and collectively, local experts assessed physical (including financial), human, and social capital and the interplay of them to arrive at a measure of capacity for a community. Selecting of experts to participate in workshops proved critical, as it determined the accuracy and quality of information obtained. In choosing workshop participants, Doak and Kusel learned that experts must understand community issues, institutions, and resources, but that effective workshops required experts who were not community boosters or overly partisan about issues.

Assessing community capacity makes it easier to understand the potential for increased opportunities for productive and rewarding involvement in a community. Though such assessment does not specify of how any single individual's well-being is affected, high community capacity itself suggests higher levels of well-being for residents. High capacity suggests, too, that expanding opportunities to meet community needs (and local well-being) is not only possible but likely.

With the continued shifting of responsibilities from state and federal entities to localities and the increased responsibility placed on locals for self-development and improvement–including communities that have long relied on federal and state subsidies for infrastructure development and maintenance–examining community capacity is an important facet of community well-being research.

CONCLUSION: FOREST COMMUNITY WELL-BEING

This review of well-being studies makes it clear there is room for considerable improvement in forest and natural resource-dependent community well-being assessment. Future studies of well-being cannot rely solely on incomplete subjective reports of well-being; exclusive reliance on measures of opulence such as income are equally limiting because they fail to address distribution issues or resource control. Additionally, if researchers are to discuss resource dependency and well-being, they must be clear about what "dependency" means and the many ways resources may be valued. The forest as a "place" embodies a diverse array of values. If a local forest, long used as a locale for the production of wood products, is reserved exclusively for recreational use or is overcut, local well-being will decline through the diminution of socially important forest values (not to mention jobs).

Researchers need to recognize that the link between resource dependency and well-being is not straightforward, nor can it be assumed; it requires empirical examination. Resource harvesting and extraction, whether they involve minerals, timber, or agricultural products, have social consequences. The manner in which these activities are organized and carried out, including who controls the resources, who manages the benefits and determines when and how locals participate in resource management, and the condition of the resources that remain have significant implications for community well-being in the short- and long-term.

Researchers also must be clear about the unit of analysis. Assessment of community well-being cannot be limited to county or regional analyses. Counties are too heterogeneous; too often jobs associated with resources comprise a small proportion of a county economy. Communities are a logical unit of study but pose methodological problems: clear identification of boundaries is often difficult and the data available within those boundaries may be limited. Well-being analysis must often strike a balance between socially meaningful units of analysis and units for which good data are available.

Communities must not only be thought of as a unit of analysis but also included as part of well-being assessment. Including a behavior category termed "commitments" broadens the concept of well-being in two significant ways. First, it acknowledges that capabilities and functionings are in part defined by the community. Communities are composed of and sustained by individuals, and individuals are shaped by their community. Viewed in this light, consideration must be given to how community–defined here as a locality-based collection of individuals–fosters or inhibits individual thinking about capabilities and individual ability to function. Hence local conditions are viewed as influencing on individual conceptualizations of well-being. A second implication of a broadened concept of well-being is that relationships within a community involve a component of responsibility to communal relationships. This involves practices of commitment that make up patterns of individual allegiance and responsibility directed towards community. These practices profoundly affect well-being.

In broad-scale ecosystem studies it is simply not possible to assess the resources each individual has and determine how they contribute to his or her functioning. An assessment of community capacity allows researchers to assess in some measure the opportunities available to residents today as well as their potential for creating additional opportunities and improving well-being. A basic assumption is that the higher the capacity of a community, the more it is likely that opportunities exist or will be created to expand individual capabilities and functioning. The very act of community capacity building is not only opportunity enhancing but itself leads to improved social well-being.

The emphasis on individuals, local community, and capacity does not mean that social and political arrangements beyond community boundaries should be ignored. They are an important–and in many cases a critical–component of capacity and local well-being. This is particularly true of forest communities where local and nearby land is owned and local jobs controlled by nonresidents. Actions originating outside of a community may contribute to or severely restrict local residents' capabilities and functionings. For example, local capabilities may be reduced by forest management decisions that neither involve local residents nor take into account local needs. Good capacity assessment will identify these arrangements. Improving a community's ability to respond positively to and influence decisions made outside community boundaries reflects on the capacity of a community and is another way of improving the well-being of community residents.

What is unique about this approach to the study of community well-being is that it involves both an analysis of factors that reduce local well-being and an analysis of how individuals and their communities respond to these factors. This assessment of capacity also encompasses an examination of how individuals and communities create opportunities or to use Sen's terminology, capabilities that expand the possible functionings or achievements of community members and improve well-being. In addition to identifying general levels of individual and community well-being, one of the significant benefits of this approach is its identification of areas with low capacities and a reduced ability to self-develop and improve local well-being. It is these areas that require the most attention and will provide the most difficult challenges for those serious about improving the health of the ecosystem and the well-being of humans and human communities.

NOTES

1. The commonly held misconception of community stability calls for a steady flow of timber products, primarily logs, to ensure stable employment in the timber industry, which, in turn, leads to community well-being. Community stability, however, was once conceived in much broader terms (see Dana 1918; Kaufman and Kaufman 1946). Beginning in the late 1920s, however, the term became inextricably linked to timber industry employment in U.S. Forest Service discussions of sustained-yield forest management (Fortmann et al. 1989).

2. This was one of three teams created by President Clinton to "identify management alternatives that attain the greatest economic and social contribution from the forests of the region and meet all requirements of applicable laws and regulations" (FEMAT 1993, ii).

3. See Lee et al. (1990) for a discussion of these concepts for forest communities.

4. Warren's (1978) observation that ties between organizations within a community (horizontal linkages) have been overwhelmed by ties to organizations and institutions outside the community (vertical linkages) is relevant here. Warren argued that the rising influence of an increasingly urban society frequently results in a decline of a community's distinctiveness, self-sufficiency, and individual interactions.

5. Machlis and Force (1988) point out that forest or timber dependency is generally determined by forest commodity production or economic measures (e.g., measures of sales by forest industries, percentage of total income from the forestry sector, and forest industry employment).

6. See Wall et al. (1998) for an exposition of alternative concepts of social capital.

REFERENCES

Albrecht, S.L. 1982. Commentary on local social disruption and western energy development, a critical review. *Pacific Sociological Review* 25:297-306.

Allardt, E. 1993. Having, loving, being: an alternative to the Swedish model of welfare research. In M. Nussbaum and A. Sen (eds.), *The Quality of Life*, pp. 88-94. Oxford: Clarendon Press.

Barlett, P.F. and P.J. Brown. 1985. Agricultural development and the quality of life: an anthropological view. *Agriculture and Human Values* 2:28-35.

Bealer, R.C., F.K. Willits, and W.P. Kuvlesky. 1965. The meaning of rurality in American society: some implications of alternative definitions. *Rural Sociology* 30:257-266.

Bellah, R.N. et al. 1985. *Habits of the Heart*. New York: Harper and Row.

Benn, S.I. 1982. Individuality, autonomy and community. In *Community as a Social Ideal*, pp. 43-62. London: Edward Arnold.

Berry, W. 1987. Does community have a value? In *Home Economics*. San Francisco: North Point Press.

Bormann, B.T. et al. 1993. A broad, strategic framework for sustainable-ecosystem management (prepared by the Eastside Forest Health Panel for presentation to the Chief of the U.S. Forest Service). Corvallis, OR: Forestry Sciences Lab.

Bradburn, N.M. and D. Caplovitz. 1965. *Reprints on Happiness*. Chicago: Aldine

Burdge, R.J. 1994. *A Conceptual Approach to Social Impact Assessment: Collection of Writings*. Middleton, Wisconsin: Social Ecology Press.

Campbell, A. 1981. *The Sense of Well-Being in America: Recent Patterns and Trends*. New York: McGraw Hill.

Chamberlain, K. 1985. Value dimensions, cultural differences and the prediction of perceived quality of life. *Social Indicators Research* 17:345-401.

Dana, S.T. 1918. Forestry and Community Development. Bulletin No. 638. U.S. Department of Agriculture, Washington, DC.

De Neufville, J.I. 1975. *Social Indicators and Public Policy.* New York: Elsevier Scientific Publishing Company.

Doak, S.C., and J. Kusel. 1996. Well-being in forest dependent communities, part II: a social assessment focus. In Sierra Nevada Ecosystem Project: Final Report to Congress, volume II, Assessment and Scientific basis for management options. Davis: University of California, Centers for Water and Wildland Resources.

Doak, S.C., and J. Kusel. 1997. Social assessment of communities in the Klamath region. Report prepared for the Klamath National Forest (project number 43-91W8-6-7077).

Drielsma, J.H. 1984. The influence of forest-based industries on rural communities. Ph.D. Dissertation, Yale University.

Erikson, R. 1993. Descriptions of inequality: the Swedish approach to welfare research. In M.C. Nussbaum and A. Sen (eds.), *The Quality of Life*, pp. 67-83. Oxford: Clarendon Press.

Etzioni, A. 1988. *The Moral Dimension: Toward a New Economics*. New York: The Free Press.

Finsterbusch, K. 1982. Commentary on local social disruption and Western energy development, a critical review. *Pacific Sociological Review* 25:307-321.

Forest Ecosystem Management Assessment Team (FEMAT). 1993. Forest Ecosystem Management: An Ecological, Economic, and Social Assessment. Report of the Forest Ecosystem Management Assessment Team. Washington, DC: U.S. Forest Service.

Fortmann, L.P., J. Kusel and S.K. Fairfax. 1989. Community stability: the foresters' figleaf. In D. LeMaster and J. Beuter (eds.), *Community Stability in Forest-Based Communities,* pp. 44-50. Beaverton, Oregon: Timber Press:.

Freudenburg, W.R. 1984. Boomtown's youth: the differential impacts of rapid community growth on adolescents and adults. *American Sociological Review* (October):697-705.

Freudenburg, W.R. and R.E. Jones. 1991. Criminal behavior and rapid community growth: examining the evidence. *Rural Sociology* 56:619-645.

Gans, H.J. 1962. *The Urban Villagers.* New York: The Free Press of Glencoe.

Gold, R.L. 1982. Commentary on local social disruption and Western energy development, a critical review. *Pacific Sociological Review* 25:349-356.

Gold, R.L. 1985. *Ranching, Mining and Human Impact of Natural Resource Development.* New Brunswick, New Jersey: Transaction, Inc.

Goldschmidt, W. 1947. *As You Sow.* Glencoe, Illinois: The Free Press.

Grumbine, E. 1994. What is ecosystem management. *Conservation Biology* 8:23-38.

Gusfield, J.R. 1975. *Community.* New York: Harper & Row.

Hazard, G.C. 1988. Communitarian ethics and legal justification. *University of Colorado Law Review* (4):721-739.

Headey, B., E. Holmstrom, and A. Wearing. 1985. Models of well-being and ill-being. *Social Indicators Research* 17:211-234.

Hester, R. 1985. Subconscious places of the heart. *Places* 2:10-22.

Hillery, G.A., Jr. 1955. Definitions of community: areas of agreement. *Rural Sociology* 20:194-204.

Hiss, T. 1990. *The Experience of Place.* New York: Alfred A. Knopf.

Interorganizational Committee. 1994. Guidelines and principles for social impact assessment. NOAA Technical Memorandum NMFS-F/SPO-16. U.S. Department of Commerce.

Jobes, P.C. 1984a. Social structure and myth: impacts of energy development in ranchland. Paper presented at the American Sociological Association meeting (August), San Antonio, Texas.

Jobes, P.C. 1984b. Disintegration of a ranching community: a longitudinal case study of the impact of coal development on gemeinschaft. Unpublished report, University of Michigan.

Johnson, D.J. 1988. Toward a comprehensive 'quality-of-life' index. *Social Indicators Research* 20:477-496.

Kaufman, H.F. and L.C. Kaufman. 1946. *Toward the Stabilization and Enrichment of a Forest Community* (The Montana Study), University of Montana. U.S. Forest Service, Region One, Missoula, Montana.

Kennedy, L.W., and N. Mehra. 1985. Effects of social change on well-being. *Social Indicators Research* 17:101-113.

Kim, K-D. 1973. Toward a sociological theory of development: a structural perspective. *Rural Sociology* 38:462-476.

Krannich, R.S., E. H. Berry and T. Greider. 1989. Fear of crime in rapidly changing rural communities: a longitudinal analysis. *Rural Sociology* 54:195-212.

Kusel, J. 1991. Ethnographic analysis of three forest communities in California. Volume II of the report, *Well-Being in Forest-Dependent Communities*. Sacramento: California Department of Forestry Forest and Rangeland Assessment Program.

Kusel, J. and L. Fortmann. 1991. Well-being in forest-dependent communities. Volume I of the report, *Well-Being in Forest-Dependent Communities*. Sacramento: California Department of Forestry Forest and Rangeland Assessment Program.

Lasch, C. 1988. The communitarian critique of liberalism. In C.H. Reynolds and R.V. Norman (eds.), *Community in America: The Challenge of Habits of the Heart*, pp. 173-184. Berkeley, CA: University of California Press.

Lee, R.G., D.R. Field and W.R. Burch, Jr. 1990. Introduction: forestry, community, and sociology of natural resources. In R.G. Lee, D.R. Field, and W.R. Burch, Jr. (eds.), *Community & Forestry: Continuities in the Sociology of Natural Resources*, pp. 3-14. Boulder, CO: Westview Press.

Lobao, L.M. 1990. *Locality and Inequality: Farm and Industry Structure and Socioeconomic Conditions*. Albany, New York: State University of New York Press.

Lobao, L.M., M.D. Schulman and L.E. Swanson. 1993. Still going: recent debates on the Goldschmidt hypothesis. *Rural Sociology* 58:277-288.

MacCannell, D. 1988. Industrial agriculture and rural community degradation. In L.E. Swanson (ed.), *Agriculture and Community Change in the U.S.* Boulder, CO: Westview Press.

Machlis, G., and J.E. Force. 1988. Community stability and timber dependent communities. *Rural Sociology* 53:220-234.

MacIntyre, A. 1984. *After Virtue: A Study in Moral Theory.* Vienna, Indiana: University of Notre Dame Press.

Marchak, P. 1990. Forest industry towns in British Columbia. In R.G. Lee, D.R. Field, and W.R. Burch, Jr. (eds.), *Community & Forestry: Continuities in the Sociology of Natural Resouirces*, pp. 95-106. Boulder, CO: Westview Press.

Manley P. et al. 1995. *Region 5 Ecosystem Management Guidebook.* USDA Forest Service, Pacific Southwest Region.

Mastekaasa, A., and T. Moum. 1984. The perceived quality of life in Norway: regional variations and contextual effects. *Social Indicators Research* 14:385-419.

Meinig, D.W. 1979. Symbolic landscapes: some idealizations of American communities (and the beholding eye: ten versions of the same scene). In D.W. Meinig (ed.), *The Interpretation of Ordinary Landscapes*, pp. 31-48. New York: Oxford University Press.

Ministry of Environment and Lands and Parks (Province of British Columbia) and Environment Canada. 1994. State of the Environment Report for British Columbia.

Moum, T. 1988. Yea-saying and mood-of-the-day effects in self-reported quality of life. *Social Indicators Research* 20:108-125.

Oakley, A. 1975. *The Sociology of Housework*. London: Robertson.

Oppong, J.R., R.G. Ironsides, and L.W. Kennedy. 1988. Quality of life in a centre-periphery framework. *Social Indicators Research* 20:605-620.

Perry, C. 1986. A proposal to recycle mechanical and organic solidarity in community sociology. *Rural Sociology* 51:263-277.

Price, M.L., and D.C. Clay. 1980. Structural disturbances in rural communities: some repercussions of the migration turnaround in Michigan. *Rural Sociology* 45: 591-607.

Putnam, R.D. 1993a. *Making Democracy Work: Civic Traditions in Modern Italy.* Princeton, New Jersey: Princeton University Press.

Putnam, R.D. 1993b. The prosperous community: social capital and public life. *The American Prospect* 13:35-42.

Quigley, T.M. and S.J. Arbelbide. 1997. *An Assessment of the Ecosystem Components in the Interior Columbia Basin and Portions of the Klamath and Great Basins*. General Technical Report, Portland, OR: U.S. Department of Agriculture, Forests Service, Pacific Northwest Research Station.

Reyna, N.E. 1998. *Economic and Social Conditions of Communities in the Interior Columbia Basin*. Part 1 of a Report Prepared by the Interior Columbia Basin Ecosystem Management Project, U.S.D.A. Forest Service and U.S.D.I. Bureau of Land Management, Walla Walla, Washington (February).

Relph, E. 1976. *Place and Placelessness*. London: Pion Limited.

Schnaiberg, A. 1986. Reflections on resistance to rural Industrialization: newcomers' culture of environmentalism. In P.D. Elkind-Savatsky (ed.), *Differential Social Impacts of Rural Resource Development*, pp. 229-258. Boulder, CO: Westview Press.

Selznick, P. 1992. *The Moral Commonwealth: Social Theory and the Promise of Community*. Berkeley, CA: The University of California Press.

Selznick, P. 1987. The idea of a communitarian morality. *California Law Review* 75:445-463.

Sen, A. 1984. *Resources, Values and Development*. Cambridge. Massachusetts: Harvard University Press.

Sen, A. 1985a. Well-being, agency and freedom (The Dewey Lectures), *Journal of Philosophy* 82:169-221.

Sen, A. 1985b. *Commodities and Capabilities* (Professor Dr. P. Hennipman Lectures in Economics, Volume 7). New York: North-Holland.

Sen, A. 1987. *The Standard of Living* (G. Hawthorn, ed.). Cambridge: Cambridge University Press.

Sen, A. 1990. Individual freedom as a social commitment. *The New York Review of Books* (June 14).

Sen, A. 1993. Capability and well-being. In M.C. Nussbaum and A. Sen (eds.), *The Quality of Life*, pp. 30-53. Oxford: Clarendon Press.

Seydlitz, R. et al. 1993. Development and social problems: the impact of the offshore oil industry on suicide and homicide rates. *Rural Sociology* 58:93-110.

Sierra Nevada Ecosystem Project. 1996. *Final Report to Congress: Assessment Summaries and Management Strategies* (volume I). Davis: University of California, Centers for Water and Wildland Resources.

Strathern, M. 1984. The social meaning of localism. In T. Bradley and P. Lowe (eds.), *Locality and Rurality: Economy and Society in Rural Regions*, pp. 191-197. Norwich, England: Geo Books.

Suttles, G.D. 1969. *The Social Order of the Slum*. Chicago: University of Chicago Press.
Swanson, L. 1988. *Agriculture and Community Change in the U.S.* Boulder, CO: Westview Press.
Tuan, Y-F. 1993. *Passing Strange and Wonderful: Aesthetics, Nature and Culture.* Island Press: Covelo, California.
Wall, E., G. Ferrazzi and F. Schryer. 1998. Getting the goods on social capital. *Rural Sociology* 63(2): 300-322.
Walter, E.V. 1988. *Placeways: A Theory of the Human Environment.* Chapel Hill: The University of North Carolina Press.
Warren, R.L. 1978. *The Community in America*. Third Edition. Chicago: Rand McNally.
Wendel, S. 1987. McCloud, California: maintaining cherished landscapes and lifestyles. Professional Report, Department of Landscape Architecture, University of California at Berkeley.
Wilkenson, K.P. et al. 1982. Local social disruption and Western energy development, a critical review. *Pacific Sociological Review* 25:297-306.
Wilson, W. 1967. Correlates of Avowed Happiness. *Psychological Bulletin* 67: 294-306.
World Commission on Environment and Development. 1987. *Our Common Future.* Oxford University Press: Oxford.

APPENDIX I

Sen (1987:28) ignores "commitments" in the calculus of personal well-being. He does so by making a distinction between actions based on "sympathy," which are included in calculations of well-being, and actions based on "commitment," which are not. Sen includes "commitments" in a category called "agency achievement" (1987:28) or "agency freedom" (1985a:203). According to Sen (1985a:203), "agency achievement" is a more inclusive category than personal well-being and includes "what a person is free to do and achieve in pursuit of whatever goals or values he or she regards as important." He points out (1987:28) that by expanding the focus of attention and including "commitments," we move from "personal well-being" to "agency achievement."

Sen defines "sympathy" as help provided to an individual that makes the helper "feel–and indeed be–better off." This increases one's personal well-being. The behavior category of "commitments," on the other hand, involves personal action (it too may be help provided to another), which Sen states, "*in the net*, [is not] beneficial to the agent himself." Sen adds, "This would put action outside the range of promoting one's own well-being."

Sen's rejection of an action because it is "in the net" not "beneficial to the agent himself" involves an evaluation of action (and its consequences) after the fact, or a prediction of its outcomes. Although Sen states (1987:28) he is concerned with effects, he does not discuss at what point this calculation should take place nor what measures should be made to determine whether a behavior is beneficial or not. Given the nature of his decision rule, Sen ignores the motivation for individual action. Both categories of behavior, "commitments" and "sympathies," may involve action for which the motivation stems from the desire to help another person. Actions to improve one's community that do not have the "effect" of contributing to one's well-being therefore are not included in Sen's well-being calculations. In this manner Sen ignores historical, social, and societal forces that not only influence action (and motivations for action) but influence value decisions implicit in the evaluation of well-being.

Etzioni (1988:40-45) states that the category of action called "commitments" is moral behavior. This is because such action is based on intentions, not consequences or effects. Intentions may also be considered the "intrinsic character" of action and are taken here to be the primary criteria by which to evaluate it, because the consequences of action may not be predictable. This valuational approach is central to a deontological social philosophy (Etzioni 1988; Hazard 1988).

Exploring Reinvestment from a Community-Based Watershed Perspective

Leah Wills
Gerald J. Gray

SUMMARY. Increasingly, urban and rural communities across the U.S. are looking at ecological systems as capital assets with values that are diminishing as they are degraded. Investments are needed so they can continue to provide ecological services critical to the environmental, social, and economic well-being of the communities.

This paper explores "reinvestment" as term encouraging the development of mechanisms for sustained and long-term investment in "natural capital," as well as the other forms of capital on which communities and society depend. It raises important questions on which there are information gaps, such as the value of ecological services, the level of investment needed to restore and maintain ecological systems, the activities to which investment dollars should be allocated, and where public and private investment dollars might come from. These questions are discussed at national, regional, and local scales, focusing on water–an ecological resource with immense value that connects ecosystems and communities at many scales. The paper presents lessons and reinvestment mechanisms from studies in California's Sierra Nevada ecosystem and community-based efforts in its northernmost drainage, the Feather River watershed. *[Article copies available for a fee from The Haworth Document Delivery Service: 1-800-342-9678. E-mail address: <getinfo@haworthpressinc.com> Website: <http://www.HaworthPress.com>]*

Leah Wills is affiliated with the Plumas Corporation, P.O. Box 3880, Quincy, CA 95971.

Gerald J. Gray is Vice President, American Forests, 910 17th Street NW, Suite 600, Washington, DC 20006.

[Haworth co-indexing entry note]: "Exploring Reinvestment from a Community-Based Watershed Perspective." Wills, Leah, and Gerald J. Gray. Co-published simultaneously in *Journal of Sustainable Forestry* (Food Products Press, an imprint of The Haworth Press, Inc.) Vol. 13, No. 1/2, 2001, pp. 385-399; and: *Understanding Community-Based Forest Ecosystem Management* (ed: Gerald J. Gray, Maia J. Enzer, and Jonathan Kusel) Food Products Press, an imprint of The Haworth Press, Inc., 2001, pp. 385-399. Single or multiple copies of this article are available for a fee from The Haworth Document Delivery Service [1-800-342-9678, 9:00 a.m. - 5:00 p.m. (EST). E-mail address: getinfo@haworthpressinc.com].

KEYWORDS. Community-based forestry, reinvestment, ecological services, natural capital, watershed perspective

Urban and rural communities across the U.S. are responding to concerns about environmental degradation, concerns often related to past and present growth and natural resource management practices. Many communities are developing strategies toward "sustainability" by identifying approaches to integrate the long-term viability of ecological, economic, and socio-cultural systems in their own local context, while recognizing links to regional, national, and even international systems. Understanding the relationship between different systems and addressing issues in an integrated fashion is a complex and, at times, overwhelming task. The President's Council on Sustainable Development (PCSD) noted the critical necessity for integrated thought and action in its 1996 report on a sustainable America:

> By recognizing that economic, environmental, and social goals are integrally linked and by having policies that reflect that interrelationship, Americans can regain their sense that they are in control of their future and that the lives of each generation will be better than the last. Thinking narrowly about jobs, energy, transportation, housing, or ecosystems–as if they were not connected–creates new problems even as it attempts to solve old ones. Asking the wrong questions is a sure way to get misleading answers that result in short-term remedies for symptoms, instead of cures for long-term basic problems. (PCSD p. 6)

One part of the sustainability challenge is the idea of maintaining the integrity of common assets. Some analysts are beginning to apply economic approaches to concerns about environmental degradation, reframing issues by thinking about ecological systems as economic assets, or "natural capital." Questions about the maintenance of natural capital are approached in the same way as maintaining financial capital (savings and investment), physical capital and industrial plant (infrastructure), and social capital (education, training, civic cohesion and community capacity, etc.). The PCSD also acknowledges the need to think of ecosystems as assets that require long-term thinking and investment to sustain their future capacity:

> A country that protects its ecosystems and manages its natural resources wisely lays a far stronger base for future prosperity than one that carelessly uses its assets and destroys its natural capital. A society that invests in its children and communities, equitably providing education and opportunity, is far more likely to prosper than one that does not make such investments and allows the gap between rich and poor to widen. (PCSD p. 6)

The term "reinvestment" affirms ecological systems' capacity, as natural capital, to produce economic value, expressing the need to maintain rather than diminish that capacity. The term describes the financial flows that are generated by an ecological system and returned to that system by those who obtain the economic benefit, often users located "downstream" or beyond the ecosystem's boundaries. The term incorporates both the need for investment in natural capital, so it will not be degraded, and the need to ensure that those who derive economic benefit from an ecological system pay their fair share in maintenance costs.

As analysts explore the idea of natural capital, it is clear there are information gaps that need to be closed. One need is to better understand the economic values of ecological systems. It is one thing to say these systems are critical to sustain life on earth as we know it, it is another to begin to quantify economic values associated with these systems and use this information effectively in policy decision making. A second need relates to the cost of maintaining the integrity of ecological systems. To date, there has been limited focus on the type of ongoing investment needed to maintain healthy, functioning ecosystems. More often, systems become degraded and require extraordinary actions and investments to mitigate the impacts and help them recover. The third need is twofold: to gain a clearer sense of who benefits from the values produced by ecological systems and to create mechanisms through which they can pay their fair share for maintaining those systems.

Economists are only now beginning to look at these information needs. One of their first efforts is quantifying the economic value of ecological services, an approach that encourages symbiotic solutions to economic and environmental issues. "Ecological services" both sustain life and enable the creation of economic value. They include such functions as water production, climate regulation, carbon sequestration, and the maintenance of biodiversity. Robert Costanza of the University of Maryland has estimated that the annual value obtained from the world's ecosystems "services" exceeds $32 trillion, a value substantially higher than the global annual Gross National Product of $18 trillion. This estimate includes significant nonmarket values and is thus seen as "soft" in the marketplace and in policy circles. The estimate is useful, however, as a starting point for dialogue and as an educational tool. It is important information for political leaders interested in heightening awareness of the need to maintain natural capital.

CHARACTERIZING REINVESTMENT: WATER IN CALIFORNIA

Reinvestment is an important policy tool when ecosystem values and the economic gains from that ecosystem are high, and when investments can sustain or enhance the ecosystem's capacity to provide these values. Areas with high carbon sequestration potential; areas surrounded by expensive recreation; areas that generate profitable biofuels, biomedicines, and florals; and water production and flood control areas all are potential candidates for reinvestment mechanisms. California's extremely high water values provide a dramatic example of how important reinvestment can be at the ecosystem and watershed scale. This paper will discuss recent information and initiatives in the Sierra Nevada ecosystem and its northernmost drainage, the Feather River watershed, as a means of exploring reinvestment.

Water supplies are becoming critical in many areas of the arid West, while in the East and South, water-quality concerns are driving new ecosystem management approaches. The California water example makes a strong economic case for maintaining the environmental health of watersheds upstream of large water and hydropower supply dams. The largely federally managed watersheds of California are critical for catching, storing, and channeling low-sediment flows of otherwise dispersed precipitation into elaborate and expensive water supply and hydropower production facilities. Combining the natural (watershed) and manufactured (engineered) infrastructure enables water and hydropower purveyors to produce billions of dollars of wealth through water and power sales. Affordable water and hydropower, in turn, support a vast array of other economic endeavors in California, most notably a multibillion dollar agricultural economy.

Most water supply and hydropower purveyors are accustomed to investing a portion of the profits from their water sales into maintaining the ditches, dams, turbines, and pipelines that harness the watershed's runoff. They are unaccustomed, however, to making similar investments in maintaining the watersheds that actually capture, store, cleanse, and channel the "favorable flows" harnessed for economic uses downstream. This is especially true in situations where the watersheds above the water supply or the treatment facilities are not owned by the water purveyors themselves.

A few water purveyors and hydroelectric utilities in California dedicate a portion of their operation and maintenance budgets to watershed maintenance–if they own a large proportion of the watershed. In other parts of the country, such as Denver, New York City, and Seattle, urban-based water authorities are experimenting with "reinvesting" in the health of watersheds in which they are relatively minor landowners. Through partnerships with other watershed landowners, these downstream water users are attempting to better conserve their water supplies through collaborative, incentive-based watershed-stewardship activities. The historic 1996 New York City watershed agreement, for example, recognizes farming and forestry as preferred land-uses for maintaining and protecting water quality in the Catskill/Delaware watersheds, which provide 90 percent of the City's drinking water. Through intense negotiations with landowner and community representatives from the upstream watersheds, city, state, and federal officials agreed to invest $1.5 billion in voluntary, incentive-based measures aimed at maintaining traditional farm and forestland uses, improving land-management practices, and updating community infrastructure, such as roads and septic tanks, in the rural watersheds. This investment in a "natural systems" approach to water-quality protection contrasts with an alternative technical approach which would have required $5 billion-$8 billion to construct a filtration plant downstream near the City (Gray 1999).

In California, however, a "tragedy of the commons" best describes the relationship between the mostly federally managed watersheds and their downstream federal, state, and private water users. Although national policy has long recognized the value of federal lands as producers of high-quality water, neither national nor state policy has developed a way to foster investment in maintaining these water-supply watersheds. The 1897 Organic Act that clarified the role of the national forests focused on two primary purposes: "securing favorable conditions of water flows, and to furnish a continuous supply of timber" for the nation (Adams 1993). Since then, a variety of mechanisms has been developed to obtain revenues from timber, recreation, forage, and other resource production activities and reinvest them in restoration and maintenance of federal lands. Examples of these mechanisms include direct payments to local jurisdictions, off-budget trust funds, yield taxes, and user fees. The cost of maintaining federal lands for water production (quality and quantity), however, has been borne entirely by U.S. taxpayers through the federal appropriations process, with little possibility of return compensation. In an era of declining federal revenues, it is appropriate to question whether federal taxpayers should continue to exempt water users from their "fair share" responsibilities for maintaining the watersheds that supply water to California's agriculture, water, and hydropower industries.

THE SIERRA NEVADA ECOSYSTEM PROJECT

The congressionally authorized and funded Sierra Nevada Ecosystem Project (SNEP) developed estimates of the economic values of natural resource activities in the Sierra Nevada ecosystem. As one of the nation's first efforts to estimate economic values for a large-scale ecosystem (20 million acres), the SNEP study broke new ground. Unlike attempts to quantify the market and nonmarket values of the "ecological services" natural ecosystems produce, SNEP estimates are based on direct economic (market-based) values produced by natural resource activities and captured by resource users and beneficiaries. The study examined and compared the value of natural resource production activities related to water, timber, recreation and tourism, and grazing and other agriculture. It found that the Sierra Nevada ecosystem produces approximately $2.2 billion worth of commodities and services annually and that water accounts for $1.5 billion–more than 60 percent of the value. Most of the water values accrue to users of hydroelectricity and municipal and agricultural water supplies, virtually all of whom are located outside the boundaries of the Sierra Nevada ecosystem. Only about 2 percent of the $2.2 billion in value generated by the Sierra Nevada ecosystem is captured by federal, state, and local governments through direct payments, taxes, and fees and reinvested in the ecosystem or local communities. Public timber production and private recreation provide the largest contributions to federal and local governments both in total dollars and as a percentage of the natural resources' total value. Water production activities generate no money to help to federal and local governments maintain watersheds or communities. The SNEP report found that "additional mechanisms to promote reinvestment are necessary to maintain and enhance the Sierra Nevada ecosystem so that it can continue to provide the socially desirable outputs" (SNEP p. 1057).

The SNEP report refers to the lack of funding for watershed maintenance as an "underinvestment problem." It suggests "a massive and directed investment of time and money" is needed to sustain and enhance the Sierra Nevada ecosystem and communities, but "compared to the size of the local communities and the value of the natural assets, the cost is small" (SNEP p. 1057).

The report identifies four primary reasons why there has not been greater investment in the Sierra Nevada ecosystem:

- Many attributes of the ecosystem are not valued in a manner that motivates investment.
- Restrictions on exchange prevent value formation for aspects of the ecosystem that generate economic benefit.
- Barriers between agencies and governments prevent efficient responses to economic values where these are known.
- Localities lack the capacity to capture economic surpluses they generate and to then invest these surpluses for ecosystem health and social well-being.

THE FEATHER RIVER WATERSHED

Each of the four reasons for "underinvestment" identified in the SNEP report suggests institutional strategies and mechanisms for attracting more investment. A number of these strategies have been tested in the Feather River watershed over the past 15 years. Various initiatives have been developed through the efforts of the Feather River Coordinated Resource Management (FR-CRM) group, a consortium of 21 local, regional, and national entities that have been cooperating together since 1985. Their initiatives strengthen local capacity, identify watershed reinvestment mechanisms and priorities, and forge alliances with groups beyond the watershed's boundaries.

The 3.2 million-acre Feather River watershed is the northernmost drainage in the Sierra Nevada; 71 percent is federally managed as national forests and national park land. The watershed delivers 3.2 million acre-feet of "firm" (e.g., "captured behind the Oroville Dam") water yield to the State Water Project (SWP), which supplies water for the municipal, industrial, and agricultural needs of 20 million Californians. The Feather River watershed delivers 4.6 million acre-feet of unimpaired flow to the famous 1,330.1 MW "Stairway of Power," currently owned by Pacific Gas and Electric Company (PGandE) and the California Department of Water Resources. PGandE serves 600,000 customers using the Feather River hydrosystem as part of its electrical supply grid. The Feather River watershed historically supported a premier timber management and woods products industry, and "Feather River Country" is increasingly attractive for recreational developments. Dr. Jeff Romm, an economist at the University of California at Berkeley, has been quantifying current economic values for the Feather River watershed's environmental services. His preliminary estimates indicate the river provided more than 3.5 million acre-feet of water to hydropower, municipal, and agricultural users in 1996, valued at more than $400 million. His current research show that, "in certain conditions, riparian and meadow restoration actually can enhance water storage more efficiently than dam augmentation" (Romm, 1999). Other analysts like Dr. Linda Bond, a consulting hydrocologist in Sacramento, and Dr. Rick Kattelmann, a hydrologist at the Forest Service's Pacific Southwest research station, estimate that investing in watershed health improvements could provide at least an additional 250,000 acre-feet or, 7 percent more useable water annually, to downstream water users (Bond 1997; Kattelmann 1987). These watershed-specific analyses mirror SNEP's earlier Sierra-wide findings and document water as the most economically important output from the Feather River watershed, yet the environmental health of the watershed continues to degrade. There is a clear need for integrated national and state policies that ensure a fair share of the profit from water and hydropower sales returns to the watershed for watershed maintenance and rehabilitation.

Local, Regional, and National Reinvestment Strategies

Reinvestment in community-based watershed management requires strategies that are inherently multi-scale and multi-issue. Watershed management can link local on-the-ground stewardship efforts with regional and national markets and governance structures. Through investment mechanisms for watershed management, water and forest conservation at the community scale can be integrated coherently within larger economic and environmental analyses and processes. Integration occurs under an overarching theme: All users of natural resources have the opportunity and responsibility to maintain natural, social, and economic assets for future generations, at the ecosystem-, community-, and personal scale. Wherever people live, work and play, watershed management means stewarding natural resources in ways that sustain and enhance them during and after each generation and evolution of uses. In remote rural areas, watershed management approaches will differ from those used in urban or suburban areas. Wealthy areas are likely to use different approaches than high-poverty areas. Ecosystem processes are even different between ecological regions; within a highly altered and urbanized eco-region, for example, the response is different from that in a pristine area.

The FR-CRM is networking to explore multi-scale, multi-issue approaches to investing in watershed management through the California Urban Forest Advisory Council, the California Brownfields Communities working group, the Cal-Fed long-term water solution process, and the Communities Committee of the Seventh American Forest Congress, a national network of organizations focused on the interdependence of forests and communities. The FR-CRM has recently worked with two regional networks, the Sierra Nevada Alliance and the Regional Council of Rural Counties, to develop the following community involvement principles which are relevant to rural and urban settings, and local, regional, and national processes (Box 1).

In 1995, the FR-CRM began to identify the most effective short- and long-term reinvestment activities for maintaining the health of the Feather

BOX 1. Principles of Watershed Community Involvement

- Watershed strategic, annual, and project planning must be open, public, and involve communities in the watershed.
- Community involvement must include a comprehensive and inclusive public education component.
- Watershed restoration and stewardship should reflect a strong component of sustainable local economies and/or revitalization of local communities implementing projects.
- Advisory and/or oversight committees must include members residing in the watershed.
- Watershed groups administering restoration projects must deposit restoration funds in institutions that actively invest in local communities and economic revitalization within the Council's jurisdiction.
- Watershed groups must adopt restoration strategies and plans of action that enhance and create local job and contracting opportunities.
- Watershed policy, restoration and stewardship plans and projects must be consistent with established principles and standards.

River watershed, and to allocate parts of a hypothetical 1995 reinvestment dollar to essential program components. Figure 1 reflects the group's years of experience with implementing site-specific watershed rehabilitation projects, sub-basin and basin-wide water-quality studies and resource-condition assessments, various multiparty investment and financing packages, and 1996-specific reinvestment priorities. It indicates the watershed management activities that need to be funded under a reinvestment program include: watershed rehabilitation projects, planning, economic incentives, critical habitat protection and enhancement, stewardship education, project effectiveness monitoring and watershed trend monitoring, resource condition assessments, job training and development, and testing and evaluation of best management practices (BMPs), with some money set aside for unallocated expenses, such as contingencies or emergencies. The percentages would change over the years but the categories would likely remain the same. The breadth of the activities and relative investment in each recognize that reinvestment efforts must integrate environmental and economic concerns by developing comprehensive assessment information at the outset of projects and comprehensive monitoring information at the conclusion.

Beyond identifying its local reinvestment priorities, the FR-CRM has also developed white papers for community forestry and watershed groups as a means of moving reinvestment strategies forward on a broader, regional basis. Its most in-depth work was sponsored by the Collaborative Learning Circle (CLC), a regional consortium of community forestry and watershed groups in the Pacific Northwest. Table 1, adapted from the 1995 CLC effort,

FIGURE 1. Feather River CRM

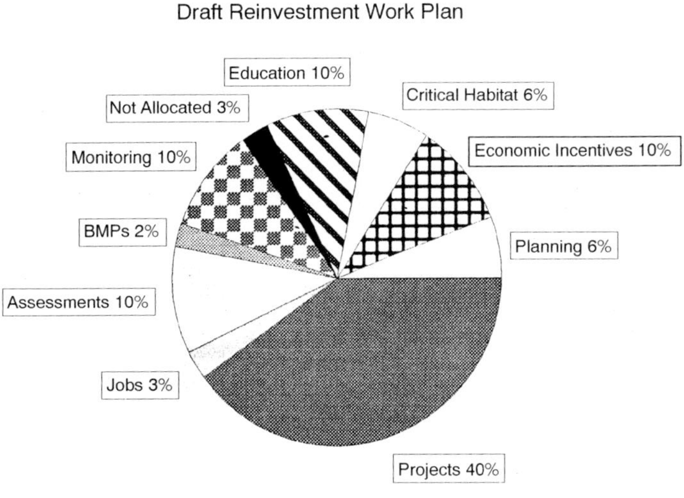

TABLE 1. Reinvestment Objectives and Activities

Reinvestment Objectives	Reinvestment Activities
Safeguarding Sustainability	All-party monitoring (ongoing at all scales) and certification.
	Developing and integrating measures and monitoring protocols for social, natural and economic capital and condition trends.
	Endowments for land trusts, acquisition of conservation easements, biotic reserves, species migration corridors, habitat conservation areas, cultural and historical heritage areas, and ecosystem-process study areas.
	Community "sense of place" heritage conservation areas and downtown revitalization and redevelopment including oral histories and photo archives of historical landscapes.
Full Life Cycle Cost Pricing and Full Mitigation of Current Negative Effects	Third-party monitoring of mitigation measures and performance standards.
	Project impact monitoring at appropriate spatial and temporal scales incorporated into performance bonds.
	Pollution trading.
	Debt for nature swaps and revenue bonds.
	Mitigation banks, habitat conservation banks, forest banks, and water conservation banks.
	Development transfers.
Watershed Maintenance	Watershed analysis: resource condition and trend assessments including resource management history and natural history including historic natural events and tenure or ownership history.
	Ensuring public access to public land management records and decision-making.
	Road and trail maintenance, rehabilitation, relocation, and closure.
	Ditch, dam, diversion, and culvert maintenance, including fish screens, fish and wildlife escapement and passage.
	Erosion control, waste management, forest fuels reduction, forest density management, restoring hydrologic function to meadows, wetlands, and floodplains.
	Toxics control, pollution cleanup and abatement.
	Flood, fire, and storm proofing, livestock management, public access management, poaching and trespass enforcement, and permit and contract enforcement.

Reinvestment Objectives	Reinvestment Activities
	Critical habitat protection, native species reintroduction, exotic species eradication, restoring habitat connectivity and refugia.
	Signage and public education.
	Coordinated implementation plans for interface areas in mixed public and private land ownerships and for interface areas between urban and rural land uses.
	Resources management and monitoring plans with performance criteria that are responsive to variable and declining budget scenarios.
Incorporating Opportunity Costs of Land Conversion to Urban Uses into Land-Use Decision making	Full life-cycle land conversion opportunity costs compared to alternative recreational, forestry, agricultural, and public-trust uses and benefits.
	Full life-cycle governmental cost- and revenue-streams including flood, fire, and pollution-hazard mitigation, as well as police and fire; public safety; hospitals, schools, and other social services; and public roads, bridges, and buildings.
Ensuring Public Benefits on Private Lands	Incorporating public benefits of land and resource stewardship into private landowner incentives, e.g., incentives for carbon sequestration, water banking, habitat conservation, cultural and historic heritage conservation, aesthetics, sense of place, open-space.
Consumer Education and Preferences	Certification of products, management plans, managers, and workers for ecosystem sustainability.
	Pricing systems to include life-cycle of waste management.
	Life-cycle maintenance of the source ecosystem is incorporated into pricing for users of ecosystem products and services.
	Pricing systems to include environmental justice impact remediation.

(Adapted from Collaborative Learning Circle, 1995).

presents a set of reinvestment objectives that might be identified in a particular region–safeguarding sustainability, mitigating current negative effects, maintaining watersheds, and consumer education, for instance–and activities that could help achieve them. County supervisors in the Northern Sierras are considering which of these and other reinvestment strategies might be useful in their region. Local officials and community-based groups generally oppose "one size fits all" policies, financial strategies, and regulatory approaches; these have often been developed without their participation and are not responsive to local contexts. Their opposition has often been misinterpreted as purely antiregulatory sentiment. What many communities are seeking, however, are flexible, broad-based reinvestment strategies, like the set of tools in Table 1, that can be used to finance specific reinvestment activities that fit unique local circumstances.

Lessons from the Feather River Watershed

The FR-CRM has 15 years of experience identifying and testing strategies for reinvestment in watershed restoration and maintenance. Its efforts have met with mixed success, but the lessons learned are instructive for policy development and important for encouraging other communities suffering from acute disinvestment. Although some circumstances are specific to the Feather River watershed, the FR-CRM's approaches to identifying and addressing reinvestment barriers are widely applicable to other areas.

While the FR-CRM has functioned as a collaborative group of organizations and interests focused on the local watershed, it has also recognized the need to form alliances with other community groups and organizations to address broader policy and market-based issues. The capacity of community-based groups to network and collaborate beyond their local boundaries is critical for addressing reinvestment issues that concern distant resource beneficiaries. One of the important dimensions of this networking involves efforts by rural communities, located in the ecosystem providing ecological services, to communicate and cooperate with urban communities that use and depend upon these services.

FR-CRM members have often discussed opportunities for water and hydropower suppliers to reinvest in maintaining the health of the Feather River watershed. Some related issues can be resolved at the community and county levels, but others require state and national involvement. It is clear that negotiations with powerful state and national interests on who reinvests in federal lands cannot be successful without leadership from the federal government. Federal public lands are often the largest holdings in the biggest water-supply and hydropower watersheds in the West. The federal government–executive, congressional, and judiciary branches–has the authority and should exercise its responsibility to encourage and enable water-supply and hydropower purveyors to "do the right thing" by paying their "fair share" for watershed restoration and maintenance. National leadership is especially important for two reasons as deregulation heightens market pressures for electric and water purveyors.

- In the absence of federal policies, individual watersheds are likely to become "stranded assets" as ecosystem restoration and maintenance compete with other investment options in increasingly competitive national markets.
- Specific reinvestment agreements will not be successful if they compromise the financial positions of potential reinvestment partners. Oth-

> er competitors in regional or national markets may not have to shoulder a comparable reinvestment responsibility. Effective reinvestment policies, therefore, must recognize the need to "maintain a level playing field" among competing hydro and water supply purveyors. Policies and mechanisms must be neutral at all market scales, from local to national, and must not penalize hydro or water purveyors for acting ethically by paying their fair share for watershed maintenance. Creating a level playing field at the national level is not possible without strong leadership from the federal government.

One economic issue often discussed by the FR-CRM is the difference between the value of the Feather River watershed as an asset and the value of the ecological services it provides. This issue is important because maintaining the ecosystem as an asset is different than optimizing resource use and values. From an economic perspective, the goals for reinvestment need to be clarified and incorporated into financial decisionmaking processes. The asset value is the ecosystem's long-term capacity to provide ecological services, some of which are then used, traded, rented, or sold in ways that are either efficient or inefficient. The value of these ecological services may fluctuate dramatically, such as the price of water during rainy and drought periods. Developing efficient resource use through effective incentive and pricing signals is important for reducing wasteful demands on the ecosystem. However, investing in efficiency alone may be inadequate to ensure long-term ecosystem health–especially in volatile and poorly regulated markets and for seriously degraded ecosystems. Several other economic considerations that the FR-CFM has identified, and that might be important to other community-based watershed efforts, are:

- A reinvestment program must learn how to respond to economic signals such as the aggregate and marginal values of improvements. There is a lot to learn about short-term improvements for specific ecological services as well as how incremental and cumulative improvements affect overall ecosystem health and productivity. These types of issues need to be monitored for adaptive learning.
- Economic evaluations must recognize the time value of water. Hydrologic improvements should be sensitive to the counter-cyclical price patterns for water runoff and storage. A water budget provides an ecological framework for evaluating the time value of water.
- Ecological boundaries and interactions should be recognized when developing economic criteria for reinvestment. Both kinds of analytical systems–the ecological and the economic–should become more integrated through adaptive learning.

- Investment analyses must include risk assessment, and reinvestment strategies must manage for risk. Risks are economic and environmental, short- and long-term, can vary at different scales, and can be reversible or irreversible over time. At this time, risk assessment is poorly developed and needs to evolve in a coordinated way with other reinvestment monitoring elements.

CONCLUSION

Increasingly, urban and rural communities across the U.S. are looking at ecological systems as capital assets with values that are diminishing as they are degraded. Investments are needed so they can continue to provide ecological services critical to the environmental, social, and economic well-being of the communities. "Reinvestment" is a term used to encourage the development of mechanisms for sustained and long-term investment in "natural capital," as well as the other forms of capital on which communities and society depend. Economists are beginning to look at critical questions concerning reinvestment, such as the value of ecological services, the level of investment needed to restore and maintain ecological systems, the activities to which investment dollars should be allocated, and where public and private investment dollars might come from. Many communities believe consistent and adequate levels of investment are needed in restoring and maintaining ecological systems to reverse cumulative environmental decline and quell the growing social and economic demand for ecological services.

Water values in areas such as New York, Seattle, Denver, and the Chesapeake Bay are driving new approaches and mechanisms that allow communities to invest in natural systems to maintain water quality and quantity. In California, ecosystem-scale studies done in the Sierra Nevada and ongoing research and community-based efforts in the Feather River watershed illustrate the importance of reinvestment. The SNEP study estimates that $1.5 billion in water values are being obtained from the Sierra Nevada ecosystem annually, yet virtually none of that value is returned to the ecosystem to help sustain it. Research in the Feather River watershed indicates similar water values and suggests investments could enhance the system's capacity to provide ecological services.

To be effective, reinvestment must function as a continuous and integrated loop where the economic values of ecological services are recognized and the beneficiaries, both within ecosystem boundaries and beyond, invest a fair share of their economic gains for ecosystem restoration and maintenance. Mechanisms for long-term financial exchange between ecosystem beneficiaries and managers must be developed and remain consistent, yet be adaptable over time. That's because the mix of profitable uses and products, and the investors and landscape managers, will invariably change over time. The investment horizon is long because processes that degrade, restore, and maintain ecosystems are long–often more than a hundred years in the Sierra Nevada. Reinvestment mechanisms must be perceived as efficient, flexible, and equitable over that long-term investment horizon. Reinvestment dollars will need to be applied to many different kinds of ecosystem stewardship activities at different spatial scales and at different times. These various, sequenced, and cumulative stewardship actions must be planned and implemented openly, and they must be perceived as lawful, necessary, and effective at restoring and maintaining ecosystem capacity. All-party monitoring and adaptive learning must be ongoing to ensure the effectiveness of stewardship actions, to maintain commitments for "fair share" reinvestment by ecosystem beneficiaries, and to build shared knowledge among the ever-changing reinvestment program participants.

REFERENCES

Adams, D. A. 1993. *Renewable Resource Policy: The Legal-Institutional Foundations*, p. 120, Washington, DC: Island Press.

Bond, L. 1997. Estimate of groundwater storage potential of the Feather River Watershed. Report to Plumas Corporation from Hydrologic Consultants, Inc., Sacramento, California.

Gray, G. J. 1999. The New York City Watershed Agreement: A grand experiment. A case study of the Communities Committee of the Seventh American Forest Congress. In press.

Kattelmann, R. 1987. Feasibility of more water from Sierra Nevada forests. Report 16. Berkeley, CA. Wildland Resources Center, Division of Agriculture and Natural Resources, University of California.

President's Council on Sustainable Development. 1996. Sustainable America: A new consensus for prosperity, opportunity, and a healthy environment for the future. Washington, DC.

Sierra Nevada Ecosystem Project. 1996. Economic assessment of the ecosystem. SNEP Final Report to Congress, Vol. III, Assessments, Commissioned Reports, and Background Information, p. 1057, Davis, California: University of California, Centers for Water and Wildland Resources.

Index

